Biology

Third Edition

John Snyder, Ph.D.
Professor
Department of Biology
Furman University
Greenville, South Carolina

C. Leland Rodgers, Ph.D.
Professor Emeritus
Department of Biology
Furman University
Greenville, South Carolina

BARRON'S

All inquiries should be addressed to:
Barron's Educational Series, Inc.
250 Wireless Boulevard
Hauppauge, New York 11788

International Standard Book No. 0-8120-1862-1

Library of Congress Catalog Card No. 94-38412

Library of Congress Cataloging-in-Publication Data
Snyder, John A., 1943–
    Biology / John Snyder, C. Leland Rodgers. — 3rd ed.
        p. cm. — (College review series)
    Rev. ed. of: Essentials of biology / by C. Leland Rodgers. 1974.
    Includes index.
    ISBN 0-8120-1862-1
    1. Biology—Outlines, syllabi, etc. I. Rodgers, C. Leland (Charles Leland), 1918– . II. Rodgers, C. Leland (Charles Leland), 1918– Essentials of biology. III. Title. IV. Series.
QH315.5.S59   1995
574—dc20   94-38412
                                                                    CIP

PRINTED IN THE UNITED STATES OF AMERICA
5678   8800   987654321

CONTENTS

PREFACE

Biology Has Grown in Scope and Importance

In the twenty plus years since the previous edition of this book, the field of biology has undergone unprecedented expansion, perhaps more so than in any other scientific discipline. Biologists have learned much about the molecular basis of heredity and embryonic development, the workings of the brain, the interactions of organisms with their environments, and the complex activities of cells, to mention just a few significant areas. Researchers have begun to manipulate organisms (including humans) at the biochemical level, raising profound ethical questions as well as giving rise to a related and vigorous new economic sector—biotechnology. We have also come to realize that certain human activities are causing the extinction of large numbers of species. This has prompted a reexamination of traditional beliefs about the absolute freedom of land use for short term enterprise in relation to modern, responsible, long-term conservation needs for the well being of future generations.

This Book Is Both Comprehensive and Comprehensible

For the reasons mentioned above, public interest in biological phenomena is at an all time high. This book was written to satisfy that interest at several levels. Biology as a field of inquiry has expanded at such an exponential rate that present textbooks have become unwieldy tomes, typically consisting of over one thousand pages. The book you hold in your hands covers the same wide range of topics but condenses them, accurately explaining their essential elements. Thus, it is valuable as a supplement and study guide in college courses or high school advanced placement biology classes, because it succinctly directs students to the most important aspects of each topic. For those who took their last biology course some time ago and need a refresher course to update themselves on the information explosion in our field, this text offers an alternative to the standard textbook. Since it touches on all of the major topics of biology, it is also a valuable study guide for those preparing to take the Graduate Record Examination subject test in biology or similar standardized tests in biology.

You Will Find Helpful Features Throughout

Since some will use the book as a reference source rather than reading through every chapter, we have provided an extensive glossary and index. Illustrations were chosen to be informative rather than entertaining. Headings within chapters are in the form of statements that summarize and emphasize major points. Each chapter concludes with a set of thought-provoking comprehension questions.

The authors thank Barron's editor Patricia Wilson for guiding this project to completion. They thank Furman University for the use of its resources. They also thank Judy Snyder and Jean Rodgers for encouragement, constructive criticism, and patience.

1
BIOLOGY AS A FIELD OF SCIENCE

What do we know about:
—*the characteristics of science?*
—*the fundamental activities of observing and performing experiments?*
—*devising scientific laws and theories?*
—*the value of skepticism when doing science?*
—*the "scientific method"?*
—*how biology is related to other scientific fields?*
—*why this can be called "The Age of Biology"?*
These and other intriguing topics will be pursued in this chapter.

SCIENCE HAS A GREAT IMPACT ON DAILY LIFE

It has been said that science is the most important agent of change in modern society. Although such a sweeping statement could be argued, it is certainly true that in this scientific age nearly every aspect of our lives is potentially affected by a myriad of scientific discoveries and their applications. Even those who have never formally studied any field of science constantly modify their daily behavior in response to scientific discoveries. Consider, for example, the current worldwide explosion of interest in various telecommunication devices made possible by the invention of orbiting satellites and of fiber optics.

Given that science changes our lives, it is obvious that its study should be part of every person's education. This book is designed to open the door to such a study for the field of science known as biology. It will also be a useful review for the reader who has done previous study and wants to bring it back to mind, to see alternative ways of explaining certain topics, or be brought up-to-date on developments that have occurred since the reader's last exposure. Throughout the book, the underlying supposition is that this portion of science is important both for its own sake and because it has great impact upon our daily lives.

1

SCIENCE HAS CERTAIN DEFINING CHARACTERISTICS

The word *science* is heard so often that almost everybody has some notion of its meaning. Its formal definition, however, is difficult for many people. The meaning of the term is confused because many endeavors masquerading under the name of science have no valid connection with it. Suppose that science is defined as *a process used to acquire knowledge about the physical universe in order to understand that universe.* Furthermore, science is characterized by including *the physical testing of each assertion about the universe,* rather than simple acceptance of it as fact. Astronomy is a field of science that adequately meets these requirements. On the other hand, astrology, which also has the stars and planets as subjects, does not qualify as a science since it does not involve the testing of its purported facts by physical means. Even in the true sciences, distinguishing fact from fiction is not always easy. For this reason, there is another aspect of science that should not be overlooked—*Any statement of fact should always be considered provisional.* That is, one should be willing to retest it for validity and to modify it if a new test demonstrates its inadequacy. There are many examples in every scientific field of statements that were sincerely believed to be true for many years, until evidence accumulated to refute them. Without a doubt, the book you are now reading contains assertions that, while currently seeming to be factual (based on our best testing), will eventually prove to need change. Scientists, like all people, develop sentimental attachment to their own ideas, but should always be willing to modify or even reject them when faced with new evidence. Without this flexibility, science cannot continue to move closer to truth about the nature of the universe.

Scientists Work by Doing Both Observation and Experimentation

Another aspect that must be mentioned is that there are two fundamental methods for scientifically studying things in nature. One is to *observe* some object or phenomenon. This is the act of acquiring information about something without altering it. If an animal's behavior is being observed, the onlooker must try to remain totally undetected to avoid influencing the animal and changing its behavior. Of course, this can sometimes be quite challenging, whether it is an organism or an atom that one is trying to observe. Sometimes observation can be greatly enhanced by the use of instruments—artificial aids to our own senses such as microphones, pH meters, or telescopes. Such devices not only increase our sensitivity, but often provide precise quantitative data to supplement our qualitative perception.

The other basic way used by scientists to understand the world is to *perform experiments.* By definition, an experiment is occurring if the object of study is being subjected to deliberate and systematic alteration. The goal here is to learn about the object by seeing how it responds to the applied change. For instance, one might learn something about the sensory ability of an organism by recording what it does in response to a flash of red light after being in darkness for some time. In this example, the light condition (absence, red, or some other color of light) is the thing that is changing; it is called a *variable*. Ideally, each experiment should involve only one variable, but sometimes this cannot be achieved (especially in complex settings such as ecosystems).

For an experiment to be of any value, one must also set up a *control*—a situation in which the object is subjected to identical conditions in every aspect except for the variable. If one is studying the response to red light, one might use continuous darkness as the control. In all ways other than light, both the experimental situation and the control situation should be as much alike as possible—same species of organism, age, sex, and conditions of sound, humidity, temperature, enclosure size, and so on. Then, if the experimental environment is changed only by the addition of a flash of red light, the scientist can make some valid conclusions about the specific effect of that light upon the organism.

Science Is Not an Appropriate Tool for Studying Some Things

Another aspect of science that must be considered is that it is always a study of real (physical) objects, not of imaginary things, or of metaphysical concepts. Of course, this is not to say that disciplines designed for studying such things as poetry, politics, music, or religion are any less important or valid than those for studying physical objects. Nor is it true that the methods of science cannot also be applied (with modification) to such disciplines.

Scientists Work to Make Laws and Theories

Most of the above is a description of the general methods of science. However, one must also consider the goals of science. Why do we observe and perform experiments? Our first attempt at a definition for science included a general statement of goals, ". . .in order to understand the universe." This is a valid assertion, but it is a bit vague. Therefore, let us become more specific at this point, by saying that the goals of scientific research are to (1) discover specific pieces of information, then (2) gather together all related data in order to make *laws*, and finally (3) link and explain the data and laws by making *scientific theories*.

What is a law in the context of science? It is *a general statement that describes a set of known facts*. For instance, a number of things have been discovered about the ways that many gases behave when temperature and pressure change. From these experimental results, certain summarizing statements have been made; they are called the gas laws. They apply only to those gases that have actually been tested and found to conform; there are some gases that do not behave in these ways. A law can be a guideline for predicting the nature of yet untested objects, but it is absolutely valid for only those that have already been tested. Since a law summarizes much information (and because much human insight is required to recognize that a law is possible) there are fewer laws than there are individual facts about nature. Why formulate laws? Because they help us make sense of nature. They pull together facts that were previously isolated and apparently unrelated. Before this topic is left, it should be noted that the idea of scientific laws is sometimes interpreted more loosely than described above, even by scientists themselves.

We come now to the ultimate goal of science, that of making scientific theories. This is the largest and most useful statement that can be devised about nature. Unfortunately, it is also the most misunderstood, especially when scientists and nonscientists try to communicate. For most people not professionally involved in science, the word theory is synonymous with hypothesis. That is, a person might have a hunch about something and say, "I have this theory about. . ." Of course, one's hypothesis might eventually be shown to be correct, but at the outset it is unsupported by rigorous testing. Indeed, scientists often use the word theory in this same way, when speaking casually. However, in formal communication, a scientific theory is quite different from a hypothesis. A scientific theory is a statement that explains a very large amount of data, and may encompass a number of laws. It is the largest, most comprehensive generalization about a given portion of the observed universe. Of course, this means that very few scientific theories have yet come into existence. Each branch of science is likely to have only a handful. As a scientific discipline matures, it should develop more theories, and an eventual goal will be to connect those theories into even larger theories.

Since a scientific theory is so important, we must consider more of its attributes. First, a theory shows previously unsuspected patterns and relationships. It connects together a number of facts that seemed unrelated to each other. In so doing, of course, a theory enables us to see nature as a whole that makes sense, rather than as an incomprehensible collection of unconnected objects and processes. Second, because a theory develops patterns of thought, it is also quite predictive. A scientist might be able to view the way previously discovered facts fit together and then know what to look for next. A rough analogy of this would be, if you string together the numbers 4, 2, 8, and 6 into the series 2-4-6-8, you are likely to notice that the pattern increases by two, and predict that the logical next number to follow 8 is 10. Every scientific theory that has been devised has been a significant catalyst—speeding the rate of new discoveries in the field because it showed what should be investigated next. Third, a theory is very robust. That is, since its basis is an accumulation of results from a very large number of already performed observations and experiments, a theory is unlikely to be quickly or easily proven false. Even if some of those are eventually shown to be invalid, there remain many more that would also have to be shown false before the whole theory would fall. This is not to say that any theory is absolutely correct, or that it will never be proven at least partially false. Science always corrects itself as new evidence is gathered, and even theories must occasionally be modified to accommodate this new knowledge. However, it is a sign of ignorance about the nature of scientific theories to say, "That theory is *only* a theory, so we shouldn't trust it until it is proven." To make such a statement is to show that one does not realize a theory's rich underpinnings of fact.

Research Can Be Basic, Applied, or Both

There is another way to look at the goals of science. That is to ask, "Why do we *want* to know about a particular thing in nature? What is our motivation?" Sometimes it is useful to say that there are two ultimate goals, and that they are illustrated by dividing all research into the categories labeled *basic research* and *applied research*. Basic research is aimed at satisfying human curiosity. Applied research is aimed at using knowledge for specific human-oriented purposes: for instance, improving health, raising standards of living, or creating new consumer products.

Sometimes practical-minded people miss the point of research in thinking only of its immediate utilization. One can see that an extraordinary amount of basic knowledge about chemistry is necessary before one can possibly understand functions of cells like respiration or photosynthesis and apply that knowledge to medical or agricultural advances. Chemists responsible for many of these discoveries could hardly have anticipated that their findings would one day result in applications of such a practical nature as those directly related to life and health. As another example, geneticists working to understand the way fruit flies pass on their genetic information could not have foreseen all the possible applications of their findings to the improvement of agricultural plants and animals through selective breeding. The host of scientists dedicating their lives to basic science are not apologetic about ignoring the practical side of their discoveries; they know from experience that most knowledge is eventually applied, and that many discoveries that now benefit our society came from basic research.

A SET OF VALUABLE ATTITUDES ENHANCES ONE'S ABILITY TO DO SCIENCE

A student of nature who approaches a problem should have certain personal qualifications that will tend to make the effort come to a successful conclusion. The person who is *inquisitive, honest, thorough, patient, objective,* and *precise* will have the greatest chance of success. Another essential attribute for a scientist is the willingness to change one's mind when faced with new evidence. Stubbornness is a valuable trait only if it does not extend to the point of holding onto an idea that others have shown to be inaccurate. Of course, each of these is a positive trait desirable in all human activities, not just those labeled scientific research.

Another positive attribute, which may seem at first to have a negative connotation, is *skepticism*. Scientists should be skeptical not only about untested hypotheses but also about what are generally recognized as facts. As discussed before, many conclusions initially accepted as facts are based on available information and may have to be revised or rejected in the light of better observations, improved equipment, and refined techniques. People make mistakes, and this includes even the so-called authorities of science.

Someone working in science should not confuse cause and effect. It is easy to fall into the fallacy that Event 2, happening right after Event 1, was therefore caused by Event 1; it is possible that this sequence was purely by a chance happening of unrelated events. Another mistake to be avoided is to treat the result as if it were the cause. An illustration is embedded in the statement, "A plant turns toward the sun *in order to* get light." Such an assertion implies that a plant knows what it is doing or that it is responding with intelligence and purpose. It would be correct to say that the plant turns toward sunlight *because of* the influence of the sun. Statements implying that activities of nonhumans are shaped by purpose are termed *teleological.* To include purpose in an explanation of a natural event is to add an attribute that is extremely complex; a simpler explanation is almost certain to suffice.

Do scientists sometimes deliberately cheat in their work, for personal gain? Unfortunately, this must be answered in the affirmative. Possessed of human ambition, a scientist may be tempted to publish a research paper that presents data not actually obtained, in order to make a hypothesis appear more acceptable. A great deal of trust is necessarily built into the process of doing science, since a worker often does experiments in isolation. However, proven cases of scientific fraud are extremely rare, for a very particular reason. Every scientist knows that his or her work will be published, widely disseminated to knowledgeable people, and eventually tested for accuracy. That testing may not be in the form of an exact repetition of the work; it could be indirectly through other experiments that are based on the assumption that one's published work is truthful. Sooner or later, erroneous or misleading material will be found out and corrections made. Persons found to have perpetrated such acts are not welcome in the scientific community.

A SCIENTIST ATTACKS A QUESTION BY USING SPECIFIC STEPS: THE SCIENTIFIC METHOD

Procedures used to solve scientific problems have been collectively referred to as the *scientific method.* At least when outlined on paper, this method is neither mysterious nor complicated but a common sense step-by-step progression toward an answer. Furthermore, its use is not confined to the realm of formal scientific study. Rather, it is valuable in many areas of in-quiry, and is often employed by ordinary citizens without formal and system-atic planning.

A scientist approaching a problem might plan the sequence of activities that is shown in Figure 1.1. He or she could begin by gathering together all of the known information relating to the problem. Much data can be gleaned by read-ing published work of other scientists. This is a major task, made easier by the use of computerized systems that can quickly scan many thousands of titles and abstracts searching for key words. When relevant papers or books are identi-fied, the scientist must find print or electronic copies of them and incorporate their information into an ever-growing mental stockpile. Another major source of information is personal communication with other scientists, especially at scientific conferences. This method avoids the inevitable time lag between a discovery and its publication in a book or journal.

From the gathered information on the topic, the scientist then attempts to make a *generalization*—a broad statement that seems to describe all of the known data. Since this statement has not yet been confirmed, it can also be called a hypothesis. Note that the process has been one of moving from spe-cific data to a broader generalization that encompasses them. Such a method of logical thinking has historically been called the *inductive method,* in con-trast to the *deductive method.* (Deduction reverses the process, by beginning with a broad general statement that is assumed to be correct, then logically deriving specific smaller statements that must also be correct. The place of de-duction in scientific investigations will be discussed later.)

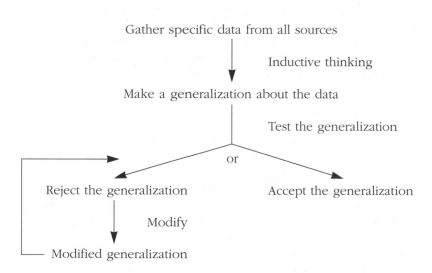

FIGURE 1.1.   *Steps taken when following the inductive version of the scientific method.*

The generalization that has come by inductive thinking may be quite close to truth (after all, it may have been based upon many previously obtained pieces of data), but its validity *must* be tested further. Indeed, this unwillingness to accept an untested generalization is at the very center of the scientific method. The scientist will use as much ingenuity as possible in devising a physical test to immediately confirm or deny the validity of the hypothesis. To the lay person, this sometimes seems strange. The scientist would seem to be attacking his or her own best intellectual work by trying to disprove it. But only by doing so will the scientist show the hypothesis to be acceptable. The actual method of testing a hypothesis can be either observational or experimental, whichever seems more appropriate, but it must be designed to provide the most unambiguous answer. Another necessary feature of this step is that a test must be repeated before one accepts its information. A mistake often made by a person just beginning in science is to accept the conclusions of a single test, in which pure chance might have skewed the results in a particular direction.

Often, testing will indicate that the original hypothesis is not quite correct. The scientist responds by modifying the hypothesis to account for this new piece of information, then devises another test to determine whether the modification has been appropriate. This cycle may be repeated many times, with each change in the hypothesis making it come closer to being a statement of fact. Eventually, the scientist feels comfortable enough with the much-modified generalization that he or she publishes it. But even then it might be subjected to further analysis by other scientists who apply their own tests. Remember that no scientific statement is considered to be forever unalterable.

What has been described above is the classical inductive approach to questions in science. However, when philosophers and historians of science examine actual records of how scientific knowledge has progressed, some claim that the inductive method is not necessarily what has been followed. The inductive approach certainly can be fruitful, but by its repetitive nature it can also be described as slow and boring. Scientists do not often describe their work as boring, so perhaps they are not always following the classical inductive method. A more interesting version of the scientific method begins with a person posing a large generalization—a big idea that might explain something about the world. Perhaps that idea has come to mind before very much specific data has been gathered. The inductive method of making a generalization on the basis of specific smaller pieces of evidence has not been followed. A person using this alternative route is likely to accept the generalization as provisional truth. That is, he or she will *pretend* that it is true, and then think about the specific things that would also be correct if it is true. It will be remembered that this sequence of logical thought is called deduction. It works in science only if one keeps in mind that the generalization is still only a hypothesis and that it has to be tested. What must happen after the smaller "truths" have been deduced is to test them. If they fail to hold up to scrutiny, not only they but also the larger generalization must then be modified. This series of thoughts and events has been labeled the *hypothetico-deductive method*, since it involves making a hypothetical generalization and then proceeding in deductive fashion. It is certainly more fun than beginning inductively, since it involves using one's imagination to leap far ahead of the known facts and to make a grand scheme that can then be tested for validity.

BIOLOGY IS THE SCIENTIFIC DISCIPLINE DEALING WITH LIFE

With an abundance of organisms around us, we invariably become curious about them. Among the obvious questions stimulating immediate interest are these: what is life, from where did it come, what is its future, what conditions are necessary to support it, and what is the significance of the great variety of living organisms? To these one may add countless others, many of which cannot immediately be answered. Biology is the science that deals with these problems and is therefore sometimes called the science of life.

Biology is classified as a natural science, because the subject matter is a part of nature. Since life is so complex and has so many still-hidden secrets, biology is sometimes said to be less exact than some other natural sciences. Such a description is misleading since it implies that there might be a difference in the precision of the biologists' methods of measurement or a laxity in their demands for accuracy and precision. If biology has historically seemed less exact, part of the reason is the difficulty of separating out small functional portions of a very complex chemical and structural organization. Analytical (rather than purely descriptive) biology had to wait until the fields of chemistry and physics developed methods and instruments that could deal with such complexity. Since the middle of the twentieth century such tools have become available and their application to living organisms has played a major role in the great advance that has been the recent hallmark of biology. This is the reason why the formal training of every modern biologist includes course work in chemistry and physics.

However, analyzing the organism as if it were a purely physical or chemical system cannot be accomplished without losing some of its fundamental nature. Truly, the living organism is an example of the old saying that the whole is more than the sum of its parts. Its complexity as a living system should make its study all the more stimulating. Ultimately, problems in biology are inseparably close to chemistry, physics, and the earth sciences, and will continue to benefit from progress made in those fields. Life itself appears to be an expression of a complex and not fully understood dynamic of physical and chemical attributes.

BIOLOGY HAS MANY BRANCHES

The field of biology is immense, if not limitless. Consequently, individual biologists specialize in phases of the science most appealing to them.

If this were not so, they would spend all their time learning what is already known without ever having time to pioneer in the realm of the unexplored.

There are many ways to divide biology into subdisciplines. One way is to emphasize the kind of organism that you are studying. If you are a generalist, rather than a specialist, you might say that you work in *zoology* (study of Kingdom Animalia), *botany* (Kingdom Plantae), *mycology* (Kingdom Fungi), or *microbiology* (Kingdoms Monera and Protista, and the viruses). Or, you could name your field of study in a way that implies greater specialization. For instance, there are biologists who call themselves *protozoologists* (studying the heterotrophic Protista), *dendrologists* (studying trees), *lepidopterists* (studying butterflies and moths), *ornithologists* (studying birds), or *mammalogists* (studying mammals). Sometimes it is more appropriate to describe one's work in a way that focuses upon the life habits of the organisms. Examples of this include *parasitology* (study of parasites, in any kingdom), or *marine biology* (study of any organisms living in salt water).

Another way to describe the subdisciplines of biology is to use names that emphasize the processes being studied, rather than the types of organisms. For instance, there are the fields of *evolutionary biology* (origins and long-term changes of organisms), *genetics* (inheritance, information management), *physiology* (short-term changes and maintenance of the body), *ecology* (interactions of organisms and their environment), and *behavior* (whole-animal actions). These large fields are sometimes divided as workers specialize. For instance, animal physiologists include *endocrinologists* (studying hormonal systems), *immunologists* (studying immune systems), *cardiologists* (studying the heart), and many others.

Sometimes a discipline's name indicates the methods of study being used, rather than the kind of organism or biological function being studied. Examples of these include *microscopy, biochemistry, biophysics,* and *biostatistics.*

Finally, it must be noted that many biologists do work that has immediate impact and usefulness with reference to humans. Of course, their activities fall within the area sometimes labeled applied science, as noted earlier in this chapter. All of the careers within the realm of *medicine* and *dentistry* (and the closely related *veterinary medicine*) are those of applied biology. So are such areas as *agriculture, food science,* and *zoo and aquarium management.*

This listing of biological career fields is far from exhaustive, but it should demonstrate that there are many facets to the subject.

BIOLOGY HAS A RICH HISTORY

The interest of humans in their natural surroundings goes back, of course, to prehistoric times. They depended on other organisms for food, clothing, and shelter, and eventually learned to domesticate some animals and cultivate some plants. It is difficult to determine where and when these first attempts at understanding biology occurred, but anthropologists and paleontologists believe that Africa and the Middle East were among the first places where agricultural practices were developed.

Undoubtedly, every society showed some degree of sophistication in naming and knowing the natural history of the organisms around it. To fail in this would mean great hardship in finding food and avoiding biological hazards such as poisons and predators. Evidence is found in the few modern nonagricultural societies that have remained isolated from formal science. When sociobiologists query members of such groups, they are often surprised by the knowledge that has been accumulated by word of mouth over many centuries of observations. A number of promising drugs for medical purposes were first brought to the attention of the pharmaceutical industry by careful listening to indigenous people far from the mainstream of modern science. It stands to reason, therefore, that ancient societies also may have had more biological knowledge than is often attributed to them.

The Greeks are often considered the originators of anything like organized study of biology. However, it should be noted that other societies may well have been doing the same sort of sophisticated thinking as the ancient Greeks, but did not leave written records for us to examine. There is good physical evidence for advanced technology in metal working and other chemical activities among ancient societies in Africa and Asia.

Greek written records beginning in the third century B.C. were complete enough to credit specific people with specific scientific speculations. Among contributors to biology (albeit, often providing flawed ideas) were Aristotle and Hippocrates. Later, around 150 A.D., Galen made important contributions to medicine, including detailed anatomical dissections of domesticated animals.

For much further progress in formal biology we must jump more than 1,000 years to the contributions of Andreas Vesalius. A Belgian living in the 1500s, he dissected human cadavers and laid the foundations of modern medicine. In the next generation, the Englishman William Harvey applied this newfound anatomical knowledge to physiological functions of the organs (especially the circulatory system).

From the 1600s forward, knowledge of biology accelerated. Listed below are some of the most important discoveries since that time, many of which will be discussed in detail within the chapters of this book. The latter half of the twentieth century witnessed a much greater growth in biology than any previous period. Of course this could be said about many fields of science, but it is perhaps more accentuated in biology than elsewhere. Much of this recent growth can be attributed to the interconnecting of biology with other sciences, especially chemistry and physics. At present, a good case can be made for the assertion that biology is growing faster (in terms of both new discoveries and number of workers in the field) than any other area of science. We live in the Age of Biology.

1600s First application of the microscope to biological study (*Malpighi* discovers capillary vessels, *Hooke* discovers cells, *Leeuwenhoek* discovers bacteria).

1700s System developed for classifying organisms, by *Linnaeus* and others.

1700s First investigations of photosynthesis, by *Priestley*.

1798 First immunization to prevent a disease (smallpox) by *Jenner*.

1839 Development of the Cell Theory of the origin and ubiquity of cells, by *Schleiden* and *Schwann*.

1859 Publication of *The Origin of Species*, first thorough description of organic evolution, by *Darwin*.

1864 Prize awarded to *Pasteur* for disproving the ancient doctrine of spontaneous generation of life from non-living materials.

1865 Public reading of *Mendel's* paper on inheritance in pea plants.

1892 Certain diseases of plants and vertebrate animals first attributed to viruses.

1900 Rediscovery of principles of heredity, previously shown by Mendel.

1902 Recognition that genes are located on chromosomes, by *Sutton*.

1907 Technique for culturing tissue apart from the organism, by *Harrison*.

1908 Equations derived for use in studying population genetics, by *Hardy* and *Weinberg*.

1910 Use of fruit fly in genetics and discovery of sex linkage, by *Morgan*.

1916 Plant succession as an important ecological concept, by *Clements*.

1920 Photoperiodism demonstrated in plants, by *Garner* and *Allard*.

1921 Concept of organizers and induction in embryos, by *Spemann*.

1922 Isolation of insulin, by *Banting* and *Best*.

1923 Ultracentrifuge invented, by *Svedberg*.

1924 Fossil remains of *Australopithecus* (early relative of *Homo*), discovered by *Dart*.

1926 Enzymes found to be proteins, by *Sumner*.

1927 First systematic study of gene mutation, by *Muller*.

1927 Concept of the ecological niche expounded, by *Elton*.

1929 First medically useful antibiotic (penicillin) discovered by *Fleming*.

1929 Discovery of ATP, by *Lohmann*.

1930	"Modern Evolutionary Synthesis" developed by *Mayr* and
–42	others, uniting Darwin's ideas with modern genetics.
1934	Interspecific competition and predator-prey relations studied systematically, by *Gause*, building on earlier work by *Lotka* and *Volterra*.
1937	Pathway of aerobic respiration (citric acid cycle), elucidated by *Krebs*.
1937	First practical electrophoresis, by *Tiselius*.
1938	Practical electron microscopes developed by several inventors.
1944	Genetic material first shown to be composed of nucleic acid, by *Avery*, *MacLeod*, and *McCarty*.
1946	Radiocarbon dating of fossils, by *Libby*.
1949	Genetic control of protein synthesis shown, by *Pauling*.
1952	Ultrastructure of the mitochondrion shown, by *Palade*.
1953	Ultrastructure of endoplasmic reticulum shown, by *Porter*.
1953	Three-dimensional structure of DNA, by *Watson* and *Crick*.
1953	First complete structure of a protein (insulin), by *Sanger*.
1954	Sliding filament model of muscle contraction, by *Huxley* and *Huxley*.
1957	Discovery of an organism using RNA as its genetic material (RNA virus), by *Fraenkel-Conrat* and *Singer*.
1957	Organization of ecological communities formally studied, by *Hutchinson*.
1958	DNA replication method shown to be semiconservative, by *Meselson* and *Stahl*.
1961	Operon concept of bacterial gene control, by *Jacob* and *Monod*.
1961	Chemiosmosis proposed as active transport method in cell membranes, by *Mitchell*.
1961	First cracking of nucleic acid triplet code, by *Nirenberg* and *Matthei*.
1962	Environmental abuses recognized and publicized in the book *Silent Spring*, by *Carson*.
1972	First recombinant DNA made *in vitro*, by *Berg*.
1972	Fluid mosaic model of the cell membrane, by *Singer* and *Nicolson*.
1976	Oncogenes (cancer genes) found to be normal components of cells, by *Varmus*, *Bishop*, and *Stehelin*.
1977	Introns discovered within eukaryotic genes, by several workers.

1977	Methods for finding the base sequence of DNA invented, by *Maxam* and *Gilbert.*
1978	First baby born who was conceived outside the womb (*in vitro* fertilization), *Edwards* and *Steptoe.*
1988	Human genome project (to map all DNA of the human) begun.
1989	First genetic disease gene (cystic fibrosis) cloned by *Tsui* and *Collins* and gene therapy clinical trials approved.

STUDY QUESTIONS

1. Why is the study of biology regarded as a science?

2. Would a scientist be satisfied with the viewpoint that the unknown cause of some phenomenon is unanswerable? Explain.

3. What is the distinction between basic and applied science?

4. What is meant by, "Any statement of fact should always be considered provisional?"

5. Compare these two ways of studying the world—observation and experiment.

6. In an experiment, what are the variables? The control?

7. What is the difference between a hypothesis and a scientific theory? Which is more robust?

8. Outline the steps of the inductive process commonly called the scientific method. Then describe how the hypothetico-deductive method differs and why it may be a more accurate description of scientific method.

2
CHARACTERISTICS AND RELATIONSHIPS OF LIFE

What do we know about:
—*how to decide whether an object is alive?*
—*the chance that life can currently arise from nonlife?*
—*the age of the earth, and of continuous life on earth?*
—*the possibility that ancient life was based on RNA, not DNA?*
—*the possibility of our possessing genes that cause aging?*
—*how the maximum human life span might be increased?*
—*defining the moment of death?*
—*looking for clues of relationships among organisms?*
—*the five kingdoms (or are there six)?*
These and other intriguing topics will be pursued in this chapter.

LIVING ORGANISMS SHOW CERTAIN CHARACTERISTICS

It is paradoxical that the quality known as life is at the heart of biology, yet explaining it is still beyond the capacity of biologists. A complete explanation will have to await the time when life can be created in the laboratory by properly organizing the chemical components of protoplasm (living matter). Biochemists have progressed far in the analysis of the constituent chemicals found in cells and of the reactions in which they participate. Nevertheless, no one can claim total knowledge of the organization and interactions that bind these into an entity that is truly alive.

Regardless of the form of life, whether a bacterium, tree, or an elephant, there appears to be a remarkable uniformity in its chemical composition. A good generalization is that organisms look more alike when examined nearer the molecular level, and more diverse when examined at organizational levels more complex than molecules. Another way of saying this is to note that there are now approximately one million recognized species on earth, but that all of these diverse types of organisms utilize identical (or very similar) molecules and do very similar things with them.

What are the functional and structural characteristics of living organisms that make them different from nonliving matter? Let us begin a list.

1. Certainly one of the most striking things about any organism is that it is *highly organized.* Furthermore, this organization is arranged in a hierarchy. A convenient way to describe this is to examine the following list, arranged from simplest to most complex.

   Levels of organization, arranged hierarchically:

   submolecular chemical structures
   molecules
   subcellular structures (organelles)
   cells
   tissues
   organs
   systems
   whole organism
   interactions among organisms

   Within this hierarchy, each component or level forms the next (more complex) component.

   Each of these components will be examined in detail in following chapters, but it is appropriate at this point to discuss more fully one of them—the cell. One of the most fundamental statements in biology is that which is usually called the *Cell Theory.* It actually consists of two related statements. First, virtually all living matter is composed of one or more cells. (Viruses are a possible exception; it can be argued that they are nonliving since they cannot exist apart from the cells in which they reproduce.) Furthermore, any cell alive today has its origin in a preexisting cell. Division of many organisms into small cellular units should not be interpreted as a separation into simple units. The cell has a complex arrangement in which there is a division of various processes among its components, as will be demonstrated in several later chapters. Cells of multicellular organisms perform processes common to all cells, plus additional processes for which they are specialized. For instance, a muscle cell carries on many self-maintaining activities that all cells must do and also uses specialized structures unique to itself that perform the activity of contraction.

2. Another characteristic of living organisms is that they *capture, store,* and *utilize energy.* For plants and plant-like microorganisms, this involves intake of pure energy in the form of light and then conversion of it to the bonds of foods that are synthesized from simpler chemicals. For all other organisms, the matter entails capturing (eating or absorbing) food

containing energy that can be ultimately traced to sunlight. Usually, energy-holding molecules are stored for some period of time after being made, either in their original form or after being converted to more stable (and usually larger) molecules such as polysaccharides or fats. For all organisms, the eventual utilization of energy involves a complex series of chemical reactions that break the bonds of food molecules and thus release the energy for use.

3. Some of the energy that an organism captures will be used in *growth and reproduction*. Overall growth of a multicellular organism often involves reproduction on the cell level—each cell in a region splitting to become two, followed by increase in size of each daughter cell. Reproduction of an entire organism can be *asexual* (splitting off or budding of a portion of a mature organism) or *sexual* (cooperative combining of components from two mature organisms to form a new individual).

4. Another characteristic of living organisms is that they are *responsive to variables in the environment.* That is, they are capable of sensing changes in their environment, then they do the appropriate thing to ensure their continued health. Obviously, organisms vary considerably in the kind and degree of response. We are most familiar with the quick interaction of nerves and muscles in animals, but there are slower responses too. For example, sunlight and gravity influence the direction of growth in trees. Animals such as the sponges, which have no nerve cells or muscles, reproduce in response to a favorable combination of environmental conditions.

   Appropriate responsiveness also occurs *within* an organism, since the internal environment constitutes a source of variability. The condition of being able to maintain optimum internal conditions is termed *homeostasis*. An example of a homeostatic mechanism is the constant monitoring of blood acidity or alkalinity (pH) in a person, with chemical changes occurring to move it in the appropriate direction when it deviates from a narrow (optimal) range.

5. The examples of response described above are all short term; that is, they are rather quickly reversible changes during the course of a single organism's life span. There is another sort of responsiveness to the environment, which involves an entire population of organisms and extends over more than one generation. This property of all organisms is termed *evolutionary adaptation*. If a population includes organisms that differ somewhat from each other, and if these differences enable some form(s) to have a greater chance of surviving and reproducing than others, then the conditions are right for the composition of the population to change as generations go by. Since the expected change makes the population (as a whole) more likely to survive and thrive, this long-term responsiveness is a useful *adaptation* to an environment.

ONLY ORGANISMS CAN PRODUCE ORGANISMS

Until it was disproved in a series of experiments beginning in the late 1600s, the general belief was that life can constantly arise from dead matter in a spontaneous fashion. Mice, insects, snails, and other forms of animal life were supposed to come from mud or putrefying matter. In 1688 Francisco Redi put pieces of meat into containers, some of which he left open, some covered with netting fine enough to exclude flies, and some covered with paper. The meat decayed in all the containers, but maggots developed only in the open ones. Flies were attracted to the meat beneath the netting but laid their eggs on the netting when they could not get to the meat. The eggs hatched into maggots on the netting. He concluded from these simple observations that maggots came from insect eggs, not from dead meat. Among others supporting Redi's views was Louis Pasteur, who with his sterilization experiments in the 1860s clinched the belief that life must come from preexisting life. He proved that bacteria had to be introduced into sterilized (nonliving) materials before decay could occur, that they did not originate spontaneously within them. In combination with Redi's work this ensured the downfall of the idea of spontaneous generation.

HOW DID THE FIRST LIFE APPEAR ON EARTH?

It is appropriate now to address the question of the *original* beginnings of life on earth. There is evidence that the ancient earth did not have nearly the same conditions of atmosphere and temperature that we see now. Therefore, the ancient appearance of living organisms from nonliving materials within that different environment cannot be ruled out.

The History of Life on Earth is Very Long

What do we know about the age of the earth, and when life first appeared? Most scientists agree that the earth formed over four billion years ago. Our way of tracking the story of life on earth rests mostly on fossilized remains of organisms and their artifacts, usually to be found in mineral formations. Over the last 50 years, several reliable methods have been developed that can provide an age estimate for rock formations in which fossil imprints are found, or for actual remains of organisms. These methods are usually based on the physics of the decay rate of radioactive atoms (see Chapter 15). With the help of these techniques, two important statements have become generally accepted. First, life on earth has existed for a very long time. The most ancient fossils currently known have been dated as being about 3.5 billion years old.

Second, those oldest materials were provided by very simple organisms, which nevertheless had shapes similar to those of some modern organisms. More specifically, the oldest known organisms were bacteria, which are the simplest of the modern types of cellular organisms. (However, recent findings suggest that single-celled eukaryotic organisms, those that are not bacterial, have ancestry going back to approximately the same time.)

We Can Speculate on Primitive Conditions for First Life

Of course, it is very difficult to know what the earth and its atmosphere were like at a time so far removed from the present. Based on knowledge of modern emissions from volcanoes, and assuming no photosynthetic activity (a property of some living organisms), the atmosphere of a lifeless earth would probably have had carbon dioxide (CO_2), carbon monoxide (CO), hydrogen sulfide (H_2S), hydrogen cyanide (HCN), gaseous water (H_2O), and gaseous nitrogen (N_2), but very little free gaseous oxygen (O_2). The surface temperature would be expected to be very hot, but cool enough for some liquid water to collect. Note that, while the hypothetical mix of materials at the ancient earth's surface is different from that of the modern earth, the fundamental rules of chemistry and physics by which the materials would interact are not presumed to be different from those that we currently see operating.

The question is, would such an inhospitable setting be suitable for a series of spontaneous events to occur that would culminate in living cells capable of reproducing and becoming the ancestors of all the life we see now? We will never know for sure whether such things actually happened. We can only devise experiments that test whether they could have happened, with certain suppositions being made. One way to do this is to set up a controlled micro-environment that mimics what we think the earth's surface might have been like, then observe what happens. Many such experiments have been done since the early 1950s. They indicate that very simple molecules can combine to make many of the complex molecules (amino acids, sugars, nucleotides, and polymers of these) that we know must be part of living organisms. Furthermore, these large molecules can sometimes be found in organized structures not unlike those found in cells.

Perhaps the most important class of molecules that would have to be formed in order to ensure a successful emergence of life is a polynucleotide used to store and release information. Although most modern forms of life use DNA for this purpose, there is much support for the idea that the most primitive life would have relied upon the closely related RNA. The major evidence for RNA is that it can catalyze chemical reactions much as protein-based enzymes do. Thus, in a primitive situation where only a minimum of molecule types would be present, RNA could do double duty by holding information for reproduction of cells and also catalyzing important reactions.

In summary, there are intriguing clues in modern settings that are not inconsistent with the idea that living cells could have formed from natural processes on the surface of a primitive earth. However, it must be said that other existing hypotheses cannot be ruled out unless physical evidence is found to do so. Among these is the long-held belief in many societies that at least some steps in the beginning of life were supernatural events directed by a Supreme Being. Another idea is that life on earth arrived from elsewhere, perhaps by way of comets or meteorites. A related, less encompassing hypothesis is that life formed on earth, but was nudged toward the critical moment by the arrival of complex organic molecules via comets or meteorites. In either case, one is still left with the question of how life or its preceding molecules might have formed on some other celestial body. It is obvious that we are far from resolving the fundamental question of the origin of life.

ORGANISMS UNDERGO THE PROCESSES OF EMBRYONIC DEVELOPMENT, MATURING, AND AGING

When uninterrupted by accidents, individual life spans pass through distinct, distinguishable phases. For multicellular organisms, the first phase is that of the embryro, which is probably the period of greatest change. During embryonic life, the organism undergoes rapid cell division (going from one cell to as many as trillions), growth in overall volume, movement of cells to form tissues and organs of proper shape and position, and specialization of cells to begin their useful functions for the body. For some organisms, such as the mammals, this period ends at birth, when the young animal becomes physically separated from its mother. For many other organisms, however, the entire embryonic development has occurred away from the mother, and the next phase is signalled by some other significant event such as hatching (some fishes, many invertebrate animals), or rapid growth of already specialized parts (many land plants, with growth of stems and roots). In all cases, embryonic life is followed by a period of maturation that culminates in the organism's becoming capable of reproducing. Finally, in many (but, intriguingly, not all) species there is an aging process that ends in death.

Any discussion of aging should begin with the reminder that most individual organisms in their natural habitat are killed long before they have the opportunity to age. For instance, it has been observed that 99 percent of wild field mice die before they are one year old (from disease, starvation, or the action of predators), even though they would be likely to live up to four years in the protected environment of a laboratory or pet store. Even though field mice can demonstrate aging, virtually none actually do so in a natural setting. Some people have speculated that aging (leading inevitably to death) is an inherited mechanism whose function is to weed out organisms that have already passed their genes to the next generation and are taking up valuable space and resources. However, the observation that aging almost never has the chance to occur in nature would seem to reduce the force of this argument, and make it less likely that there are aging genes or death genes.

Some Organisms Seem to Avoid Aging

Some organisms never seem to age, even if given the chance. Many trees continue growth and reproduction for their entire lives, with only the ravages of physical and biological attack causing death. Bristlecone pines are estimated to live longer than 4,000 years, with no apparent natural limit. As an example in the animal realm, some fish species appear to have no innate age limit. Single-celled organisms, such as bacteria and protozoans, do not age if given the chance to reproduce. The only reason that a population of bacteria does not increase forever is that its members eventually run out of nutrients and space to occupy.

Many Animals Have Characteristic Life Spans

On the other hand, vertebrate animals demonstrate maximum life spans that are characteristic of their species, when given the chance to avoid rigors of the wild. Among long-lived vertebrate animals are tortoises, known to live over 200 years; carp, 150 years; falcons, 160 years; eagles, over 110 years; parrots, 100 years; and crows, 70 years. Among domesticated animals, dogs may live more than 20 years (average 10); cats, more than 30 years (average 14); cows, 30 years; and horses, 50 years. The typical length of human life in developed countries has increased dramatically the last 100 years, a tribute to our increasing knowledge of nutrition, disease control, and surgery.

In 1900, it was estimated that 63 percent of caucasian American males who could have reached the age of 40 had actually done so. In 1940, 85 percent had succeeded in living to that age. However, there is no indication that the *maximum* human life span has increased. Almost no survivors exist past the age of 100 years, and the very small fraction doing so is gone before many more years have passed. (Currently, no one has been proven to have lived over the age of 121 years.) Over the past century (when reliable records were first widely available), this maximum has remained steady. It would appear, therefore, that many organisms have a natural aging process and a resulting maximum life span that are heritable along with all of their other species-characteristic shapes and functions. What is not known at this time is the exact cause of this inevitable loss of ability.

There Are Multiple Hypotheses on the Cause of Aging

Within an organism some cell types continue to undergo reproduction throughout life, and in some ways retain a youthfulness that nonreproducing cells gradually lose. For instance, within a tree the growing stem and root tips and cambium (the growing layer that increases the diameter of stems and roots) remain perpetually capable of reproducing. In the human, actively dividing unspecialized cells like those in the lower layer of epidermis and blood-forming tissue do not show the usual signs of aging that eventually appear in nondividing cells.

When normally nondividing cells are transferred to the artificial environment of a culture vessel, and given unlimited nutritional supplies and room to grow, they often revert to a relatively undifferentiated condition and go into rapid reproduction just like their embryonic ancestor cells. However, even here there is often a change that eventually occurs—many cultured cell populations stop reproducing when about 50 cycles of mitosis have occurred, after which the cells begin to show signs of aging.

Some feel that aging is intimately tied to metabolic rates—the rates at which the multitude of chemical reactions are occurring within cells. During the summer, when honey bees are most active, the workers live from three to six weeks; yet they survive much longer periods during the winter when they are relatively inactive. In temperature-regulating animals, including the birds and mammals, there seems to be some correlation between speed of living, size, and length of life. During the day the tiny hummingbird reaches a peak of metabolic activity among the highest known for any animal. So much energy is expended that it feeds almost continuously to keep from starving. The bird would die of starvation at night were it not to go into a state of hibernation during which its metabolic rate is strikingly reduced. Hummingbirds live only a few years. On the other hand, the long-lived elephant has a relatively low metabolic rate.

Perhaps related to the metabolism hypothesis is the observation that rats with significantly reduced access to food tend to live as much as 30 percent longer than those given unlimited amounts. The most prominent physiological effect of semistarvation in this species is the prolongation of the juvenile period of life (retardation of sexual maturity); it is this extension that leads to a longer total life span. Thus, this study may be a stronger link between aging and reproduction than between aging and metabolic processes.

Another possible critical factor in cellular aging is the accumulation of toxic materials that inevitably occurs in a very active cell. Such molecules could act directly on vital cell processes, or indirectly by increasing the rate of mutations in vital genes. This is an attractive idea, but no specific occurrence of such accumulation has been documented.

Currently, the most popular model for cellular aging is based on accumulation of errors in genes. Rather than postulating the existence of genes whose function is to cause aging, this is an assertion that all genes necessary for normal life gradually become defective because they undergo attack during the course of life. Errors accumulate and genes become less efficient in directing cell activities. It is well known that a variety of physical and chemical insults occurs at the gene level, and that they can come from outside the organism (e.g., X rays, asbestos), or from within (free radicals produced in every cell). If errors in genes can occur at any time, and if they are sometimes kept as long as the genes are present, then it is logical that more and more would accumulate as the cell grows older. It is probable that a number of defects have to be present simultaneously before a cell or a whole organism shows the classical signs of aging and then dies.

Can cells detect gene errors and correct them? There are several mechanisms in a normal cell for this important activity. It is possible that these maintenance and repair mechanisms gradually lose effectiveness, and that errors eventually accumulate. This idea can be linked to the concept of natural selection. It would be in the best interest of a species to attempt to repair errors up to the time of reproduction, but there would be no usefulness for a repair mechanism to continue vigorous activity after genes have been passed along to new individuals, since the continuity of the species has already been ensured by successful reproduction. Thus, according to this hypothesis, natural selection would favor an error-correcting mechanism only up to the point of successful reproduction. Then there would be no selection operating for organisms to continue protecting their genes after that point. If such a view is correct, a strategy to elongate the normal human life span would involve finding ways to decrease the damage to genes, or to increase the efficiency of repair mechanisms, or to prolong the length of time that repair mechanisms continue to work well, or a combination of all of these. Of course, the sociological, psychological, and demographic consequences of significantly increasing the life span of humans would be matters of much debate, should such biological methods come into existence.

DEATH IS A DIFFICULT EVENT TO MONITOR

Drawing a sharp line between life and death is a problem. The glib response when asked to define death is that it is the cessation of those activities that occur in a living organism: gaining and using energy, reproducing, responding to environmental changes, and the like. However, there are many exceptions. Would a sterile person be considered dead, just because of an inability to reproduce? Would a portion of a body be considered dead just because it could no longer respond to nerve messages?

On the level of the individual cell, the moment of death is difficult to document, because a cascade of several events is likely to occur. If, for instance, a specific poison enters the cell and causes cessation of vital energy-releasing reactions, it may take some time for consequences to appear. Minutes or hours later there will be important changes, such as lack of movement and breakdown of intracellular structures. Eventually it becomes quite apparent that the cell has died, but death came by a series of events over a period of time.

A multicellular organism's death may be even more difficult to document. Large portions of a tree may be dead when examined on the level of individual cells, but enough of it might be operating to allow energy conversion, growth, and reproduction to continue indefinitely in certain portions. Progressive loss of function might continue for decades before the entire plant is truly dead.

Of course, the definition of human death is very important to us. For many centuries, the critical signs to be looked for were cessation of heartbeat and breathing. We now know that a person can survive several minutes in the absence of these activities, and there are documented cases in which a drowned person revived after more than twenty minutes without breathing (helped by lowered oxygen need in extremely cold water). It is now apparent that various portions of the body do not die simultaneously under the same conditions. At normal body temperature cells of the brain (requiring much more oxygen than most other cells) begin to die within a minute after being deprived, while muscles (including heart muscle) can survive for much longer periods. Because the human brain controls so much of the rest of the body's functions, its irreversible death has come to be considered the end point of a person's functional life, even if other organs can be artificially maintained alive.

SYSTEMATISTS STUDY HOW ORGANISMS ARE RELATED TO EACH OTHER

Throughout this chapter there has been ample evidence that all living organisms have a number of features in common. But of course we also see many strikingly different characteristics, by which we distinguish them and apply names for identification. The scientific process of placing organism types into appropriate categories, and using that information to understand relationships and ancestries, is carried out within the discipline called *systematics*. Many other fields of biology depend greatly on the work of systematists, who provide a grand scheme of the diversity of life within which various specialized studies must find their place.

Even at first glance it appears that systematizing the great number of organisms living on this earth is an impossible task. Indeed, we now believe that there are many more kinds of organisms out there than we ever believed before, and that it may not be feasible to catalog all modern species. Nevertheless, the rewards, both theoretical and practical, of gaining as much knowledge as possible about biodiversity make the attempt worthwhile.

Structural and Functional Similarities Are Clues in Deciding Relationships

The best way to proceed is to see which organisms share characteristics that might reveal some degree of kinship. One can quickly see that some organisms resemble each other more than others. A horse is obviously more closely related to a cow than to a snake, just as a peach tree is obviously more closely related to a cherry tree than to a daisy. These similarities make possible the lumping of large numbers of organisms into relatively few categories based on major characteristics that they have in common.

Systematists do not agree totally on all aspects of classification, but as time passes they are gradually constructing a natural system based on relationships. They replace artificial groupings based on superficial resemblances by natural groupings based on genuine similarities arising from common origins. This branch of biology takes its clues from many sources, such as gross anatomy, cellular and molecular structures, embryological development, and fossil records.

The easiest way to begin is to look at obvious anatomical structures of adult specimens. In many cases, however, visual evidence of actual kinship is obscure, so biologists are often forced to turn to procedures in physiology, embryology, and genetics to clarify some of the less apparent relationships. An example of one that puzzled zoologists for a long time is the true position of the strange-looking sea squirts that grow attached to surfaces in the oceans. At one time they were classified with clams and snails. Finally someone studied their development and discovered that the larval stage has a temporary supporting rod identical to the notochord in the chordate animals. Then they were correctly placed with humans, fishes, and other chordates.

A pitfall to avoid is drawing conclusions from superficial appearances. The whale may appear similar to a fish, or the bat may appear similar to a bird, but both whales and bats are mammals. Their body forms are adaptations to the environment in which they live. The Spanish moss of the southern United States may be mistaken for a true moss until a closer examination reveals it to be a flowering plant belonging to the pineapple family. There are numerous cases in which animals are mimics, such as the insect that looks like a rose thorn and a type of fly that looks like a stinging bee.

When studying seed plants, one may find temporary use for artificially grouping them into such categories as trees, shrubs, and herbs, or annuals, biennials, and perennials, but these are based on some conspicuous characteristics with no regard for their natural positions. These groups are useful to describe body form or duration of life but are fallacious indicators of any common ancestry. Sometimes, subtle features are the most accurate indicators of relationships. For instance, the locust or mimosa trees, by virtue of their size and general appearance, would seem at first to be close relatives of oak or maple trees, but they have entirely different kinds of flowers, and are properly placed in the pea family with such un-treelike plants as the clovers and beans.

THE FORMAL CLASSIFICATION OF ORGANISMS IS A PRODUCT OF SYSTEMATISTS

Very little was done toward finding a workable method of classification until Linnaeus (1707–1778), a Swedish naturalist, laid the foundation of a new system. Until his day, attempts at placing organisms into categories were almost entirely artificial. Even Linnaeus did not have the ideas of origins and kinships as they are understood today. Nevertheless, his system of using four groups—*class*, *order*, *genus*, and *species*—provided a nucleus for a really workable system. His four categories, plus some additional ones, are still in use. A major contribution of Linnaeus was the establishment of a system of giving to each living organism a two-part Latin name, a *binomial*.

The first word in a binomial is the organism's genus, and the second is its species. For instance, a common North American butterfly is in the genus *Danaus*, along with many related butterflies, and its species name is *plexippus*. The butterfly's binomial is, therefore, *Danaus plexippus*. Its genus name indicates relationship with other butterfly species that share many characteristics, and the species name provides a unique designation for this member of the genus. Perhaps you know this butterfly by its common English name, the Monarch. In everyday usage among English-speaking North Americans that name suffices, but someone outside North America may recognize another butterfly by that same common name. A Spanish speaker in Mexico, where the butterfly is common, will not be familiar with the English name, and a number of Native American languages undoubtedly have their own words for this butterfly. Upon reflection, it becomes obvious that assigning a unique internationally recognized binomial to this animal greatly enhances communication, at the same time that it provides clues to relatedness.

The following outline shows the major categories (*taxa*; singular, *taxon*) used in the modern system of classification. The outline also includes the classification of one common North American plant and one animal as illustrations.

CATEGORIES	WHITE OAK	HUMAN
Kingdom	Plantae	Animalia
Phylum (pl. Phyla)*	Tracheophyta	Chordata
Class	Angiospermae	Mammalia
Order	Fagales	Primates
Family	Fagaceae	Hominidae
Genus (pl. Genera)	*Quercus*	*Homo*
Species (pl. Species)	*alba*	*sapiens*

*The term *division* is used for plants.

The largest numbers of individuals are included in the upper categories because they need to have fewer characteristics in common. Those characteristics are considered of major importance and apply to all individuals in lower categories. For example, the fundamental characteristics of a kingdom are found in all groups within that kingdom. Or to express the relationship differently, all the lower groups are included as a part of a higher group because they have certain characteristics in common. Many systematists consider that Kingdom Animalia is composed of more than fifty phyla; Kingdom Plantae is composed of ten divisions. At the opposite end of the scale are the smallest categories, where individual types of members are fewer but where their members have more characteristics in common.

One might be wondering why not all systematists agree on, say, the number of animal phyla. It should be kept in mind that we do not yet know exact relationships, and that the same evidence might be interpreted differently by two people. One expert might consider a particular characteristic as being very important and assign all organisms carrying it to a unique phylum. Another person, looking at the same characteristic, would consider it fundamental enough to delineate a class but would not give it a separate phylum designation. Keep in mind the idea that science is a dynamic area of human endeavor, which is constantly being corrected as additional evidence and insights accumulate.

Organisms Are Either Prokaryotes or Eukaryotes

An important set of characteristics has been used to divide all cellular organisms into two categories, *prokaryotes* and *eukaryotes*. These are not part of the formal hierarchy shown in the list above, but help to place organisms into that system. The cell of a prokaryote *lacks* two important characteristics that are found in any eukaryotic cell—a membrane surrounding its genetic material (a nuclear membrane) and a set of other structures that are also membranous (mitochondria, endoplasmic reticulum, etc.). These and other less easily observed differences lead biologists to conclude that modern prokaryotes and eukaryotes should not be considered to be very closely related. Indeed, while most think that prokaryotes preceded eukaryotes, there is some recent evidence that the eukaryotic line has just as long a history on earth, and may not be derived from the prokaryotes.

Kingdoms Are the Largest Classification Categories

The highest ranking formal category of classification is the kingdom, which encompasses many organisms sharing very significant characteristics. Even at this level it is difficult to decide how to place organisms in a way that accurately depicts their degree of similarity and their lineage. Historically, the first way to do this was to use just two kingdoms, Plantae and Animalia. Eventually, systematists realized that more kingdoms would have to be designated in order to accomodate those organisms that were clearly different from either plants or animals. Since 1969, a five-kingdom system has been the one most used by biologists, but many now argue for the addition of at least one more kingdom. These six kingdoms are introduced below.

1. The smallest kingdom is *Archaea*, composed of bacteria-like single-celled organisms (lumped with the true bacteria in a kingdom called Monera if a five-kingdom system is used). These and the true bacteria

are the only known prokaryotic organisms. Most members of Kingdom Archaea are found living under extreme conditions—high heat, low pH, or great saltiness. Molecular data published in 1994 indicate that modern Archaea may be more closely related to the eukaryotic kingdoms than are the true bacteria.

2. If one prefers a six-kingdom system, the true bacteria are placed in Kingdom *Bacteria*. These are the more common prokaryotes, of which about 4,000 distinct species have been identified (although many more are probably awaiting our discovery).

3. Kingdom *Protista* includes the simplest modern eukaryotic organisms. However, it should be pointed out that even these are far more complex than the prokaryotic organisms. One indicator of this is that a typical algal cell, a Protistan, contains about ten times as much DNA as a typical bacterium (although both phyla show much variation from these averages). There is some controversy as to whether certain organisms belong in Kingdom Protista. Certainly all single-celled eukaryotic organisms are Protista. This includes many kinds of algae and all of the organisms called protozoa. Organisms about which there is difference of opinion include the multicellular algae (often placed in Kingdom Plantae) and some fungus-like organisms (sometimes placed with multicellular organisms of Kingdom Fungi). Well over 60,000 species of Kingdom Protista have been identified.

4. Kingdom *Plantae* includes all of the multicellular organisms that carry on photosynthesis to capture energy (but see the controversy over multicellular algae, above). Of the approximately 265,000 identified plant species, fully 235,000 are the familiar ones that reproduce using the structure called a flower. Nonflowering plants include mosses, ferns, and others.

5. Most of the multicellular organisms that cannot perform photosynthesis, but must rely on gathering food molecules from outside the body, are in Kingdom *Animalia*. In addition to their method of capturing energy, animals differ from plants in a number of other features. This indicates to biologists that the ancestries of animals and plants diverged a very long time ago.

6. Kingdom *Fungi* includes a wide variety of organisms that were once considered to be plants. Some of its members are mushrooms, yeasts, and molds. The fungi were originally lumped with plants because of their immobility and plant-like external morphology. Closer examination revealed an inability to perform photosynthesis and many other biochemical differences that demanded a unique kingdom for these organisms. Still, many biologists considered them to have descended from the plant kingdom. In the early 1990s several independent studies at the biochemical level led to a rather startling conclusion: the fungi appear to be more closely related to members of Kingdom Animalia. According to the best data now available, fungi and animals may

share a common Protistan ancestor, and the fungi can no longer be considered as degenerated or specialized offshoots from Kingdom Plantae or from photosynthetic Protista. Over 100,000 fungal species have been classified.

The Species Is the Most Fundamental Classification Unit

The taxonomic category called *species* (singular and plural) is the narrowest. To the nonscientist, it is usually equivalent to what is meant when the word kind is used. It may come as a surprise to learn that this most fundamental of all taxa is also one of the most difficult to define, and is the subject of much debate among biologists. Historically, the species has most often been defined as a group of similar individuals, usually differing little from one another and differing distinctly from members of other species. Under this definition, species have generally been named and described on the basis of clearly visible external characteristics and (less often) internal anatomy. However, more recently a biological (rather than anatomical) criterion has become the norm. The biological definition of a species is *a group of natural populations whose organisms can successfully reproduce among themselves, and cannot successfully reproduce with organisms of other groups*. The term *successfully reproduce* means that reproduction occurs to bring an appropriate number of offspring into existence, *and* that those offspring themselves are able to reproduce. Some closely related but distinct species can reproduce (producing a hybrid organism), but this hybrid will be sterile. Some systematists hesitate to place two groups into a single species if reproduction occurs but produces an abnormally small number of offspring compared to the number resulting from mating within a group.

From the foregoing paragraph, it becomes obvious that the biological definition of species has potential for being more clear-cut, since it provides a test that can be put to any groups—the test of reproduction. However, this method may be very expensive in terms of time and money. Therefore, it is not surprising that the great majority of newly described species is still defined by the less rigorous criteria of anatomy.

The species may sometimes be divided into *subspecies, variety, form, breed, race*, or *strain*. These divisions are not always clearly definable and may sometimes refer to the same thing. The subspecies is used by some biologists to separate variants of a species having a distinct geographical distribution. Variety often means the same thing as subspecies, but it is also used in special senses for discontinuous variants of an interbreeding population or for variants horticulturally produced. Breed is used in connection with dogs and other domesticated animals; strain in connection with crops having physiological differences such as disease resistance; and race in connection with humans. All categories below that of the species are subject to much interpretation and are to be avoided unless their use clearly helps in identification or in showing relationships.

STUDY QUESTIONS

1. What are the two general statements that comprise the Cell Theory?

2. List the recognized levels of organization found in a living organism, beginning with the smallest (submolecular structures) and progressing toward the largest.

3. What are good examples of quick responses to the environment, and slower ones? Can responsiveness of living organisms extend beyond individual life spans?

4. According to current scientific estimates, how long did the earth exist without any form of life?

5. Why do some molecular biologists suggest that this was once an RNA-based world of life?

6. Has the typical human life span increased during the past century? Has the maximum life span done so?

7. Describe the hypothesis relating cellular aging to accumulation of genetic errors. Do cells have ways of correcting such errors?

8. How does a natural system of classification differ from an artificial one?

9. From what fields of biology has information been obtained to clarify relationships among organisms? From what branch of biology was information obtained to clarify the taxonomic position of sea squirts?

10. What naturalist was responsible for laying the foundation of the system of classification in use today?

11. For every identified organism, a binomial name is given. What are the classification categories shown by the binomial?

12. How could one distinguish a eukaryotic cell from a prokaryotic cell? Which kingdom or kingdoms include(s) prokaryotes?

13. Why is the number of kingdoms still controversial among systematists?

14. Kingdom Fungi was once part of Kingdom Plantae. Why was it necessary to split it away?

15. What is the biological definition of species? Why is this definition sometimes hard to apply to real organisms?

3
PHYSICAL AND CHEMICAL FOUNDATIONS OF BIOLOGY

What do we know about:
—*the fundamental components of life—atoms and molecules?*
—*the relative abundance and importance of various elements in organisms?*
—*some fundamental rules for atoms bonding together to make molecules?*
—*the great importance of hydrogen bonds in building biologically important molecules?*
—*water's properties that make it an almost uniquely versatile molecule for life?*
—*the concept of pH, and the effect of pH upon cell activities?*
—*how carbohydrates, lipids, proteins, and nucleic acids are shaped?*
—*the roles of vitamins in maintaining health?*
These and other intriguing topics will be pursued in this chapter.

CELLS AND ORGANISMS FOLLOW PHYSICAL AND CHEMICAL LAWS

A living organism has some striking properties that make it different from the nonliving environment around it, as has been described in the preceding chapter. However, all organisms and the cells of which they are composed are built from the same fundamental physical objects as are all other materials in the universe. Furthermore, the behavior of these objects is everywhere governed by the same set of rules. Thus, all life scientists must be good synthesizers, understanding and using the concepts of the fields of physics and chemistry in addition to those unique to biology. This chapter provides some necessary information about elements and how they interact to form the molecules necessary to support life.

ELEMENTS AND ATOMS
ARE THE FUNDAMENTAL UNITS
OF CHEMICAL STRUCTURES

At the present time there are 109 known *elements*, several of which appear to be made only with human intervention. Some of these elements—for instance oxygen, carbon, gold, calcium, or uranium—are very familiar to us. As the word implies, any such entities are elementary substances and cannot be broken into simpler materials by chemical means. However, they can interact with each other to form innumerable combinations.

The smallest units of elements are *atoms*. An atom consists of a *nucleus* around which are *electrons* at various energy levels. The nucleus of the atom is composed of positively charged particles called *protons* and uncharged particles called *neutrons* (not present in hydrogen); the electrons outside the nucleus are negatively charged. Each of these three subatomic particles can be further subdivided into smaller units, but those most elementary particles of all are not often the concern of biologists and will not be discussed here.

An atom of the simplest element, hydrogen, consists only of one proton and one electron. Others have additional protons, neutrons, and electrons. As one looks through the periodic table (a chart listing the elements in a logical and orderly arrangement), each succeeding element has one additional proton in its nucleus, therefore one additional positive nuclear charge. The nucleus of each element larger than hydrogen also contains at least as many neutrons as it has protons, but it may sometimes contain more. The different forms of any particular element, which are the result of their having a different number of neutrons in the nucleus, are *isotopes* of the element. For example, the most common isotope of oxygen has eight protons and eight neutrons; another (much rarer) isotope of oxygen has eight protons and nine neutrons.

In a neutral atom the number of (negative) electrons is the same as the number of (positive) protons. An atom can gain or lose one or more electrons, in which case it is no longer electrically neutral. Such a charged atom (or a group of atoms of which it is a part) is called an *ion*.

Elements are usually symbolized by capital letters that are the first letters of their names. Examples are O for oxygen, C for carbon, N for nitrogen, P for phosphorus, H for hydrogen, and S for sulfur. Since the names of several other elements begin with the same letters, two letters of a name, often the first two, are used; for instance, Ca for calcium, Cl for chlorine, or Si for silicon. In still other cases, an abbreviation of the Latin or Greek name for the element is used—for example, Hg for mercury or Ag for silver.

Of all the elements known, only a few comprise the bulk of atoms normally found in living cells. Four elements—carbon, hydrogen, oxygen, and nitrogen—comprise about 95 percent (by weight) of the typical cell. Two others, phosphorus and sulfur, are found in much smaller amounts but are also very important within living cells. Together, these six elements can be remembered by the acronym CHNOPS, made from their chemical symbols.

One should keep in mind that the major importance of these elements (and many others found in trace amounts or in specialized tissues) is that they can be combined to make structures; they often have no biological usefulness by themselves. However, it is useful to know the essential elements in order to supply them when they are deficient. The practice of fertilizing crops and taking minerals in pill form are examples of providing for deficiencies. However, such supplements cannot always be supplied as pure elements but must sometimes be in combination with other elements. Pure nitrogen cannot be used by green plants; it must first be combined with hydrogen in the form of ammonia (NH_3) or with oxygen in the form of nitrites ($-NO_2$) or nitrates ($-NO_3$). Similarly, human anemia due to iron deficiency cannot be corrected by taking pure iron; it must come from certain compounds of iron.

CHEMICAL BONDS CONNECT ATOMS TO MAKE MOLECULES AND COMPOUNDS

Most atoms cannot exist singly; they must combine with one or more other atoms in order to be stable. When two or more atoms (same or different elements) chemically combine to become stable, the resulting group is called a *molecule*. (Technically a single atom of the sort of element that can exist independently, such as helium, is also a molecule.) If a molecule is composed of two or more different elements, such as H_2O, it can also be called a *compound*.

Electrons Are the Most Important Atomic Particles for Making Bonds

Since molecules, rather than individual elements, are the fundamental building blocks of cells and organisms, it is necessary to carefully examine how they are put together. Most interactions between atoms involve their electrons, so it is necessary to examine the arrangement of electrons around an atom's nucleus. The exact position of any electron at any moment cannot be determined, but they tend to place themselves at somewhat predictable distances from the nucleus. A simplistic, but useful, way to imagine the arrangement of electrons is to consider that they occupy regions called shells, which are arranged concentrically around the nucleus (Fig. 3.1). Electrons occupying outer shells carry more energy than those orbiting closer to the nucleus. If the number of electrons of an atom is two or less, they occupy the same shell, quite near the nucleus. An atom with three through ten electrons places two in this innermost shell and the remaining electrons in a shell farther out. The third and subsequent shells, even farther from the nucleus, can also hold up to eight

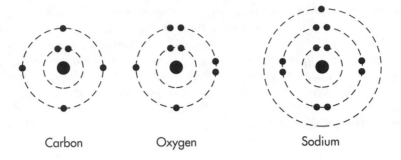

Carbon   Oxygen   Sodium

FIGURE 3.1.   *The arrangement of electrons in atoms of some elements important to living organisms.*

electrons each. Thus, as shown in Figure 3.1, an atom of the element chlorine places two of its seventeen electrons in the shell of lowest energy, eight more in the next shell outward, and the remaining seven in the outer shell, where they possess the most energy.

The atoms of nearly all elements have their outermost occupied shell unfilled. They are not stable when this is their condition and they tend to undergo reactions leading to a change in this condition. An atom whose outermost shell has four or more electrons tends to react with another atom in such a way that it fills the shell to capacity. For instance, an atom of chlorine reacts with another atom in some way that leads to chlorine's gaining the single electron that will fill its outermost shell. Alternatively, an atom can become stable by losing any electron(s) that it carries in its outermost shell. This action converts the full shell that had been next closer to the nucleus into becoming the outermost occupied shell. Such an action is likely to be taken by an atom whose outermost shell is occupied by fewer than four electrons. Whatever the method, when two atoms interact to obtain filled outermost shells, the result is a *bond* between them.

Ionic Bonds Involve Transfer of Electrons

Some atom pairs interact by *moving* an electron from one atom to the other (Fig. 3.2). Since an electron carries a negative charge, the donating atom becomes positively charged and the receiving atom becomes negatively charged. That is, they both become ions, of opposite charge. Since opposite charges attract, the two atoms tend to draw toward each other and remain close together.

This attraction is called an *ionic bond,* and the resulting molecule is called a *salt.* Some salts involve ionic bonding among three or more atoms. For instance, the salt calcium chloride is symbolized $CaCl_2$ to indicate that each of two chlorine atoms forms an ionic bond with one calcium. Each chlorine gives one electron to the calcium, thus filling its outermost shell

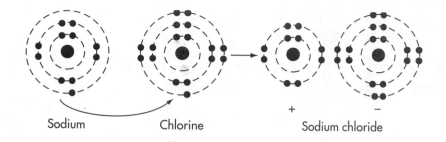

Sodium   Chlorine   +   −
                                                                  Sodium chloride

FIGURE 3.2.   *An ionic bond between sodium and chlorine, brought about by transfer of an electron from sodium to chlorine.*

(which initially contained six electrons). Ionic bonds are relatively stable, as shown by the ability of table salt (NaCl) to remain unchanged over long periods.

Covalent Bonds Involve Sharing of Electrons

An alternative way that atoms interact to provide stability is that of *sharing* electrons, rather than donating or accepting them (Fig. 3.3). A bond that forms because of two atoms sharing one or more electrons is called a *covalent bond.* As with ionic bonding, the involved electrons are those of the outermost occupied shell of each atom. In most cases of covalent bonding, each atom has this shell filled exactly half of the time, since each shared electron can be orbiting only one atom at any moment. If a pair of electrons (one from each atom) is shared, this bond is said to be a single covalent bond; if two pairs are shared, it is a double covalent bond. In structural diagrams of covalently bonded molecules, where there is no need to show each electron, a single covalent bond is symbolized by a straight line (——→) between the involved atoms and a double covalent bond is symbolized by two parallel straight lines (⚌⚌ . Examples of such diagrams

a.   b.

FIGURE 3.3.   *Two examples of molecules held together by sharing electrons (covalent bonding. (a) Molecular oxygen, O_2. (b) Methane, CH_4.*

occur in many portions of this book. As was stated for ionic bonds, covalent bonds are relatively strong.

Polar Covalent Bonds and Hydrogen Bonds Involve Unequal Sharing of Electrons

Some elements exert more attraction toward electrons than do many others. If an atom of such an element becomes involved in a covalent bond, it is likely to hold the shared electron(s) near itself; the sharing will be unequal. Since electrons are negatively charged, this slight inequality will lead to the attracting atom being slightly negative, and the other atom being slightly positive. A covalent bond of this sort is called a *polar covalent bond*, and the molecule in which this occurs is called a *polar molecule*.

Two elements that are very common in molecules of living cells, nitrogen and oxygen, exert a very strong attraction for electrons; therefore, a number of biologically important molecules are polar. In most such situations, the positively charged atom of the polar pair is likely to be hydrogen. If there are two polar molecules quite near each other, the positively charged hydrogen of one can become attracted to the negatively charged atom (oxygen or nitrogen) of the other molecule. When this happens, and the two molecules stick to one another, the bond between them is called a *hydrogen bond*. The most common occurrence of this is the bonding that holds adjacent water molecules together (Fig. 3.4). It is also possible for separate regions of a single large molecule to join by hydrogen bonds within the molecule. An example of this is found within a nucleic acid, shown in Chapter 12. Hydrogen bonds are much weaker than covalent or ionic bonds (on the order of only 5 to 10 per cent as strong) but, as will be demonstrated, they play very important roles in biological settings. In structural diagrams, a hydrogen bond is symbolized by a dotted line.

Molecules Are Either Inorganic or Organic

Molecules are classified as *inorganic* or *organic*. Organic molecules are generally thought of as most molecules containing carbon and inorganic mole-

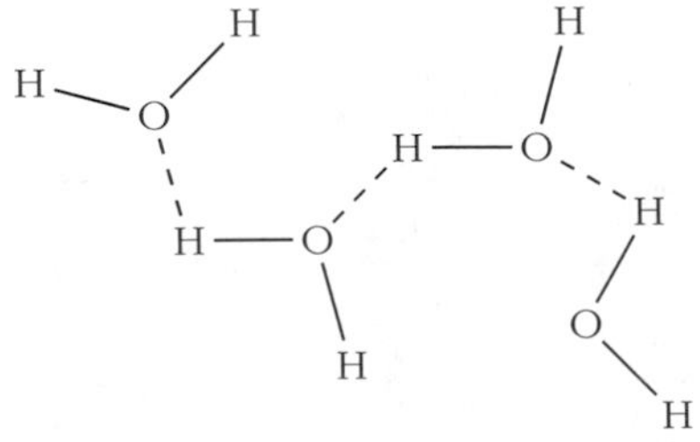

FIGURE 3.4.   *Water molecules are polar and clump together by hydrogen bonding (shown by dotted lines).*

cules as those of other elements and a few carbon compounds similar to earth-like substances, such as calcium carbonate ($CaCO_3$). Organic compounds are so named because of the original belief that they can be made only by living organisms. However, since the early 1800s many thousands of organic compounds have been synthesized in laboratories. Both organic and inorganic compounds are necessary to life.

WATER IS A HIGHLY VERSATILE MOLECULE FOR LIFE

Water, an inorganic compound, is uniquely useful in sustaining life on earth. The following paragraphs provide only some of the reasons for such a statement.

The most prevalent compound in cells is water. The water content varies so much, however, that it is impossible to give an average figure. It is estimated that the entire human body, including the hard materials of bone and teeth, is 65 to 70 percent water by weight. Of course, this varies from tissue to tissue. Dentine of teeth is estimated to be 10 percent water and skeletal muscles 75 percent water. Old or inactive tissue contains less water. Dormant plant tissue may average about 30 percent water. In extreme cases, such as the relatively inactive cells of dry seeds, the water content may be below 10 percent.

Cellular water is important in many ways. It is the medium in which the various subcellular structures are dispersed. Water's polar nature gives it the ability to form a coat over charged molecules, pulling them away from each other. That is, water is a very good solvent, causing a very large number of substances to dissolve. The movement of foods, gases, minerals, and by-products of metabolic activity in or out of cells is dependent on their being dissolved in water. Two important chemical processes, photosynthesis and digestion, depend directly on water as an initial reactant. Hydrogen of water combines with carbon dioxide in the photosynthetic process, thus forming simple organic combinations out of which carbohydrates, fats, proteins, and vitamins are made. Conversely, it combines with complex foods during digestion, thus breaking single large, unabsorbable molecules into smaller absorbable ones.

In addition to being an important constituent of the cell, water occupies spaces between cells as in tissue fluid, blood plasma, and lymph. In some organisms it provides the vehicle for the movement of materials from one place to another. It sometimes acts as a lubricant to reduce friction when parts rub together. In almost all organisms it is a necessary medium for the movement of the sperm to the egg. In all animals, whether they live in water or on land, water is a necessary medium for the normal development of the embryo.

Water protects against extremes of temperature. It has a large capacity for absorbing heat without the characteristic of a corresponding increase in temperature. This is directly related to the hydrogen bonding that occurs among water molecules. As heat enters a body of water, some of it must be used to break these bonds before any increased movement of water can occur. Thus, the water's temperature, which is a measure of the average speed of movement of the molecules, does not initially rise even though heat has been transferred to the water. However, water cools more slowly than many other substances do. The moderating effect of water is beneficial for organisms living in lakes and oceans, and also for organisms living on land near those bodies of water. On a smaller scale, this same tendency of water to warm and cool slowly lends stability to an organism that is mostly water inside.

If any substance evaporates from a body's surface, it lowers body temperature because the molecules that are leaving are those with the greatest energy (heat). Water's possession of hydrogen bonds means that evaporation is slowed and that any water that succeeds in leaving a surface will carry with it a larger amount of heat. Many animals utilize evaporative cooling (induced by sweating, panting, or other actions) to survive at high environmental temperature and to bring body temperature down to normal after exertion. Evaporative cooling is also important in reducing the temperature of tropical bodies of water, making them more hospitable for life.

THE CONCENTRATION OF HYDROGEN IONS DEFINES ACIDS, BASES, AND PH

An important concept in both chemistry and biology is that some molecules *release* H^+ ions into a solution as they dissociate (break bonds). A molecule that is capable of doing this is called an *acid*. Another type of molecule is capable of *accepting* H^+ ions and therefore taking them out of a solution; such a molecule is called a *base* (or alkali). Some bases accept H^+ directly, while others cause the binding of H^+ without themselves doing so. An example of the latter sort of base is sodium hydroxide, NaOH. This molecule dissociates, releasing OH^- ions into a solution. These ions in turn bind to H^+ ions, to form water.

Water itself can dissociate into two ions, H^+ and OH^-. This is extremely rare, and the resulting concentration of H^+ is made even lower by the constant re-forming of bonds to make water molecules again. Nevertheless, there are always some H^+ ions in any body of water. When dissociation of H_2O and its reassociation have reached a point of equilibrium, the actual concentration of H^+ can be reliably measured as 10^{-7} Molar. It is convenient to convert this to a logarithmic value: the log of H^+ concentration in pure water is -7 Molar. One more manipulation provides a universally used unit to describe this same H^+ concentration. This is the *pH unit*, which is defined as the negative log of the concentration of H^+ in a solution. For water, of course, this conversion is from -7 to 7. The pH of pure water is 7.

This pH is sometimes referred to as the point of neutrality—it is neither acidic nor basic. If additional H^+ ions enter a water solution, from whatever source, the pH becomes lower than 7. Such a solution is said to be *acidic*. Conversely, anything that lowers the concentration of H^+ in a water solution (such as addition of OH^- ions that then sop up H^+) will cause the pH value to move above 7. Such a solution is said to be *basic*.

Many cellular processes are very sensitive to pH. For instance, certain enzymes cannot do their catalytic work when in an environment that is too basic or acidic. A major task of any organism is to maintain the optimum pH of the fluids inside and outside its cells. One way to help in this is to place certain molecules into the fluid. These molecules, called *buffers*, accept H^+ (taking it out of solution) when it is in excess, and releasing it back into solution when it is in abnormally low concentration. Buffers in human blood keep its pH at about 7.4, unless overwhelmed by extraordinary conditions.

ORGANIC MOLECULES ARE COMPLEX AND NECESSARY FOR LIFE

Something as complex as a living organism would be expected to be built from a very large repertoire of molecules, some of them very complex. It is not surprising that most of these molecules are carbon-based. That is, they are organic molecules. Since an atom of carbon has four electrons in its outer shell, it can form covalent bonds with as many as four other atoms. An example of this is the molecule called methane (CH_4), shown in Figure 3.3. Some or all of the four atoms bonded to a carbon can themselves be carbon, each of which adds as many as three more sites at which even more atoms can be bonded. An example of the ability to make complex molecules by linking together several carbon atoms is the sugar glucose (Fig. 3.5). This molecule illustrates another useful feature of organic molecules—it can twist itself into a ring-shaped conformation.

Although glucose is a large molecule when compared to most inorganic molecules, it is a rather small organic molecule. Much larger ones can be made by *polymerization* of smaller units. That is, many hundreds of molecules such as glucose can be covalently bonded to form one huge chain. Some polymers have great variety in the units that form them, making an almost infinite number of possible molecules. This is the sort of complexity and diversity on the chemical level that in turn confers complexity and diversity upon living organisms in which the molecules reside.

There are four major structural categories of organic molecules in living cells: carbohydrates, lipids, proteins, and nucleic acids. Each will be introduced below and discussed in many places throughout this book.

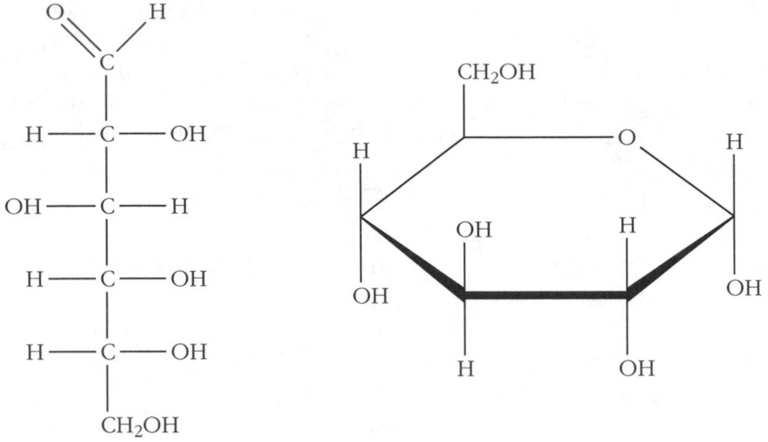

FIGURE 3.5.   *(a) Glucose in its linear form. (b) Glucose in its ring form. These are inter-convertible, but the ring is the more commonly occurring form.*

CARBOHYDRATES ARE MADE OF CARBON, HYDROGEN, AND OXYGEN

Carbohydrates are characteristically composed of carbon, hydrogen, and oxygen in the proportion of CH_2O. They are the sugars and the polymers made by stringing together sugars. The most common simple sugars are those with five carbons ($C_5H_{10}O_5$) and those with six carbons ($C_6H_{12}O_6$). Simple sugars of the types just mentioned are known as *monosaccharides* because they cannot be broken down into any simpler sugar molecule. Three of the most abundant six-carbon monosaccharides in the human diet all have the same formula, but arrange their atoms in slightly different ways. These are glucose (see Fig. 3.5), fructose, and galactose. Glucose and fructose are made in many plants and stored in fruits and nectar. Galactose is released into the intestine when milk is digested (see below).

Oligosaccharides Are Small Polymers of Sugars

Carbohydrates that are small polymers (two to six monosaccharides covalently bonded in a chain) are known as *oligosaccharides*. The most common oligosaccharides are the *disaccharides*, each of which can be broken into two molecules of a simple sugar. Common table sugar, or sucrose, is a disaccharide formed by bonding a glucose to a fructose (a five-carbon sugar). Two others are maltose and lactose. Maltose (two units of glucose bonded together) is formed when large carbohydrate polymers are digested during the sprouting of grain. This, of course, is a natural process of making food available to developing embryos in germinating seeds. Lactose is the sugar found in milk, composed of galactose and glucose. Oligosaccharides larger than disaccharides do not usually exist free; they tend to be attached to proteins (glycoproteins) in such places as cell surfaces.

Polysaccharides Can Store Energy or Make Structures

When a carbohydrate molecule is a polymer of more than six monosaccharide units, the substance is a *polysaccharide*. Generally, polysaccharides contain hundreds or thousands of these units and are among the largest known molecules. Better known polysaccharides are starches, glycogen, cellulose, and chitin. In each of these cases the exact number of monosaccharide units linked together is variable. Starches manufactured in plants are chains of glucose molecules. Glycogen, which is found in animals and in some algae and fungi, is also a chain of glucose molecules, but they are arranged in a slightly different fashion than that of any starch. Both starch and glycogen are *storage polysaccharides*; that is, they function to hold glucose until it is needed to provide energy for the organism. Cellulose is a larger polymer than either of the other two polysaccharides mentioned here. It is the major component of plant cell walls; as such, it is an example of a *structural polysaccharide*. Cellulose is probably the most abundant organic compound on earth. The paper pages of this book, manufactured from wood materials, are largely cellulose. The animal counterpart of cellulose is chitin, which is used by insects and other Arthropods to build the tough exoskeleton.

LIPIDS ARE DEFINED BY THEIR INSOLUBILITY IN WATER

The second most abundant organic group in most cells is the lipids, which vary from 3 percent in muscle to 15 percent in brain tissue. Lipids are defined more by one of their properties than by their molecular shape. That property is insolubility in water. It will be remembered that water is a very good solvent for many molecules, but it fails to dissolve lipids. Most of the bonds in lipids are nonpolar; this makes the molecules unlikely to have much affinity for the polar water molecule. The insolubility of lipids is a key to understanding several of their functions in organisms. Structurally, there is much diversity among the lipids, although all are rather complex molecules that consist wholly or largely of the elements carbon, hydrogen, and oxygen. Four classes of biologically important lipids will be introduced here: fats, phospholipids, waxes, and steroids.

Fats Include Long Chains of Carbon and Hydrogen

A fat consists of four portions: a unit of glycerol plus three units of fatty acids (Fig. 3.6). Each of the three carbon atoms of glycerol bonds to a fatty acid. Fatty acids are long chains of carbons (4 to 24 carbon atoms, depending on the fat), with their attendant hydrogens.

Most or all of the carbons of a fatty acid are *saturated* with hydrogen (Fig. 3.7a). That is, they are bonded to the maximum possible number of hydrogen atoms. *Unsaturated* fatty acids, however, have one or more carbon pairs that double bond with each other and therefore carry fewer hydrogens. Such a C—C bond produces a bend in the fatty acid chain (Fig. 3.7b). Perhaps it is this structural difference that makes unsaturated fats more easily digestible in humans and less likely to contribute to build-up of unhealthy deposits in arteries.

The most common use of fats in organisms is as storage depots for energy. Their long fatty acid chains can be broken into two-carbon pieces, which can

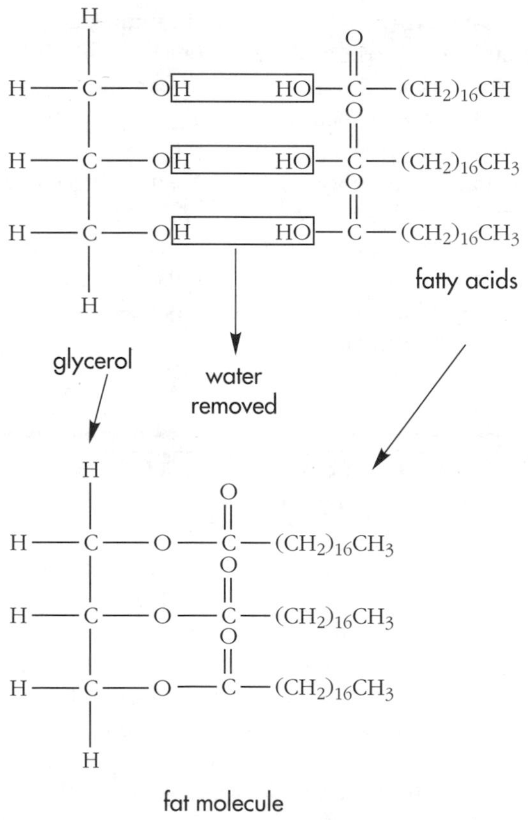

FIGURE 3.6.   *A fat molecule is synthesized by linking three fatty acid molecules to one glycerol molecule. Three molecules of water are released as this occurs.*

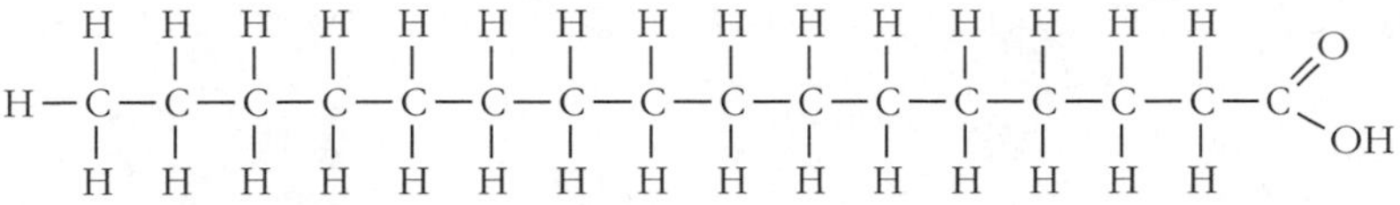

a.

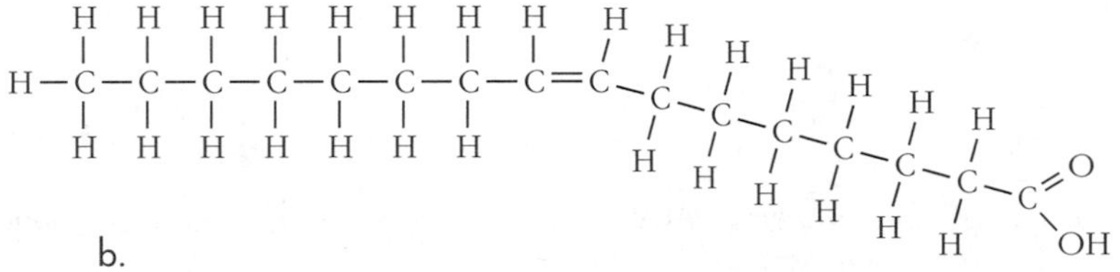

b.

FIGURE 3.7.   *(a) A saturated fatty acid, palmitic acid. (b) An unsaturated fatty acid, oleic acid.*

then be metabolized to yield energy. As a food, fat yields two and a quarter times as much energy as the same weight of a carbohydrate. In mammals, fats for this purpose are stored in specialized *fat cells* scattered in the connective tissue. In some mammals, large deposits of fats under the skin also serve to insulate against heat loss during cold weather. Plants often store fats in large quantities in reproductive structures like nuts, cotton seeds, peanuts, or flax seeds.

Phospholipids Are Modified from Fats

The substitution of a phosphate ($-PO_4^=$) and other groups for one of a fat's fatty acids produces a molecule called a phospholipid (Fig. 3.8). Such a molecule has some very different properties from those of a fat, largely because it is a partially polar molecule. The portion that includes the phosphate group has both positive and negative charges on it. The two remaining fatty acids constitute the nonpolar portion of a phospholipid. The dual qualities of polar and nonpolar in the same large molecule lead to biologically useful interactions when many such molecules are placed in contact with water. The nonpolar portions are *hydrophobic*, meaning that (like fats) they clump together and push away from watery areas. The polar end of each phospholipid, in contrast, interacts readily with water (it is *hydrophilic*). When many phospholipids are surrounded by water, they spontaneously arrange themselves to make the double layered configuration shown in Figure 3.9. It is this conformation that is a major part of every membrane of a cell. In that place, the double lipid layer acts effectively to build important structures that might be dissolved away by surrounding water if composed of other molecule types. (Remember that all lipids, including phospholipids, cannot be dissolved in water.)

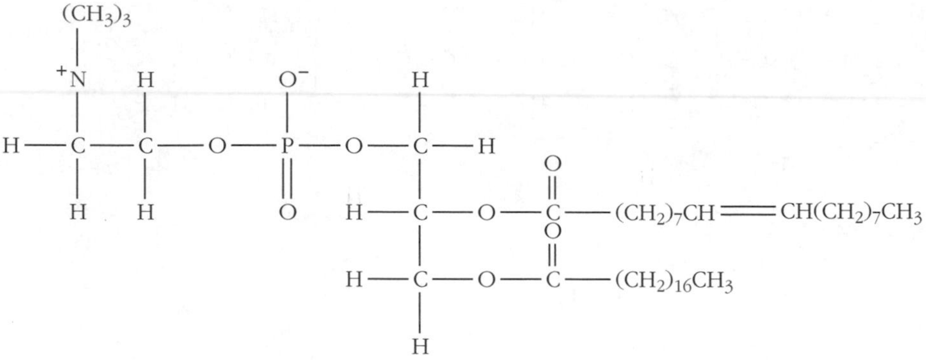

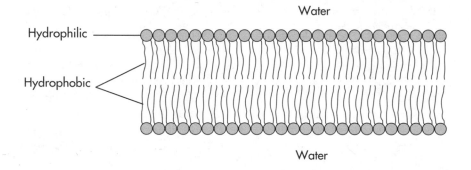

FIGURE 3.8. *A typical phospholipid (phosphotidylcholine, a membrane component of many organisms).*

Water

Hydrophilic ———

Hydrophobic ———

Water

FIGURE 3.9. *A bilayer of phospholipids, with hydrophobic fatty acids inward and hydrophilic regions outward.*

Waxes Help to Block the Passage of Water

Waxes also have long-chain fatty acids as their major component. Here, a pair of fatty acids are joined more directly than those of fats (Fig. 3.10). Most biological uses of wax are as waterproofing, playing on their hydrophobic nature. Wax covers most of a leaf's surface, drastically cutting down the amount of water lost there. The same function is played by the wax covering the surface of insects, a feature that helps explain why some insects can thrive in places where they rarely or never find water to drink. Waxes also serve to keep skin pliable.

Steroids Are Lipids Composed of Rings Rather Than Chains

The steroids are strikingly different from all other lipids, since they are composed of a series of four rings (Fig. 3.11). The most common steroid in a typical

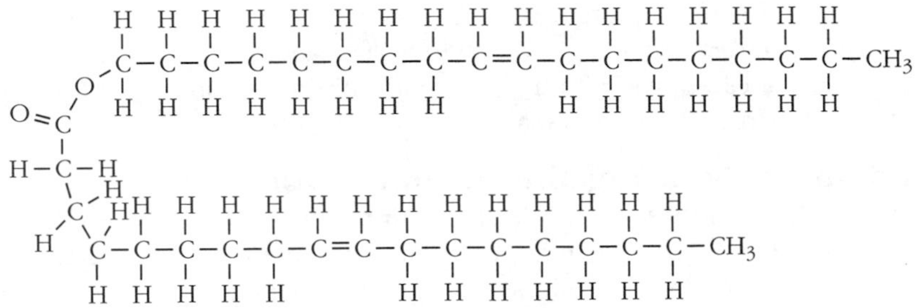

FIGURE 3.10. *A typical wax, composed of two fatty acids joined by an ester bond.*

animal is cholesterol, which is found interspersed among the phospholipids of membranes. Cholesterol is also the molecule from which another class of steroids is made—the steroid hormones such as testosterone, estrogen, and progesterone. These three are involved in determining and controlling sexual characteristics in some animals, but there are steroid hormones in other animals (such as insects) that are not related to sex. Vitamin D, needed for normal bone formation in children and calcium and phosphorus metabolism at all ages, is also a steroid.

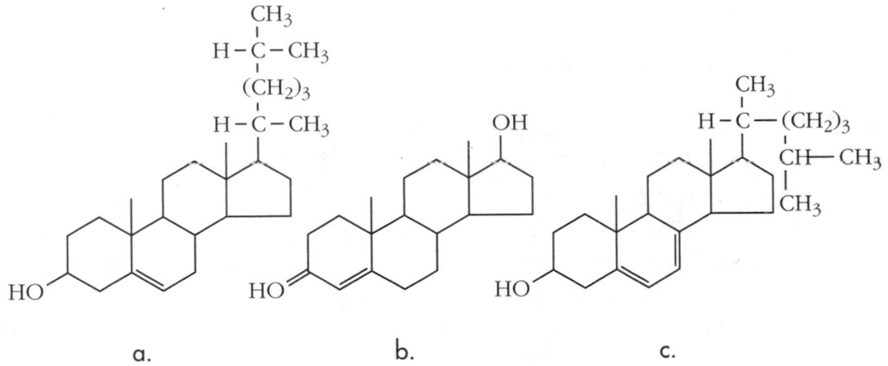

FIGURE 3.11. *Several common steroids. (a) cholesterol. (b) testosterone. (c) a form of vitamin D.*

PROTEINS ARE AMONG THE LARGEST MOLECULES

The most abundant large molecules in any organism are the proteins. If all of the water is removed from a cell or organism, about half of the remaining weight is accounted for by proteins. Some cells contain thousands of different proteins at any time. A protein is both very large and very complex. Fortunately for our understanding, it is a polymer of smaller molecules called amino acids. Thus, we can begin with these simple units and work toward the complexity of the structure of proteins.

Amino Acids Link Together to Make Proteins

The backbone of these molecules is a pair of carbon atoms and a single nitrogen atom (Fig. 3.12). The nitrogen, with accompanying hydrogens, comprises the *amino* end of an amino acid. The terminal carbon, at the opposite end from the amino group, is part of a conformation called a *carboxyl* group (−COOH). The carboxyl portion is capable of relinquishing its hydrogen, so it is the acidic portion of an amino acid. The remaining carbon always has one hydrogen attached to it, but its remaining bonding position can be filled by a wide variety of components. In organisms, there are more than twenty different amino acids that differ only by what is attached to the backbone at this position (examples: Fig. 3.13).

Free Amino Acids Can Have Biological Activity

Although the major use of amino acids is to become part of proteins, free amino acids have some value in cells. They can become sources of energy when fed into certain metabolic processes (see Chapter 6). Some amino acids act as neurotransmitters, facilitating communication between brain neurons. The amino acid thyroxine is an animal hormone, initiating metamorphosis in frogs and controlling metabolic rate in mammals.

Linked Amino Acids Are Peptides

Amino acids can be linked through bonding between the carboxyl end of one and the amino end of another (Fig. 3.14). Note that a molecule of water is produced each time such a linkage is formed. The bond between two linked amino acids is called a *peptide bond* and the resulting molecule is a *dipeptide*.

This same bonding can continue to make tripeptides, tetrapeptides, and the like. The general name for such molecules is *polypeptide*. Proteins are composed of one or more large polypeptides, but there are also biologically active polypeptides too small to be called proteins. An example is the category of brain-produced pain suppressers called endorphins, which are chains of

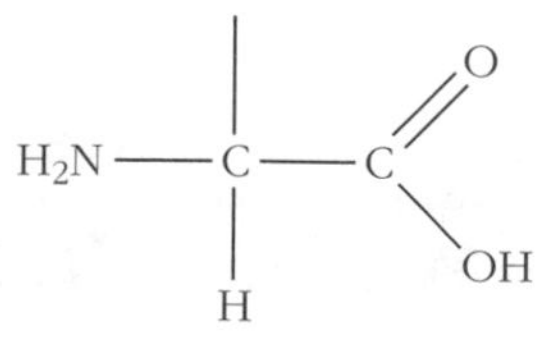

FIGURE 3.12.   *The "backbone" portion of any amino acid. The hydrogen of the carboxyl end may be reversibly removed and an additional hydrogen may be attached to the amino end, depending upon environmental conditions.*

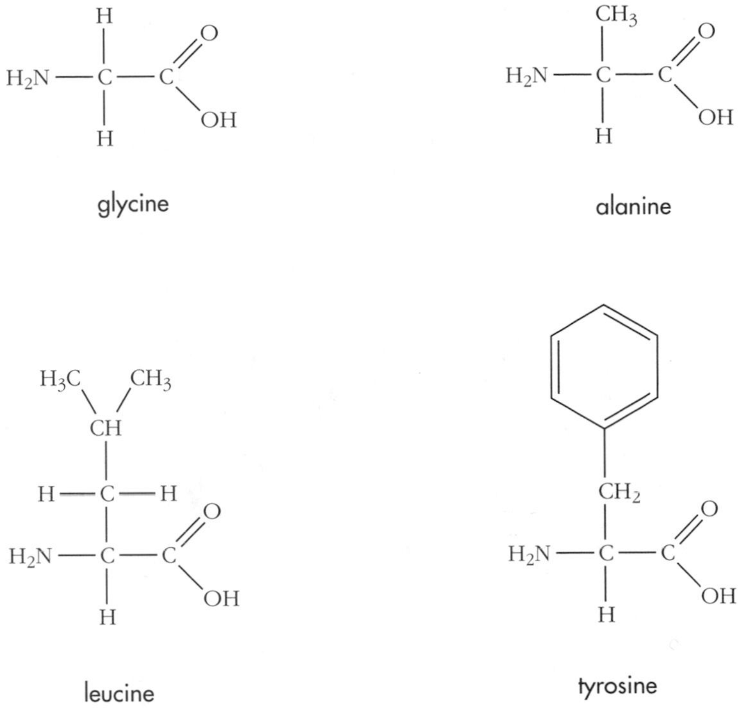

FIGURE 3.13. *Several naturally occurring amino acids. Notice that they differ only by what is attached to the middle carbon atom of the "backbone."*

about 32 amino acids. A second example is oxytocin, a nine amino acid polypeptide hormone that induces labor and milk release in mammals.

Proteins Must Have Precise Three-dimensional Shapes

Larger polypeptides (or combinations of several polypeptides) are called proteins. Some cells contain well over a thousand different proteins simultaneously, each performing an essential chemical or structural role. One can visualize the enormity of possibilities for constructing proteins when it is remembered that they are combinations made from about 20 different amino acids. Keep in mind that any one of the 20 amino acids can take up any one of the 100 or more positions of a typically sized protein. It is safe to say that all of the different species of organisms that have ever existed or will exist on earth have more than enough different proteins theoretically available from which to make themselves.

If it is to be functional in an organism, a protein must be more than a long chain of amino acids; it must also fold into a complex three-dimensional shape. Depending on the protein, there are either three or four levels of structure-forming that a protein must successfully negotiate before it becomes active. The *pri-*

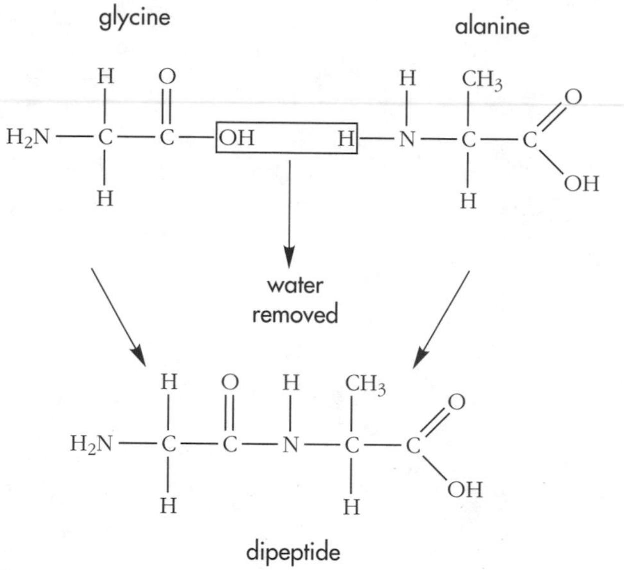

FIGURE 3.14.   *Synthesis of a dipeptide from two amino acid molecules.*

mary structure is accomplished by stringing together the amino acids in proper sequence. Then, some regions of the chain must coil into a helical conformation (or, in some proteins, a nearly flat sheet-like conformation). This is called the *secondary* structure of the protein. After this is accomplished, the polypeptide twists into a much more compact three-dimensional shape. This tertiary structure is often globular (ball-shaped), but some proteins are elongated or fibrous. Smaller proteins are correctly shaped for biological functions when their tertiary folding has been accomplished. However, many proteins are formed by combining two or more polypeptides. It is only when these subunits have successfully linked that the *quaternary* structure of such large proteins has been formed. A good example of a protein made from several polypeptides is the oxygen-carrying protein of blood, hemoglobin. This protein consists of two identical copies of an alpha subunit and two of a different one, the beta subunit.

Secondary, tertiary, and quaternary structures of a protein are held in place by a combination of weak hydrogen bonds and stronger covalent bonds among their amino acids. If the cell makes an error in placing the amino acids of a polypeptide (making the primary structure), it is possible that this error will cause incorrect three-dimensional folding and thus loss of biological function.

Proteins have many functions. Most are enzymes (biological catalysts), as described in detail in Chapter 6. Some proteins form physical structures, such as the moveable elements of muscle and the tough connective materials of tendons and ligaments.

NUCLEIC ACIDS ARE VERY LONG POLYMERS OF NUCLEOTIDES

Another category of very large and complex biologically important molecules is that of the nucleic acids. Like proteins, these molecules are polymers of much smaller units. These smaller molecules are the *nucleotides*.

A Nucleotide Includes Three Functional Groups

Figure 3.15 shows that a nucleotide is composed of three portions. Between a phosphate group ($—PO_4$) and a complex structure called a nitrogenous base is a five-carbon sugar. The sugar can be either ribose or deoxyribose, the latter being different only by substituting a hydrogen atom at one of the places where a hydroxyl (—OH) group is found in ribose. A nitrogenous base is attached to one of the carbons of the sugar. Most nucleotides have one of five different bases at this position; since they are very important in the genetic function of nucleic acids, their names and structures are presented here (Fig. 3.15). Notice that three of the five bases are single-ring structures (called pyrimidines) and the other two are double-ring structures (called purines).

Polynucleotides (Nucleic Acids) Form by Specific Bonds among Nucleotides

Symbolizing some portions of the nucleotide molecule will help demonstrate the manner in which the parts are linked together. Let NB mean Nitrog-

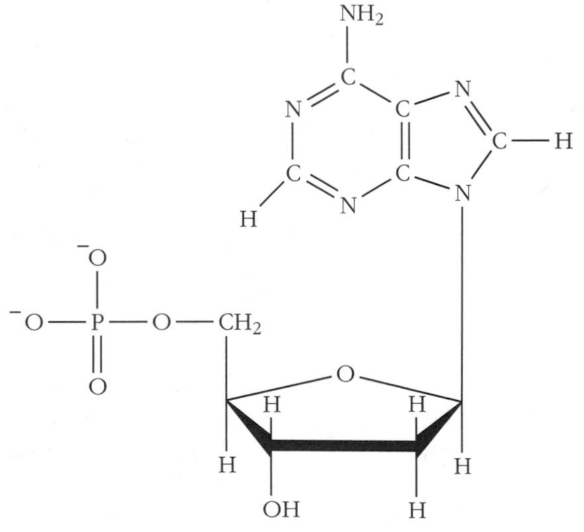

FIGURE 3.15. *A typical nucleotide. This one is deoxyadenylic acid, so named because its sugar is deoxyribose and its nitrogenous base is adenine.*

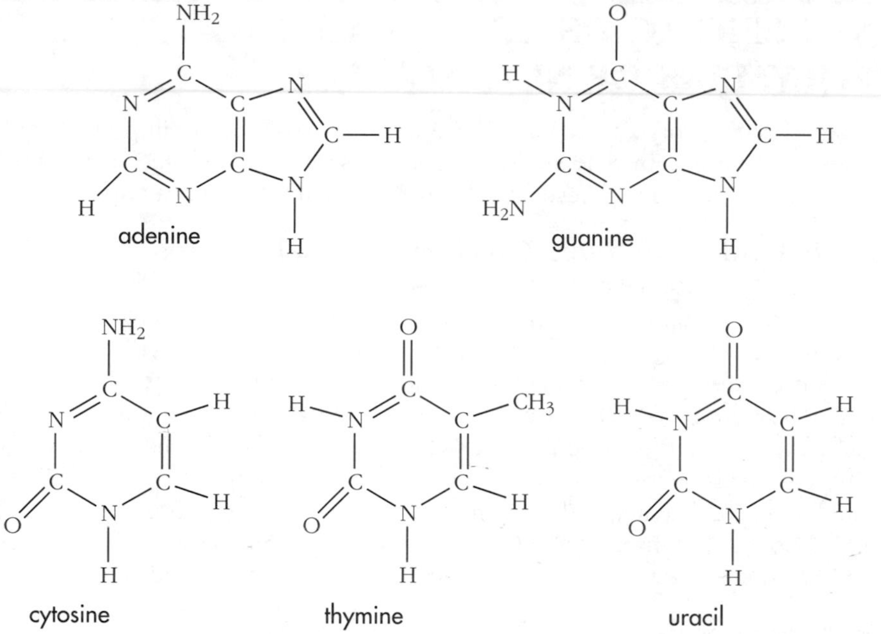

adenine

guanine

cytosine

thymine

uracil

FIGURE 3.16.   *The five nitrogenous bases most commonly found in naturally occurring nucleotides. Thymine is found only in nucleotides of DNA; uracil is found only in those of RNA. The remaining three bases are incorporated into both DNA and RNA.*

enous Base attached to the first carbon of the ring and a circled P mean Phosphate group attached to the fifth carbon of the ring.

By substitution of the first letter of each nitrogenous base for the NB used above, there will be the following four symbols for the types of nucleotides that can join to make the polynucleotide called *deoxyribonucleic acid (DNA)*:

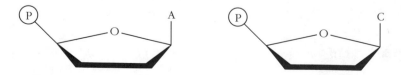

They can be linked in any order to make a chain. The nucleotides are linked together between their sugar and phosphate components as shown below. The length of chain formed in this manner can be very long; DNA consisting of hundreds of thousands of nucleotides is common.

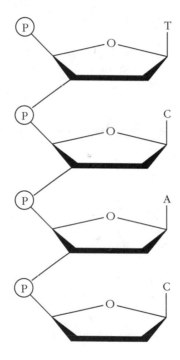

Ribonucleic acid (RNA) is also a polynucleotide, with the same kinds of linkages among its nucleotides. Both RNA and DNA are also sometimes called *nucleic acids.* RNA is usually a smaller chain; some functioning forms contain only 75 to 90 nucleotides. Both DNA and RNA function to store and release genetic information. Chapter 12 describes these activities in some detail.

VITAMINS ARE ESSENTIAL IN FOODS

The last category of organic molecules to be considered here is one that is defined by function rather than by chemical structure or chemical behavior. The vitamins are a diverse group of molecules that are:

essential to the well-being of the organism in question,

not able to be synthesized within that organism, and must therefore be imported,

needed in very small quantities at any time (and are therefore not food).

Most (but not all) vitamins act as *coenzymes*. This means that they are essential in helping enzymes do their work of catalyzing chemical reactions. Some of the common vitamins for humans are not actually in the form in which they help us; they must be chemically modified by our cells after they are ingested. There is some current evidence suggesting that ingestion of certain human vitamins (such as Vitamin C and Vitamin E) in amounts far above those needed as vitamins may help protect against certain cellular changes linked to aging, cancer, and heart disease. However, it is yet too early to know whether claims in this area are valid. Retinoic acid, a form of Vitamin A, is also believed to play an important role in directing the differentiation and shaping of some organs in developing embryos.

STUDY QUESTIONS

1. Define the following: atom, element, isotope, and ion.

2. When is a molecule also a compound? Name a biologically important molecule that is not also a compound.

3. If one atom of an element has a total of 11 electrons, how are they expected to be arranged around the nucleus? Think about rules for filling shells.

4. Consider the atom described in question 3, above. If it approaches an atom with seven electrons in its outermost shell, what is likely to happen to create a bond?

5. How does ionic bonding differ from covalent bonding?

6. What are the properties of water that make it so useful (necessary) for supporting life on earth?

7. Compare nonpolar covalent with polar covalent bonds. Why are both important for biologists?

8. Define pH. Why is any organism concerned with its internal pH?

9. Why does evaporation of water (as sweat) from a person's skin cause cooling? Would nonpolar molecules do as well?

10. What is distinctive about organic compounds?

11. Name several polysaccharides and describe their usefulness for organisms.

12. What are the major categories of lipids? What property do they share that makes them all lipids?

13. Explain how one molecule of glycerol combines with three molecules of fatty acids in the synthesis of one fat molecule.

14. How does a saturated fatty acid differ from an unsaturated one? Which is considered to be more healthful in one's diet?

15. What is the structure of an amino acid? What is the name of the —NH_2 group? Of the —COOH group?

16. What are the major categories of protein, distinguished by their functions in organisms?

17. The nucleotide, a unit of nucleic acids, is composed of what three smaller units?

18. What nucleic acids are important in cells? How do they differ?

19. How is the word *vitamin* defined? Are all vitamins similar in structure?

4
BODY ORGANIZATION: CELLS, TISSUES, ORGANS, SYSTEMS

What do we know about:
—*how our knowledge of cells is tied to the history of microscopes?*
—*what the internal structures of a cell are and what functions they perform?*
—*the great importance of the plasma membrane surrounding a cell?*
—*why normal cells must remain very small?*
—*how cells make exact copies of themselves?*
—*how normal cells control their reproduction and cancer cells escape this control?*
—*the specialized cell divisions that make sexual reproduction possible?*
—*the levels of organization that occur above the cellular level, to make tissues and organs?*
These and other intriguing topics will be pursued in this chapter.

ORGANISMS ARE HIGHLY ORGANIZED

One of the most fascinating aspects of life is that it is richly detailed, in both structure and function. As will be described, the history of biology has been a discovery of ever more complexity as instruments have been developed to look more closely at organisms. It is gratifying to discover that most of the details of organisms' structures can be closely associated with the activities that are necessary for carrying on life. To borrow the words of the great American architect Louis Sullivan, "form follows function," even down to the level of structures within cells. This chapter introduces the reader to subcellular, cellular, tissue, organ, and systemic organization.

DEVELOPMENT OF OPTICAL INSTRUMENTS PARALLELED ADVANCES IN CELL BIOLOGY

Following the early recognition that organisms are composed of cells, scientists studied them intensely as the most fundamental units of life. From the time that Robert Hooke first used a primitive light microscope to see the cell wall (1655) to the present generation, the trend has been for scientists to focus their interests on ever smaller physiological and structural entities within cells. Soon evident was the fact that cellular organization is essentially universal and that the living part of the cell is its internal contents rather than its wall. (Indeed, it was soon discovered that not all cells have a wall.) In the 1830s, Brown and Purkinje published descriptions of the contents of cells, including the fact that each cell has a rather large central object, the nucleus. This period was also the time of Schleiden's and Schwann's publications, which clearly presented the Cell Theory (described in Chapter 2).

In the late 1800s the refinement of light microscope lenses and the use of specific stains made possible the study of more structures within cells. Mitochondria were described (1857), chromosome behavior was followed during cell reproduction (1879), and specific types of bacteria were first identified (1882). In the 1930s the development of instruments called interference microscopes and phase-contrast microscopes enabled workers to see structures within living cells, without the need for killing and staining. In 1941 some cellular components were first made visible by attachment of fluorescent antibodies, making possible the technique now generally called fluorescence microscopy. In the 1980s, the combination of lasers, fluorescence, and special optical techniques produced an instrument called a confocal scanning microscope. This microscope allows analysis of the three-dimensional structure of objects with excellent detail.

Much of the usefulness of any microscope can be described in terms of its ability to show that two objects lying close together are really separate objects. This resolving power (measured as the distance between the two objects) of an optical device is limited by the wavelength of the illuminating medium. With visible light, as used in any light microscope, the closest objects that can be resolved are about 0.2 micrometer apart (a typical cell is about 20 micrometers in diameter). A logical way to get past this limitation is to replace visible light with some other medium having a shorter wavelength, such as a beam of electrons. This is the basis for the greater resolving power of the electron microscope. From the 1930s onward, this instrument was refined to the point that it became a highly valuable tool for studying the smallest details of structures within cells. With a theoretical resolving limit of about 0.002 nanometer, an electron microscope could be 10,000 times better than the best light microscope. In practical use with biological materials, electron microscopes provide about 100 times better resolution than the best light microscopes. From the 1950s onward, electron microscopists provided detailed views of such structures as membranes, muscle filaments, and the internal features of chloroplasts and mitochondria. In the 1960s the first scanning electron microscopes became commercially available, providing detailed three-dimensional views of the surfaces of biological objects.

THE CELL IS A STRUCTURAL UNIT

To define a cell properly and make the definition all-inclusive is not easy. In general terms, it is a membrane-bound compartment containing water and many materials either dissolved or suspended in the water. This definition rules out viruses as cells, since they are not membranous. It does not rule out the organisms in Kingdom Protista (Protozoa, algae, etc.), but they are so specialized and independent of other cells that they are sometimes called acellular organisms.

A special difficulty in defining terms exists when a cell, tissue, or body is multinucleate or has no nucleus at all. Ordinarily a cell has a boundary and a single nucleus; but osteoclasts, which play a part in the replacement of embryonic cartilage by bone, are large cells with multiple nuclei. Likewise, a skeletal muscle fiber is a multinucleate unit not separated by cell membranes (although its embryonic precursors were individual cells that became fused together). The same is true of slime molds, thread fungi, and some algae. Special names, like *syncytium* or *coenocyte,* are given to such tissues and organisms respectively. At the other extreme are cells without nuclei. All prokaryotic organisms are of this sort. The precursors of human red blood cells, found in bone marrow, are nucleated but the functioning cells that are released to circulate and carry oxygen do not include a nucleus.

The remainder of this chapter focuses on eukaryotic cells with the common basic parts (*organelles*).

A TYPICAL CELL'S ORGANELLES ARE SUSPENDED IN A FLUID

A cell (Fig. 4.1) is the smallest unit of organized matter having characteristics of life. A eukaryotic cell's contents (held within an outer limiting membrane) are a central nucleus and a complex area called the *cytoplasm.* The cytoplasm is composed of many specific structures (organelles) and a semiliquid material (the *cytosol*) in which the organelles are suspended. Cytosol is mostly water, but includes a variable array of dissolved molecules—proteins, salts, sugars, hormones, waste materials, and so on. Because of the presence of these dissolved materials, cytosol has the property of a *colloid.* That is, it is interconvertible between being very liquid (easily flowing) and being more

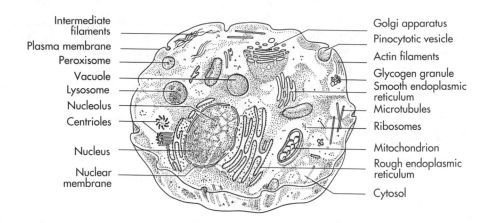

FIGURE 4.1. *A generalized animal cell. (From* Biology *(6th edition) by Kimball. Copyright © 1994. Reprinted by permission of Wm. C. Brown Communications, Inc., Dubuque, Iowa.)*

rigid. Depending on such factors as temperature and ionic concentration, cytosol of a cell can divide the interior of the cell into more or less liquid compartments. Some of the dissolved materials in cytosol can move across membranes, a topic that will be pursued in detail in Chapter 8.

THE NUCLEUS CONTAINS UNITS OF INHERITANCE

The boundary of a cell's nucleus is a membranous structure called the *nuclear envelope*. Within the confines of the envelope are chromosomes, the sites of genes (see Chapter 11). The volume of a nucleus is somewhat proportional to the volume of cytoplasm surrounding it. A closer look at the nucleus reveals the complexity of its structure and function. The nuclear envelope is really two membranes separated throughout most of their extent. It is clearly porous when viewed under the electron microscope. Doubtless it regulates the passage of materials as other cellular membranes do.

The viscous material contained inside the nucleus is the *nucleoplasm*, within which the chromosomes float. How the chromosomes look depends on what is happening in the cell. If the cell is not dividing and if proper observation techniques are used, they appear as fine, twisted, and tangled. As cells begin to divide, the chromosomes coil, shorten, and thicken to the point where they are easily seen by use of ordinary light microscopic techniques. Nuclei of nondividing cells may contain one to several spherical bodies known as *nucleoli* (singular: *nucleolus*), which are temporary congregations of RNA. This RNA was made under the direction of nuclear genes and soon moves into the cytoplasm to direct the building of proteins (see details, Chapter 12).

MANY ORGANELLES ARE POSITIONED IN THE CYTOPLASMIC PORTION OF A CELL

Outside the nucleus is the cytoplasm. The electron microscope reveals an extremely complex ultrastructure. Let us examine each of the organelles to be found there.

The Endoplasmic Reticulum Provides Surfaces and Spaces for Enzyme Activity

Throughout the cytoplasm run flat sheets of membranes, collectively called the *endoplasmic reticulum*. Close examination (using the electron microscope) reveals that all of these membranes are connected to each other and that they are continuations of the outer membrane of the nucleus. These membrane surfaces are sites of much chemical activity, as indicated by the presence of many enzymes embedded in them. Studded over the outer surface of some (but not all) of the endoplasmic reticulum membrane are thousands of tiny ball-like structures called *ribosomes*. The portions that include ribosomes are called *rough* endoplasmic reticulum and portions lacking ribosome are called *smooth* endoplasmic reticulum. Proteins that are newly manufactured at ribosomes (for details, see Chapter 12) can be brought into the endoplasmic reticulum for further processing and eventual transport to other parts of the cell. Proteins destined to be used elsewhere in the cell (or exported from the cell) are packaged by the pinching off of small portions of the endoplasmic reticulum to form *transport vesicles*. These vesicles then move into other parts of the cell, carrying the proteins within them.

The Golgi Complex Further Processes Materials for Export

One of the destinations of transport vesicles is a separate set of membranous structures, the *Golgi complex*. Each complex (some cells have more than one) consists of a loose stack of about four to six plate-like sacs. As proteins are brought to the Golgi region the transport vesicles carrying them become fused with the membrane of one of the outlying sacs. Then, the proteins travel from sac to sac (again, carried by small vesicles), undergoing modifications along the way. For instance, some proteins are not functional until small chains of carbohydrates are attached; this can occur as each molecule progresses through the Golgi complex. After molecules have been transported from one side of the Golgi complex to the other, they are packaged once more into small vesicles, which pinch off from the sac lying at the outermost edge of the complex. Depending on the function of their enclosed molecules, these vesicles may move to the cell surface or stay suspended in the cytosol.

Cells secreting large amounts of material to their exterior (such as mucus-producing cells lining the breathing passages) have very active Golgi complexes that constantly produce secretory vesicles. *Lysosomes* are Golgi-produced vesicles that contain digestive enzymes for use within the producing cell. Some lysosomes release their enzymes to destroy foreign materials (such as bacteria) that have entered the cell. Others do not become active until after the cell dies, when it is necessary to digest the entire cell to make room for its replacement and to release its molecules for re-use.

Peroxisomes Are Specialized Vesicles

Some cells contain membranous bags filled with the enzyme called peroxidase. These *peroxisomes*, about the same size as lysosomes, carry out the important process of breaking down hydrogen peroxide (H_2O_2), which is a potentially dangerous by-product of metabolic activity. Some peroxisomes are also active in breaking down ethyl alcohol, the toxic material in alcoholic beverages. In animals (including humans), peroxisome activity is limited to cells of the liver and kidneys.

Vacuoles Are Large Vesicle-like Structures

Vacuoles are membranous bags much larger than the vesicles described above. They are more likely to be within plant cells than other cells and may occupy up to 90 percent of the internal volume of a mature plant cell. Ordinarily in a plant cell a vacuole functions in maintaining osmotic balance and stores sugars, organic acids, salts, plant pigments, and other substances. Some plants manufacturing noxious chemicals to detract potential predators store them in their cells' vacuoles.

Special cases exist in freshwater Protozoa (Kingdom Protista) where the vacuoles accumulate water that continuously diffuses into the cell. Periodically they contract and expel the excess water to the outside. Such cells would burst if they did not have a mechanism for expelling the water. It is interesting that both internally parasitic and marine Protozoa do not have these *contractile vacuoles*. The absence of a contractile vacuole in such environments makes sense, for the osmotic condition of the environment around parasitic and marine Protozoa is much nearer to that of their cytosol than is the case of freshwater Protozoa.

Many of the Protozoa, as well as some cells of higher animals, incorporate food into themselves in what is known as a *food vacuole*. A food vacuole is formed as a cell engulfs its prey in a movement called *phagocytosis*. Enzymes are then secreted into the vacuole via lysosomes that fuse their membranes with that of the vacuole. Digestion occurs inside the food vacuole, after which individual food molecules pass through the vacuole's membrane into the cytosol.

Mitochondria Are Energy-converting Organelles

Mitochondria (singular: *mitochondrion*) are in almost all cells and are in largest numbers within those that are most active in making energy accessible for use. They contain an ordered system of enzymes that works to wring energy out of certain food molecules, a subject to be discussed thoroughly in Chapter 6. Structurally, mitochondria vary in shape from spheres and rods to filaments and are of sufficient size to be seen with the light microscope, although the details of their interior require the better resolving power of the electron microscope (Fig. 4.2). Mitochondria are bounded by two membranes with the inside one folded in such a way that the surface area is greatly increased. Many critical enzymes and structures for metabolic processes are found on that innermost membrane.

Chloroplasts Resemble Mitochondria and Are Also Involved in Energy Conversion

Chloroplasts are the organelles specialized to carry out the conversion of light energy into the energy of chemical bonds—the process called photosynthesis. Of course, they are not found in cells of animals, fungi, or the Protista that are incapable of photosynthesis. A typical above-ground plant cell contains 20 to 100 of these organelles. Chloroplasts resemble mitochondria in general size and external shape (fig. 4.3). Like a mitochondrion, a chloroplast has a pair of membranes, one internal to the other. However, the most complex and functional portion of a chloroplast is a third membranous region that takes up most of the interior. This area consists of a series of membranous disks called *thylakoids*, arranged into stacks. Each stack is called a *granum:* (plural: *grana*). The membranes composing the grana, called *thylakoid membranes*, are the sites of the apparatus used to catch the energy of sunlight and convert it into bonds of molecules. The portion of a chloroplast not occupied by grana is a fluid-filled compartment called the *stroma*. Enzymes located in the stroma are involved in certain steps of photosynthesis. The details of thylakoid membranes and the photosynthetic process are in Chapter 6.

Both Mitochondria and Chloroplasts May Have Once Been Independent Organisms

These two cytoplasmic organelles are (as far as we know) unique in that they contain DNA, the molecule associated with storing and releasing genetic information. Both mitochondria and chloroplasts have many (but not all) of the instructions on how to build copies of themselves, independent of the cell nucleus. This feature, plus their size and general shape, has led many biologists to speculate that these organelles were once independent bacteria or bacteria-like organisms. This idea is called the *endosymbiont hypothesis*, because it postulates that bacteria took up residence in eukaryotic cells and that a mutually advantageous or symbiotic relationship evolved into a permanent coexistence.

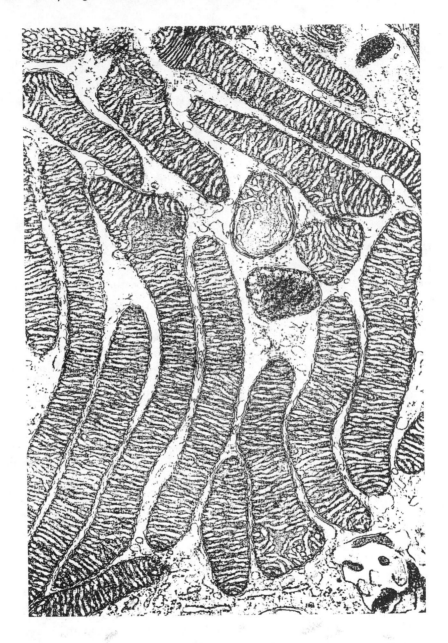

FIGURE 4.2.   *Electron microscope view of many mitochondria in an adrenal cell of a hamster. Internal finger-like projections are cristae. (From* A Textbook of Histology *by Fawcett, Chapman and Hall, 1994.)*

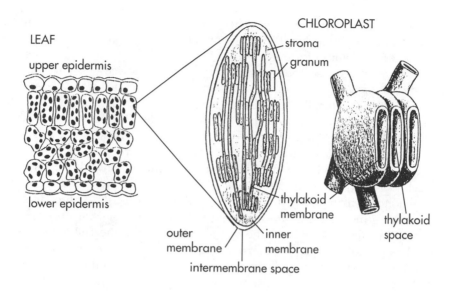

FIGURE 4.3. *Chloroplast location and structure. (From* Molecular Biology of the Cell *by Alberts et al., Garland Publishing, Inc., 1994.)*

THE CYTOSKELETON AIDS IN MOVEMENT AND PROVIDES SUPPORT

Throughout the cytoplasm there exists an intricate network of rods and tubes made of proteins. Falling into three categories, this cytoskeleton has multiple functions.

Microtubules Are Associated with Movements and Shape Changes

One category of cytoskeletal materials is that of the *microtubules.* They are very thin hollow rods, constructed by linking subunits of a protein called *tubulin.* When a cell is not dividing, an arrangement of microtubules is found at a particular place in the cell called the *cell center.* This region can be seen in animal cells and in most algal and fungal cells, but not in plant cells. It is located just outside the nucleus and helps prepare a cell for division, during which time the cell center undergoes changes. In the nondividing cell, the cell center consists of a pair of cylindrical structures known as *centrioles,* which are surrounded by a rather clear spherical zone. Each centriole consists of nine sets of three short microtubules, arranged to form a tube. As a cell approaches the time for dividing, new microtubules are quickly assembled, radiating out from each centriole. These become the major components of a structure called the *spindle* (Fig. 4.11), which is essential in moving chromosomes during cell division (see later in this chapter). The centrioles in animal cells appear to be active in organizing the spindle, but plant cells (lacking centrioles) must have some other mechanism for this.

Microtubules are also found within two types of device used by cells for movement: *flagella* and *cilia*. Both of these are projections from the surface of a cell, flagella generally being longer and less numerous. In each case, the internal structure is an intricate array of microtubules (Fig. 4.4). In a remarkably consistent pattern among cells of quite diverse organisms, the flagellum or cilium will have a set of nine pairs of microtubules running its length in a cylindrical pattern, with two more microtubules running through the center of the cylinder from one end to the other. These microtubules are able to contract (shorten) rhythmically, thus changing the shape of the entire structure and enabling the characteristic whipping motion that pushes against the surrounding medium and enables the cell to move. (Or, if the cell is firmly anchored as part of a tissue, the surrounding medium is moved by the beating action.) At the base of each cilium or flagellum there is a centriole-like object called the *basal body*, closely resembling a centriole. This structure is active in forming a cilium or flagellum, but the mechanism is not well understood.

Microtubules are also involved in changing the shape of cells. If a cell is to become elongated, one of the first changes that it undergoes is to develop a set of parallel microtubules oriented in the direction of elongation. Such a shape change occurs at numerous regions of a developing animal embryo, helping them to take on appropriate three-dimensional configurations.

Microfilaments Help Determine a Cell's Shape

A second category of cytoskeletal material is a type of solid rod called microfilaments. These are also composed of polymerized proteins, mostly *actin*. Microfilaments are associated with outreaching edges of cells that are in mo-

FIGURE 4.4.   *Cross section through a cilium or flagellum, showing nine pairs of microtubules surrounding two central microtubules.*

tion or changing shape. They undoubtedly provide internal support for the cell and its organelles. When combined with another protein, *myosin*, the actin of a microfilament can cause contraction of the rod. During division of animal cells, a loop of microfilaments forms around the perimeter of the cell (just under the cell's surface), then it contracts to constrict the cell down the middle.

Intermediate Filaments Provide Internal Support

The last category of cytoskeleton elements, intermediate filaments, is a complex lacework of solid proteinaceous threads that is found throughout the cytoplasm. Their proteins are tough fibrous molecules, appropriate for building semipermanent skeletal rods that support the cell and anchor its organelles. Intermediate filaments seem to be found only in animal cells.

THE PLASMA MEMBRANE IS THE OUTER BOUNDARY OF A CELL

The most important part of a cell for interaction with its environment is the outermost limiting membrane, the *plasma membrane* (also called the cytoplasmic membrane or the cell membrane). It acts both as gatekeeper and address proclaimer for a cell. If its structural integrity is compromised, cell death almost inevitably follows.

The Plasma Membrane Fits a Fluid Mosaic Structural Model

The current hypothesis on the structure of the plasma membrane (and of most other membranes of a cell, such as the endoplasmic reticulum), is that it is a very dynamic and complex combination of lipids, proteins, and carbohydrates (Fig. 4.5). As was indicated in Chapter 3, phospholipids can interact with each other to form a bilayer when surrounded by water (Fig. 3.9). It is this bilayer, insoluble in water, that comprises most of the plasma membrane. Interspersed among the phospholipids are individual molecules of cholesterol, acting to stabilize the bilayer when it is at temperature extremes.

Proteins are found floating in and upon the lipid bilayer, not unlike pieces of wood floating at a lake's surface. It is this combination of a very fluid lipid bilayer and individual protein molecules interspersed in it that led to the name fluid mosaic for this model of the membrane. Many specific types of proteins are associated with the membrane. Some of them are only on the outside surface, others are characteristically found facing the cytoplasm, and still others penetrate through the lipid bilayer with portions both outside and inside. Some of the proteins that are on the inward side of the plasma membrane act as attachment points for the cytoskeleton.

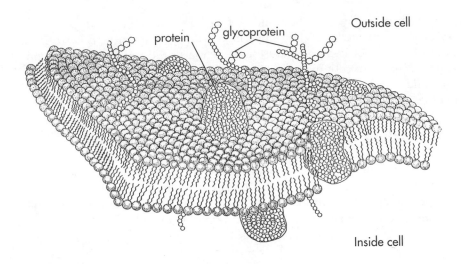

FIGURE 4.5.   *Diagram of a cell's membrane, according to the fluid mosaic model. Proteins (some with attached carbohydrate chains) float in a bilayer of phospholipids. (From* Zoology *by Mitchell et al., Benjamin/Cummings Publishing. Co., 1988.)*

Some of the outward-facing proteins are *glycoproteins*; that is, they have short chains of carbohydrates attached to them. It is the combination of proteins and these carbohydrates that acts to mark the surface of a cell for identification. As cells move during embryonic development, they slide over each other and read their neighbor's surfaces as indications of where they are and whether they should continue moving. The cells of an animal's immune system use these same addresses to determine whether any cell they touch is part of the body or if it is foreign. If the latter is detected, the proper response is to destroy the cell since it is potentially harmful to the organism. This same recognition system is used by immune system cells to detect surface changes that are characteristic of a person's cells becoming cancerous. If the immune system is not compromised in some way, it can recognize and destroy cancerous cells before they cause harm (an action called *immune surveillance*).

Some of the proteins that are large enough to stick out from the both the inside and outside surfaces of the plasma membrane can act as channels for transporting materials into and out of the cell. Although some molecule types can pass through the lipid bilayer without the necessity of finding a "hole," most are blocked by the bilayer. It is only via specially shaped membrane proteins that they can pass through. These proteins act as gatekeepers since they must recognize specific molecules or ions by their shapes before they react to provide an opening. The topic of how materials enter and leave a cell is explored more thoroughly in Chapter 8.

MANY CELLS ARE SURROUNDED BY A NONLIVING COATING

A number of cell types, in all kingdoms, produce an extracellular material of some sort. The best known of these is the *cell wall* formed by the cells of many nonanimal organisms. Of course, this is not to be confused with the plasma membrane, which is the actual limiting boundary of the cell and which lies interior to the cell wall. Most plant cells secrete walls of nonliving substances around themselves. These materials reinforce the plasma membranes and may be considered primarily protective or supportive. Unlike the extremely thin plasma membrane, the wall is clearly visible under the light microscope. Plant cells deposit walls consisting of a framework of the polysaccharide cellulose in which other substances like cutin, suberin, pectin, hemicellulose, and lignin may be added. Primary walls, the first to be deposited, commonly contain pectin (a polysaccharide) and the secondary walls, the last to be deposited, contain lignin (a phenolic compound). Together, cellulose and lignin give rigidity to the plant body. Cells located on the outside of organs deposit waterproofing cutin (a wax) within the outside walls and on the exposed surface of the epidermis (where it is called *cuticle*) and another wax, suberin, within the walls of cork cells. As a result of these materials being present the cells are protected, especially against desiccation. Plant cells are held together by an intercellular layer known as the *middle lamella,* which is composed at first of pectin but later modified.

Among nonplant organisms, cell walls are found in the Kingdoms Archaea, Bacteria, and Fungi. Bacteria and Archaebacteria secrete walls composed primarily of specific polysaccharides and very different chemically from those of plants. Most fungi produce walls of the polysaccharide chitin, again quite distinct from the walls of Kingdom Plantae.

In animals, bone and cartilage cells secrete matrices (singular: matrix), not called walls, that give support to the body. Other materials that are secreted by animals are keratin in the skin of mammals and the chitinous exoskeleton of Arthropods (insects, crustaceans, etc.).

THE SIZE OF A CELL GIVES IT IMPORTANT ABILITIES

Most cells are easily visible under the light microscope. A few are even large enough to be seen by the naked eye. The yolk of the hen's egg is a single cell with a diameter often exceeding two centimeters. The largest cell, of course, is that of the ostrich egg. Some nerve cells may be very long but usually have microscopically small diameters. Processes of a single nerve cell extending from the foot to the spinal cord of some large animals may be over a meter in length. Plant fibers, with only their walls persisting, are likewise elongated. Ramie fibers are as long as 50 centimeters, and cotton fibers may be four centimeters. Nevertheless, the typical cell is quite small. About 20 average-sized eukaryotic cells could be placed side-by-side along a one-millimeter line, and ten times as many prokaryotic cells would occupy that same length.

Size is just one aspect related to function. The largest cells, the yolks of birds' eggs, are large because they contain quantities of stored food. The yolk (egg) has a nucleus and a concentration of cytoplasm near the surface at one side. It differs from the human egg (0.15 mm in diameter) in having a large amount of reserve food. This adaptation is necessary for survival because bird embryos are entirely isolated from outside food sources until they hatch.

The size of cells definitely affects their ability to function. The small-sized cell has an advantage in that a relatively large surface area is exposed to the environment. The smaller the cell, the greater the surface exposed in proportion to the volume of cytoplasm contained within it. Consider some simple geometry. If a cell were a cube in shape, shrinking in half the lengths of its sides would cause its surface area to become one-fourth as large, but its volume would decrease even more, becoming only one-eighth as large as in the original (Fig. 4.6). That is, the ratio of surface to volume would be increased as the size of the cell was decreased. Now, the volume of a cell is an indicator of its *needs* (food, oxygen, etc.) that must be supplied from outside. On the other hand, the surface area represents the *ability* of the cell to provide for those needs; the amount of plasma membrane available dictates how much material can be delivered to the interior in a given time. So, the organism that chops up its internal volume into a very large number of small units (cells), each with its own membrane for transfer of materials, is the organism with a distinct advantage.

The same size consideration comes into play at the organismal level. For instance, mammals living in hot climates tend to have long ears. This provides

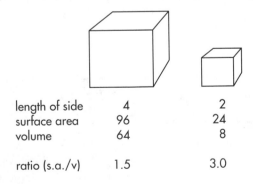

length of side	4	2
surface area	96	24
volume	64	8
ratio (s.a./v)	1.5	3.0

FIGURE 4.6.   *Consider two cubes, with the dimensions shown. The ratio of surface area to volume is much higher for the smaller cube.*

a larger surface area across which heat can be dissipated. Conversely, related species living in very cold areas tend to have much shorter ears, providing less surface and retaining more heat.

THE SHAPE OF A CELL IS OFTEN RELATED TO ITS FUNCTIONS

No matter what variation in shape the cells of multicellular organisms may assume, they originate from almost spherical cells. At first these developing cells deviate from a sphere by flattening slightly where they are in contact with adjacent cells. As development proceeds, further deviation in shape may be a consequence of cells dividing only in certain planes or being subjected to certain stresses and strains due to the growth of adjacent cells. Although the manner of development and its consequent effect on shape is set by the heredity of the organism, the precise way such development is accomplished is still unknown. As described above, the orientation of cytoskeletal elements within a cell can be a determining factor in changing its shape.

Some cells have shapes or other characteristics that increase the surface in proportion to the volume. Extremely elongated cells with small diameters, like nerve or muscle cells, expose a large surface in proportion to their volume. Any elongating, flattening, or branching produces the same effect. Large plant cells like those of the water plant *Elodea* (Fig. 4.7) accomplish the same result with their central vacuoles that occupy most of the volume, in effect forcing the protoplasm to a peripheral position where chances for exchanges are enhanced.

Even unicellular organisms are usually specialized enough to vary from the generalized spherical shape. Some, like the *Amoeba,* change their shapes con-

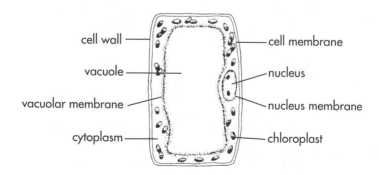

cell wall — — cell membrane

vacuole — — nucleus

vacuolar membrane — — nucleus membrane

cytoplasm — — chloroplast

FIGURE 4.7.   *A cell of* Elodea, *a plant found in freshwater ponds and lakes. Note the very large central vacuole, forcing metabolizing portions to spread to the periphery of the cell, close to the plasma membrane.*

stantly as they move from place to place (Fig. 4.8). Others have a fixed shape. In addition, they may bear specialized structures like cilia or flagella.

An early embryonic cell is one that is not overtly specialized in either shape or function. It is likely to be more or less spherical and performs all metabolic processes necessary for survival. When the cell becomes specialized (differentiated), it becomes most efficient in one or a few functions such as protection, secretion, contraction, message transmission, or support. Specialization in function usually means that there is a corresponding alteration in shape or size.

Cells that conduct materials or messages can do their jobs more efficiently with elongated shapes. The fewer membranes or intercellular barriers that have to be crossed, the faster will be transmission from place to place. Fluid-conducting cells of plants may be considerably elongated. For instance, tracheids, the hollow cells that line up to form the xylem tubes of vascular plants, can be as long as a millimeter. Conduction in nerve cells is also performed best by elongated cells, because elongation means fewer synapses for the impulse to cross (see Chapter 10). The same shape is advantageous to muscle fibers in making possible their operation over longer distances. Guard cells around pores (stomata) in leaves have shapes that can open or close the pore, depending upon the amount of water that they absorb. Root hairs, the most efficient absorbing cells of roots, are elongated and thus are in contact with more soil. The flat shape of the red blood cell makes gas exchange much faster between it and the surround blood plasma.

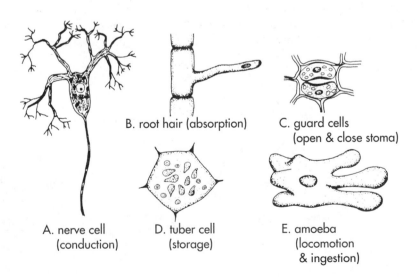

B. root hair (absorption)

C. guard cells
(open & close stoma)

A. nerve cell
(conduction)

D. tuber cell
(storage)

E. amoeba
(locomotion
& ingestion)

FIGURE 4.8.   *Relation of some cell shapes and functions.*

DIFFERENTIATION (SPECIALIZATION) OF CELLS OCCURS DURING THE EMBRYONIC PERIOD

Soon after an embryo has produced a sufficiently large number of cells by repeated divisions, they begin to differentiate toward specialized shapes and functions. The mechanisms involved in becoming differentiated will be discussed in Chapter 14. As differentiation proceeds, many cells lose certain vital functions, relying on others to work for them. For instance, cells that move deeply within the organism can no longer directly exchange important gases (oxygen and carbon dioxide) with the outside environment; they come to rely on cells that are part of transport systems for this function.

Some cells become so specialized that they lose their ability to divide. Such is the case with muscle cells, most nerve cells, and red blood cells. Inability to divide means, of course, that if they are killed, as heart muscles are during a heart attack, they cannot be replaced. The dead cells are replaced with scar tissue in the hearts of those who survive the attacks. Since scar tissue is composed of rather unspecialized cells that cannot contract like muscle cells, the heart's efficiency is impaired. The same is true for nervous tissue. The best a nerve cell can do is to regenerate its elongated portions if one or more have been severed, but the cell cannot ordinarily divide to replace lost nerve cells. (There is evidence that brain cells of a few vertebrate animal species can reproduce.) Red blood cells wear out quickly as they circulate (typical lifetime in humans: four months) and are destroyed. However, they are constantly replaced by division of the *stem cells* in bone marrow. Each time a stem cell divides, one of the two daughter cells remains a stem cell and the other one differentiates into a non-reproducing blood cell.

Some cells are specialized to an even greater degree. Not only do they lose the ability to reproduce themselves, but they die in the process of becoming specialized. Such is the case of fibers, cork, and some conducting cells of plants as well as the outer epidermal cells of the human skin (Fig. 4.9). Waterproofing of cork cells is efficient in stopping the loss of water, but these cells cannot survive after their own water supply is blocked. Plant fibers afford better support because they have thick walls, but when their walls become thick, the cytoplasm is crowded out.

The most efficient water- and mineral-conducting system of plants is composed of highly specialized tubular cells or of tubes made by linking a series of tubular cells. Of course, to be tubular the center of a cell must be hollow. This hollowing is accomplished through the removal of the cytoplasm and nucleus contained therein. The functions of support and conduction are combined in primitive tracheophytes (plants with conducting tissue) like ferns or gymnosperms in cells called *tracheids*. They are elongate, tubular, and thick-walled. Flowering plants are specialized further by having the functions of

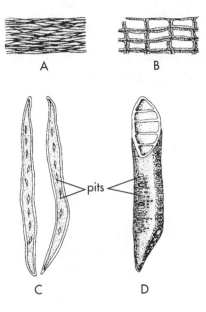

FIGURE 4.9.   *Specialized dead cells. A: outer epidermis of skin. B: cork. C: wood fiber. D: wood tracheal cell.*

support and conduction separated (Fig. 4.9 C and D). Support is accomplished by *fibers* that have thicker walls and smaller cavities. Conduction is accomplished by *tracheae* (tubes) that are composed of cells with much larger cavities whose end walls have disappeared.

No living cells, even in the most complicated body, are ever specialized to the extent that they perform only one function. In addition to their specialized function, they must carry on energy-yielding reactions, repair damaged portions, and sometimes divide. One may summarize by saying that specialization results in the loss or impairment of the ability to carry on some generalized functions; once a cell has reached a certain degree of specialization, there is no reversal or reacquiring of generalized functions previously lost.

CELLS REPRODUCE BY DIVIDING

From the simplest unicellular organisms to the most complex multicellular ones, there is a duplication of cells. If duplication is in a unicellular organism, the process is called *asexual reproduction* (Fig. 13.1 B); if duplication is in a multicellular organism, it is called *cell division*. Prokaryotic cells, having only a single circular chromosome and lacking a nucleus, go through much simpler reproductive actions than do cells of eukaryotic organisms, but even these must be very precise in duplicating and dividing the genetic information.

Mitosis Occurs in Eukaryotic Organisms

Mitosis is the method of cell division in all eukaryotic organisms. Strictly speaking, mitosis is only one of three sequential activities that together accomplish the making of two cells from one. The first activity is the exact replication of the DNA composing each chromosome within the nucleus. The details of how this is accomplished are presented in Chapter 12. Figure 4.10 shows that, before chromosomes are redistributed in mitosis movements, the cell is initially inactive in DNA production (a gap phase symbolized as G_1). Then it goes through a synthesis phase (symbolized as the S phase), followed by a second period without DNA synthesis (the G_2 phase). Only after the G_2 phase is the cell ready to perform mitosis movements.

Following the complicated movements of chromosomes that comprise mitosis (described in detail below), the cell completes its reproductive cycle by constricting down its middle and eventually becoming two distinct cells. This third and final stage is called *cytokinesis*.

As a consequence of cell division the two daughter nuclei are qualitatively and quantitatively the same as the nucleus of the mother cell from which they were derived. Therefore, all body (somatic) cells of a particular organism have exactly the same number and kinds of chromosomes. This is critical, since any cell might have to use any of the many thousands of genes that are positioned on the chromosomes.

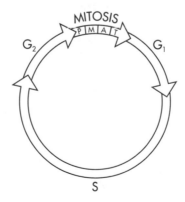

FIGURE 4.10.   *A complete cycle of cells going through division. Mitosis, in a formal sense, is only a small portion of the cycle. G_1 and G_2: gap portions of interphase. S: DNA synthesis portion of interphase. P, M, A, T: phases of mitosis (see text).*

Mitosis Is a Series of Movements to Distribute Chromosomes Evenly

Let us follow a typical plant cell as it goes through the series of events that together comprise mitosis. Then we will compare this with mitosis in animal cells. An nondividing cell (in *interphase*) has an intact nuclear envelope. During interphase, the chromosomes are loosely coiled threads (Fig. 4.11). These threads are reduplicated during the portion of interphase called the S phase (Fig. 4.10), but this doubling is not evident when the nucleus is examined by microscope. The nucleus also contains one or more nucleoli, which disappear during mitosis and reappear in the new cells.

Upon the initiation of cell division, the doubled chromosomes become more tightly folded upon themselves, condensed enough to be seen with the light microscope after DNA-binding stain is applied. This signals the first phase of mitosis itself, called *prophase*. As prophase continues, it becomes visually apparent that each chromosome has a characteristic length and that the cells of a particular species all have the same number and kinds of chromosomes. It also becomes apparent that each chromosome is actually doubled, because

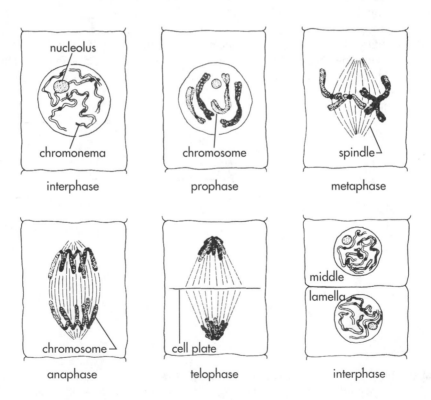

FIGURE 4.11. *Plant mitosis and cytokinesis.*

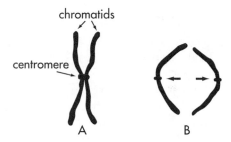

FIGURE 4.12. *A: A chromosome in either prophase or metaphase, composed of two chromatids held together by a centromere. B: At early anaphase the centromere has divided and each chromatid (now a chromosome) moves independently toward a pole of the cell.*

it consists of two parallel *chromatids*, each held to the other at a junction called the *centromere* (Fig. 4.12A). The centromere will be important in later events of the process. Each chromatid is destined to become an independent chromosome toward the end of mitosis. Simultaneous with the condensation of the chromosomes is the disappearance of the nucleoli and nuclear membrane and the formation of a structure known as the *spindle* (Fig. 4.11). This is a biconical set of microtubules, with the two points referred to as *poles* and the wide part across the middle as the *equator.*

The distinguishing characteristic of the next mitosis phase, *metaphase*, is the migration of all of the chromosomes to the middle of the cell, where each becomes attached to a spindle fiber at the equator of the spindle. It is at this time when the chromosomes are most condensed. Therefore, if the shapes and number of chromosome are objects of study, it is best to take photographs of the cell during metaphase. *Karyotypes*, or analyses of the chromosomes in an organism's cells, are usually made from metaphase cells.

Immediately following metaphase is a period of rather rapid movement of chromosomes toward the two spindle poles. This period, known as *anaphase*, begins with the splitting of each chromosome's centromere into two, each associated with one of the two chromatids (Fig. 4.11, 4.12). As soon as each has its own centromere, it can be called an independent chromosome. This sometimes leads to confusion on the part of a person trying to decide exactly what a chromosome is. By definition, each chromosome has one centromere. Thus, a pair of attached chromatids during metaphase constitutes one chromosome and a little later (in anaphase) each of those now-single chromatids with its centromere is a chromosome. Chromosomes continue to move until they reach the two poles at opposite ends of the cell. The mechanism by which chromosomes move during anaphase is not yet well understood, but it seems to involve both a pulling and a pushing action by spindle fibers.

It is very important to note that (a) each chromatid of a metaphase chromosome is identical to the other with which it shares a centromere and (b) these two identical units move away from each other in anaphase, heading for opposite poles of the spindle apparatus. Unless something very abnormal happens, two identical chromatids never move in the same direction. Thus, the two opposite ends of the cell form two congregations of chromosomes and each of those groups contains exactly identical numbers and kinds of chromosomes. Since the chromosomes are the sites for holding genetic information, mitosis ensures that the two daughter cells will have identical and complete sets of this information.

Upon reaching the poles, the chromosomes begin losing their compact shapes. A nuclear envelope forms around each set of chromosomes. From the time the chromosomes reach the poles to the reformation of the nucleus, the process is called *telophase*.

Cytokinesis Is the Physical Dividing of a Cell After Mitosis

Simultaneous with the events occurring in telophase, the machinery is in operation for the division of the cytoplasm, which is a separate event known as *cytokinesis*. It is also called *cell cleavage*, particularly when occurring in a developing embryo. In plants, the first sign of cytoplasmic division is a thickening of the spindle fibers (microtubules) at the equator during telophase. The fibers participate in forming a partition (cell plate), in conjunction with polysaccharide wall-forming material transported in vesicles from the Golgi apparatus. Gradually, the cell plate enlarges toward the periphery of the cell. The completed partition is called the *middle lamella*. Each new plant cell eventually deposits the polysaccharide cellulose on either side of this partition, to complete the building of its own cell wall.

ANIMAL AND PLANT CELLS REPRODUCE IN SLIGHTLY DIFFERENT WAYS

Animal cells have a *cell center* near the nucleus in interphase. As described earlier, it contains two microtubular structures known as centrioles (Fig. 4.1). Visible radiations form around each of the centrioles as they migrate to positions on opposite sides of the nucleus. Each centriole with all of its radiations is an *aster*. Each aster is in a polar position within the cell that is about to divide. Radiations between the two asters (collectively known as the spindle) are longer than radiations toward the peripheral region of the cell. The centrioles are usually duplicated during late telophase, so that each of the two daughter cells has a pair for use when it later divides.

Another important difference in animal cell division concerns the method of cytokinesis. Animal cells are bounded by flexible membranes rather than rigid walls. They constrict across the equatorial region of the cell, using a contracting ring of microfilaments, as described earlier.

CELL REPRODUCTION INVOLVES A SEQUENCE OF SIX PERIODS

The described events appear to be accurate for the majority of cells, but some variations should be expected. A summary of highlights in each phase follows:

Interphase (before mitosis begins)
Chromosome material like a tangled net of threads when viewed microscopically.
During S phase of interphase, all chromosomes double longitudinally, as DNA replicates.
Nuclear membrane present.
Nucleoli present.
Cell center beside nucleus in animal cells.

Prophase (first phase of mitosis)
Chromosomes condensing. Each seen to consist of two chromatids, attached at a centromere.
Spindle forming.
Nuclear membrane disappearing.
Nucleoli disappearing.

Metaphase
Chromosomes arranged in equatorial plane of spindle, with centromeres attached to spindle fibers.

Anaphase
New chromosomes (previous chromatids) moving toward opposite poles of spindle.

Telophase (last phase of mitosis)
Chromosomes in two bundles at poles and in the process of transforming to the interphase pattern.
Nuclear membrane reforming around each bundle of chromosomes.
Nucleoli reforming in each nucleus.
Membranes form at cell's equator.

Cytokinesis
Plasma membrane completely splits the original single cell into two cells (begun in latter portion of telophase).

THE CYCLE OF CELL DIVISION CAN BE QUITE RAPID

So many factors influence the rate of cell division that, it is impossible to give an average time taken for its completion. Some bacteria are known to divide in 15 to 20 minutes after their previous division. (However, it should be noted that bacteria reproduce in a more direct fashion than eukaryotic cells, without the elaborate production of spindles.) In the early stages of the fruit fly embryo, the time required for cell division has been measured at as little as 10 minutes, a rate that must be near the minimum for all of the steps involved. More typically, cells of the early frog embryo divide every 45 minutes to an hour. The early cleavages of the human embryo are quite slow, proceeding at about one cleavage per day for the first several days.

UNDERSTANDING WHAT CONTROLS CELL DIVISION IS CRUCIAL IN COMBATING CANCER

An oversimplified definition of *cancer* is that it is cell division occurring at the wrong time and place. In higher animals, only a relatively few types of cells (e.g., blood, skin, intestinal epithelium) should continue reproducing after the body reaches its mature size, but mitosis is quite normal and necessary during embryonic life and the growth period following it. If we could understand how most cells normally inhibit their own reproduction after adult body size is reached, we would be well along the path toward understanding how to prevent or reverse cancer.

A number of genes are involved in the control of cell division (see Chapter 12, oncogenes). Some of them cause the production of certain proteins known to interact in a cell's cytoplasm and to induce the various steps of cell division. It appears that nondividing cells also possess some sort of inhibitor of mitosis in their cytoplasm, which overrules the action of those inducing proteins even if they are present. If cytoplasm from a nondividing cell and the nucleus from a cell that has been rapidly dividing are artificially placed together, the resulting hybrid cell is almost always unable to reproduce. Thus, the normal control of reproduction could be both at the gene level (keeping oncogenes turned off) and at the cytoplasmic level (inhibiting the products of oncogenes).

Cancer involves much more than inappropriate cell division, of course. As a cancerous mass grows, it invades healthy tissue, causing dysfunction. It uses up many body resources, starving other cells. Perhaps most insidiously, individual cancer cells change their surfaces in such a way that they no longer adhere to each other as cells of normal tissues do. Therefore, cancer cells easily break free, to be washed passively to other places in the body by normal fluid movements (in lymphatic and blood vessels). Wherever such a cancer cell becomes lodged it can divide to form a new tumor. This spreading ability is called *metastasis* and is a major factor in many cancer cases becoming fatal.

REGENERATION AND WOUND HEALING REPAIR DAMAGE

If regeneration should be defined as replacement of parts, then it is possibly a universal phenomenon of living organisms, for bodies replace some cells continuously. Humans constantly replace blood, epidermal, and intestinal cells. Regeneration, however, is usually used in the sense of replacing organs or even larger portions of a body. Some invertebrate animals, such as the sponges, hydras, planarians, starfishes, sea cucumbers, earthworms, and crustaceans, routinely replace lost body parts.

In vertebrate animals, regeneration is not nearly as spectacular in extent. If a particular tissue cannot repair itself with its own kind of cells when it is injured, it is repaired with fibrous connective tissue. This is the material of scars. In humans, some of the more specialized cell types can be regenerated. Among these are portions of the liver and the kidneys. Of course, virtually all animals have some ability to close off recently cut surfaces. This phenomenon, called wound healing, is a rapid division of skin cells adjacent to the damaged area. The champions of vertebrate regeneration are salamanders, which can regenerate entire limbs. Additionally, some frogs can regenerate lost eye lenses.

Higher plants regenerate parts in several ways just as the animals do. The sealing of wounds or the growth of roots on cuttings are well known examples.

Whatever the organism in which it occurs, regeneration is a complex set of activities. There must first be a recognition by neighboring cells that a loss has occurred. Then, there must be generated a supply of cells to replace the lost ones. Although it is still somewhat a point of debate, it is most likely that these cells are produced by de-differentiation of specialized cells, followed by rapid division and re-differentiation into the various types needed to make the replacement part. This means that the normal and necessary inhibition of cell division that has previously been in place must be temporarily suspended. As newly produced cells accumulate, they must be arranged to make the shape of the lost part and differentiated according to their position in that newly developing shape. Finally, there must be recognition by the reproducing cells that the final size has been achieved, followed by cessation of further mitosis.

All of these activities are the same sort that occurred in creating the original anatomy during embryonic life. Indeed, many scientists in this field believe that regeneration is nearly identical to embryonic development except that it is confined to a particular portion of the body. There is much to be learned about the mechanism of regeneration, but the reward of understanding it might be an ability to induce regeneration in humans who have suffered accidental loss.

MEIOSIS IS SPECIALIZED CELL DIVISION TO MAKE SEXUAL REPRODUCTION POSSIBLE

Sexual reproduction (as contrasted to asexual) involves combining cells from two organisms to form a third organism. Cells that participate in this union are sometimes called sex cells, but their more formal name is *gametes*. Gametes made by a male are usually called sperm cells, or simply *sperm*. Gametes of a female are *eggs*.

Both Gametes and Spores Must Be Haploid

Of course, eggs and sperm are highly specialized in their shapes and particular roles played in the sexual reproduction event (see Chapter 13). Their shapes make them easily distinguishable from all other cells of the body. However, they are also specialized internally, particularly within their nuclei. In order to approach this topic, we must examine the number of chromosomes found in a normal non-dividing cell that is not a gamete. (These cells are termed *somatic* cells.) Such a cell almost always has two copies of each of its chromosomes; that is, it is *diploid* (the prefix *di* means two). In almost all cases, if a somatic cell has more than two copies of each chromosome (for instance, having three copies—being *triploid*), that cell becomes quite abnormal in many ways. If two diploid cells acted as gametes and became joined together, the resulting cell would have four of each type of chromosome (*teteraploid*) and would not be viable. (Some plants are exceptional on this point and tolerate their cells being greater than diploid.) Since sexual reproduction necessarily involves joining together two cells, it is necessary to do something that will avoid this problem. The obvious solution is to reduce the number of chromosomes in the gametes from two of each type to only one. Such a cell is said to be *haploid*. Then, when two haploid cells join, the resulting cell and all of its descendants that build the multicellular organism have the normal diploid number.

Reducing the nuclear chromosomal condition from diploid to haploid is accomplished during the cell divisions that make gametes. These special divisions are together called *meiosis*.

All of the above has concerned haploid cells that unite rather soon after their production, to make a diploid organism. Another use of meiosis occurs in the life cycle of some organisms that are haploid (rather than diploid) much of the time, such as algae and fungi. In these organisms, when two gametes unite, the resulting diploid organism rather soon goes into meiosis to produce haploid cells again. These cells, called *spores*, quickly undergo asexual reproduction to form (in the case of fungi, for instance) a multicellular haploid organism that represents the major portion of the life cycle. The difference between a spore and a gamete is this: the former is the cell that is the first cell of a haploid organism and the latter is one of two cells that must join to become a diploid organism. When an organism is one that makes spores by meiosis, it is called a *sporophyte.* All members of Kingdom Plantae include a sporophyte form, but the spores that most of them make are functionally equivalent to gametes, since plants are diploid through the bulk of their life cycle and their spores are not considered to be (or to produce) independent haploid organisms.

The First Cell Division of Meiosis Produces Haploid Cells

Meiosis occurs only in certain organs. In animals these are the *testes* (in males) and the *ovaries* (in females). In either organ, only certain cell populations undergo meiosis; others are supporting cells for these *germ cells.* In plants, the germ cells are also called *spore mother cells,* located in reproductive tissue of the sporophytic form of the plant. In any organism meiosis begins with a germ cell dividing in a process that superficially resembles mitosis, to produce two cells. However, this division differs significantly from mitosis in that it reduces the number of chromosomes of each type. That is, the two daughter cells are each haploid, whereas the original germ cell was diploid. Therefore, this cell reproduction event is termed *reductional division.* It is also sometimes called *meiosis I,* to distinguish it from a second portion of meiosis that also involves a cell division. As Figure 4.13 shows, reductional division is preceded by a replication of chromosomal material, causing each chromosome to be composed of two chromatids. The ensuing cell division involves the production of a spindle and chromosome movements much like those of mitosis. However, there is a very significant difference. In meiosis I, the two pairs of chromosomes (one from each parent) that are essentially identical in size, shape, and type of genetic information find each other during prophase and become aligned along their lengths. Another way of saying this is that pairs of *homologous* chromosomes enter into a condition of *synapsis.* Then, during anaphase, each chromosome (still consisting of two chromatids) separates from its homologous chromosome and goes to an opposite pole. Contrast this with anaphase of mitosis, where each chromosome splits its two chromatids apart, with each chromatid moving toward an opposite pole. In meiosis I, the result is that each daughter cell receives only one of the two homologous chromosomes (although each of the homologs consists of two chromatids). Thus, a cell that had two of each type of chromosome has produced two daughter cells, each of which has only one chromosome of each type. The number has been *reduced* from two to one.

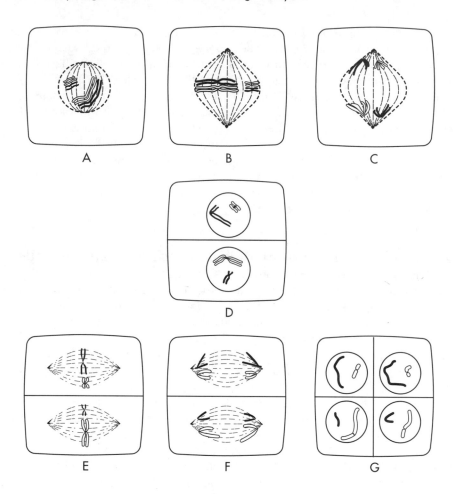

FIGURE 4.13. *Meiosis resulting in the production of spores. Dark chromosomes are from one parent, and the corresponding light ones are from the other parent. A: meiosis I prophase of spore mother cell (chromosomes in synapsis). B: meiosis I metaphase. C: meiosis I anaphase (each chromosome still composed of two chromatids). D: meiosis II prophase. E: meiosis II metaphase. F: meiosis II anaphase (individual chromatids migrating). G: set of four haploid spores in interphase.*

However, there is an important feature of these haploid cells that makes it necessary to perform another maneuver before gametes are formed. Although each chromosome exists singly, it carries two copies of each gene (one on each chromatid). If two such cells united in sexual reproduction, the resulting cell and its descendants would be technically diploid but would have four copies of each gene. Since this condition is likely to lead to gross abnormalities, it must be avoided. Meiosis II performs this service.

The Second Cell Division of Meiosis Yields Cells with Only One Copy of Each Gene

The essential difference between meiosis I and II is that the latter is not preceded by DNA replication. Each cell produced by meiosis I builds a spindle apparatus and lines its chromosomes up along its equator. These chromosomes, one of each type, then split during anaphase and each chromatid moves to an opposite pole. After cytokinesis, each of the resulting cells (four of them, since two cells move from meiosis I into meiosis II) has only one chromosome of each type and that chromosome consists of only one unit (previously called a chromatid). Thus, each cell has only one of each gene. Since the cells at both the beginning and end of meiosis II are haploid, this phase of meiosis is often called *equational division.*

After meiosis, the final activity necessary to yield functional gametes in animals is a maturation (differentiation) into either egg or sperm. Sperm cells must be specialized for movement. Therefore, each loses much of its cytoplasm and develops a tail. The sperm tail includes contractile microtubules arranged in a pattern nearly identical to that of a flagellum.

The maturing animal egg, by contrast, is likely to gain much cytoplasmic volume. This certainly includes a large quantity of the cytoplasmic components that are lacking in a sperm, so that the cell formed when these two gametes unite (called a *zygote*) will have a normal complement of these materials. Helping this to happen is a highly unequal division of the cytoplasm during the cytokinesis at the end of both meiosis I and II. In each case, one of the cells receives a nucleus but very little cytoplasm. It is not able to become a gamete, but sits at the edge of the viable cell (Fig. 4.14). It is called a *polar body.* The results of polar body production are (a) only one egg cell is produced from a germ cell that began meiosis and (b) that cell contains essentially all of the cytoplasm of the original cell.

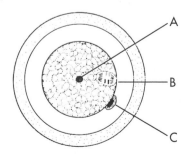

FIGURE 4.14. *A stage in* Ascaris *meiosis. In this roundworm, the egg's meiosis activity does not begin until a sperm cell has made contact. A: sperm nucleus. B: chromosomes in anaphase of meiosis II (results in elimination of second polar body). C: first polar body. Compare with Fig. 4.13.*

Depending on the species, the egg may also have a considerable amount of a nutrient material, the *yolk*. Yolk is a complex of proteins and lipids that provide energy and materials for building embryonic structures before the embryo can gain food from an external source. Obviously, a species whose embryo is isolated within a protective shell (such as birds and some reptiles and fishes) is the sort that must provide a very large amount of yolk.

Meiosis in Plants Is Somewhat Different

The production of *meiospores* in plants is the result of the same reductional procedure (nonplant spores are produced by mitosis). Meiospores are a part of most plant life cycles. The spore mother cell, which is diploid, is the precursor of the spores and is produced by the sporophytic form of the plant. As in the production of animal gametes, there is a series of two divisions with the resultant four haploid cells. An interesting observation is that in some life cycles of seed plants one spore is larger and becomes the only functional one, while the other three degenerate. This production of two sizes of spores is reminiscent of the production of one egg and two or three polar bodies in animals. Despite the similarity in how they are formed, eggs require fertilization for further development whereas spores do not.

TISSUES ARE EFFICIENT GROUPS OF SIMILAR CELLS

The concept of organization as a characteristic of living organisms may be seen not only in the manner in which chemicals are organized into subcellular structures and these are organized into cells, but also in the organization of cells into tissues. Any association of cells having a similarity of shape and activity and acting as a unit is a *tissue*. The term is loosely applied to include tissue with a high degree of uniformity of cellular structure such as muscles as well as to those composed of variable cells and intercellular materials that function rather closely as a unit like blood or complex tissues of plants.

One way to classify tissues is into the categories of *reproductive, embryonic,* and *mature*. Tissues are considered reproductive when they are gametes (or spores) or have the ability to develop into them. They are embryonic when they are in the process of developing into mature body tissues. Embryonic tissues (termed *meristematic* in plants) are noted for the active division of their cells and their lack of complete differentiation. Mature tissues are fully differentiated.

Animals Have Five Types of Tissue

Mature body tissues in higher animals are *epithelial, nervous, connective* (bone, cartilage, etc.), *vascular* (blood and lymph), and *muscular* (Fig. 4.15).

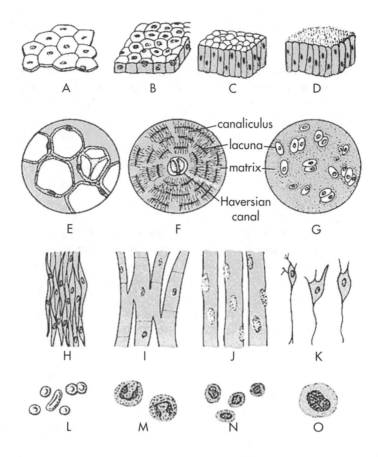

FIGURE 4.15. *Mature animal tissues. A: squamous epithelium (surface view). B: cuboidal epithelium. C: columnar epithelium. D. ciliated epithelium. E: adipose (fat). F: bone. G: cartilage. H: smooth muscle. I: cardiac muscle. J: skeletal muscle. K: nerve cells (neurons). L: erythrocytes (red blood cells). M: granular leukocytes. N: lymphocytes. O: monocyte. M, N, and O are all white blood cells.*

These types of tissue are distinguished by both the anatomy of their cells and their functions.

Epithelial tissues are found in thin sheets that cover surfaces and line cavities, or in masses that comprise the bulk of some glands. Where they are on surfaces or line cavities, they may protect, secrete, absorb, circulate materials, or act as limiting membranes. Where they are the main part of glands, they secrete or excrete. Thin sheets of epithelial tissue that cover surfaces and line cavities are nonvascular (i.e., not supplied with blood). With respect to the number of layers of cells, epithelial tissue may be classified as *simple* (one layered) or *stratified* (more than one layer). With respect to shape of cells, they may be classified as *squamous* (flat and thin like a scale, Fig. 4.15 A), *cuboidal* (like a cube, Fig. 4.15 B), or *columnar* (like a column, Fig. 4.15 C). There are other varieties that will not be considered, however. Epithelium may be ciliated (Fig. 4.15 D) as in a part of the respiratory tract of humans or in the Fallopian tube of humans. Cilia keep the surface clean in respiratory tracts and transport eggs in the Fallopian tubes. Squamous epithelium makes up the inner lining of blood vessels, lining of body cavities (peritoneum, pleura, and pericardium), the outer layer of skin, and respiratory surface of lungs. Cuboidal epithelium makes up the secreting cells of the thyroid gland, salivary glands, pancreas, liver, and sebaceous (oil-producing) glands of skin, as well as cells of small tubes like the nephrons in the kidneys. Columnar cells line much of the digestive tract.

Nervous tissue is made of cells called neurons (Fig. 4.15 K). While they vary a great deal in shape, they have distinct polarity and transport messages in only one direction (Chapter 10). Neurons sending messages toward the central nervous system are called *sensory* neurons and neurons sending messages from the central nervous system to muscles or glands are called *motor* neurons. Usually there are central nervous system neurons that connect sensory neurons with motor neurons (or with other neurons within the central nervous system). These connecting neurons are called either *association* neurons or *interneurons.* A neuron has extensions called *dendrites* and *axons.* The former transport messages into the cell body and the latter away from it. The axons of vertebrate animals may or may not be insulated with sheaths (called *myelin sheaths*), depending upon where they are located. Nerve messages are transmitted from cell to cell from the axon of one to a dendrite of another across tiny gaps known as *synapses.*

Nerves should be defined so that there will be no confusion of them with neurons. They are cords composed of neuron extensions (dendrites, axons, or both) lying parallel to one another and bound together with connective tissue. Nerves are easily visible to the naked eye, and some of the larger ones have been given anatomical names.

Connective tissue contains a variety of cell types. They are distinct from other kinds of tissue in having cells that are rather scattered through different kinds of intercellular substances such as fluids, flexible fibers, firm protein-carbohydrate material, or stony deposits. Sometimes vascular tissues are included along with supporting and connective tissues, but they will be considered separately in this arrangement.

Some types of connective tissue bind the other tissues of the body together. The cells are branching or spindle-shaped and are scattered among intercellular materials that are of watery to firm consistency. Most of these tissues have intercellular fibers that also vary in elasticity and density. The two most common types of fibers are white *(collagenous)* and yellow *(elastic).* White fibers predominate in tougher tissues such as ligaments and tendons, yellow fibers predominate in structures that stretch such as the walls of blood vessels. Connective tissues occur in sheets, cords, bands, loose entanglements, or masses. They are found in such diverse places as the vitreous humor of the eye, lymph nodes, spleen, bone marrow, tendons, ligaments, nerves, walls of blood vessels, skin, and between various tissues and organs.

Some types of connective tissue act to support and protect other tissues. These connective tissues, cartilage and bone, are much firmer than those described above. However, they are similar in having much intercellular substance, including fibers. *Cartilage* (Fig. 4.15 G) is found in rings in the trachea, on surfaces of joints, and at the ends of ribs, nose, external ear, and intervertebral discs. There are three types: the *hyaline* as in the tracheal rings, *elastic* as in the external ear, and *fibrous* as in the intervertebral discs. (All types are fibrous, but the latter is considerably more fibrous and tougher than the other two.) Cartilage tissue is unusual in that it does not have a blood supply.

Both cartilage and bone cells are located in cavities (*lacunae*) and separated by intercellular substances known as *matrix*. The matrix of bone is about 85 percent calcium phosphate deposited within a fibrous framework. The arrangement of bone cells within this matrix is quite intricate (Fig. 4.15 F). Bone cells are arranged in several concentric circles (*lamellae*) around a central canal (*Haversian canal*). The Haversian canal contains a small artery and vein as well as a nerve supply. The lacunae have small canals (*canaliculi*) radiating from them so that there are connections between lacunae and between the lacunae and the Haversian canal. These small canals facilitate movement of material to and from the blood vessels of the Haversian canal.

Blood and lymph are fluids that comprise the category called vascular tissue. Blood cells are of two general types. The *erythrocytes* (red blood cells, Fig. 4.15 L) are by far the more numerous, there being around four to five million per milliliter of circulating blood. Although they originate from nucleated cells, they have no nuclei at maturity. They are coin-shaped but thin in the middle, a form often called biconcave. They contain the pigment *hemoglobin,* which combines quickly and reversibly with oxygen. Attached in this manner, oxygen is carried through the system of blood vessels. It eventually reaches the thin-walled capillaries, where it is released to enter the tissues.

Leukocytes (white blood cells) number about eight to ten thousand per milliliter of blood. As the name implies, they are colorless. They are nucleated and have some ability to perform amoeboid motion. They may be grouped as *granulocytes* (Fig. 4.15 M), *lymphocytes*, (Fig. 4.15 N), and *monocytes* (Fig. 4.15 O), distinguishable from each other by size and internal structures. They are a part of the body's defense system and at least some of them engulf bacteria and other foreign cells. Some diseases are diagnosed by finding disproportionate numbers of some types of leukocytes. The whiteness of pus is due largely to accumulations of dead leukocytes at any area of infection.

Blood also contains fragments of cells called *platelets*. These carry materials necessary in the forming of clots to close gaps in damaged blood vessels. Platelets originate from the same mass of cells within bone marrow that produces erythrocytes and leukocytes.

The intercellular substance of blood is *plasma,* largely water in which various materials are dissolved. Some plasma and leukocytes escape from the vessels and are recollected in a system of lymphatic vessels that returns these materials to the circulatory system. This recollected material is *lymph.*

Muscular tissue is responsible for movement of an animal's body. The cells are elongated and contain contractile fibrils within them. There are three types of muscles: *skeletal, cardiac,* and *smooth.* Skeletal muscles (Fig. 4.15 J), the bulky ones usually connected to the skeletal bones and functioning in the moving of certain parts of the body, are composed of multinucleate *muscle fibers.* A number of embryonic cells (each with the normal number of one nucleus) join together to form a muscle fiber. As they do so, the plasma membranes between them disappear so the fiber comes to include a number of nuclei but only a single membrane binding together the cytoplasm of all of the original cells. The general name of such a composite is *syncytium.* The nuclei occupy a peripheral position in a fiber, enabling it to contract without hindrance. Skeletal muscle fibers have alternating light and dark bands (not shown in the figure) that give them a striped (striated) appearance. Therefore, skeletal muscle is sometime also called *striated muscle.* Because most skeletal muscles in a human are under the control of the will, they are also sometimes described as *voluntary.*

Cardiac (heart) muscles (Fig. 4.15 I), unlike skeletal muscles, have distinct cells with only one nucleus apiece and are involuntary. They are, however, striated. Their nucleus occupies a central rather than peripheral position. Cardiac muscles are unique in having transverse markings called *intercalated discs.* When examined carefully, each disc is found to be a very tight junction that helps to hold two adjacent cardiac cells together in a fashion that allows quick and efficient contraction. In addition, intercalated discs have tiny porelike areas that facilitate transmitting the message to start contracting.

Smooth muscles (Fig. 4.15 H) are composed of uninucleate, spindle-shaped cells and are involuntary. They are the slowest to contract after being stimulated (skeletal muscles are the fastest). They are the most primitive muscle types in the sense that their cells look most like non-muscle cells. Smooth muscles are part of the walls of tubular structures such as the digestive tract, reproductive tubes, and blood vessels.

Higher Plants also Possess Several Tissue Types

Plant tissues that include actively dividing, unspecialized cells are *meristems* (adj., meristematic). Meristematic tissue is located in the apex (tip) of roots and stems where it is known as *apical meristem,* or in lateral layers in roots and stems where it is known as *cambium* (Fig. 8.9). Growth and maturation of apical meristems lengthen the roots and stems. Tissues maturing from apical meristems are *primary tissues.* Growth and maturation of cambium increase the diameter of roots and stems in many plants. Vascular cambium located between the xylem and phloem produces xylem toward the center and phloem toward the outside. Functions of these structures are described below and in

Chapter 8. Cork cambium, usually located in the cortex, produces cork toward the outside and cortical cells on the inside. Tissues derived from cambium are *secondary tissues*.

Mature tissues of higher plants can be classified as either *simple* or *complex* tissues. The simple ones are *epidermis, cork, parenchyma, collenchyma,* and *sclerenchyma*. The complex tissues are *phloem* and *xylem*.

Epidermis (Fig. 4.16 A) is ordinarily one layer thick and present on the outside of young stems and roots as well as on leaves, flowers, and fruits. The cell shape varies. Epidermis is always alive and covered on the outside with a cutin (wax) of different thicknesses. Ordinarily epidermis contains very little or no chlorophyll.

Cork (Fig. 4.9 B) is a tissue that replaces epidermis in older stems and roots. It varies in color and texture, from cork that peels in thin sheets as in the pine to cork that crumbles into grains as in oaks. Cork cells are block-shaped and arranged in layers of a few to many cells in thickness. All cork cell are dead at maturity.

Parenchyma (Fig. 4.16 B) is composed of thin-walled living cells of various sizes and shapes. Some of them contain chlorophyll and capture energy by photosynthesis. Others may store food, as with the flesh of an apple.

Collenchymatous cells (Fig. 4.16 C) are somewhat elongated and have walls that are unevenly thickened, often in the corners where cells are adjacent. The

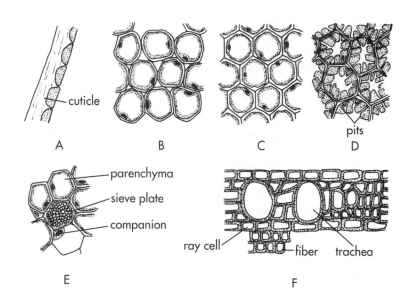

FIGURE 4.16.   *Some mature plant tissues. A: epidermis. B: parenchyma. C: collenchyma (sunflower stem). D: sclerenchyma (stone cell of wall of hickory nut). E: phloem (cross section). F: xylem (cross section).*

cells are thus mechanically strengthened. They are commonly found in the outer cortex of herbs and in veins of leave where they give a measure of support. Collenchymatous cells are usually alive. They are much less common than parenchyma.

Sclerenchymatous cells are uniformly thick-walled and most are dead at maturity. Tough cells that generally conduct and support, they vary in shape from the nearly polyhedral cells *(stone cells)* that make up the hull of a nut (Fig. 4.16 D) or the grains in the flesh of a pear to the elongated fibers (Fig. 4.9 C) and tubes (Fig. 4.9 D) of stems, roots, or leaves. The walls are thickened by additional deposits (secondary walls) of cellulose and lignin that may be variously sculptured. Many of them have conspicuous pits (pores) through which cytoplasm of adjacent cells were connected when the cells were alive.

Phloem (Fig. 4.16 E) is one of the complex tissues, composed of different kinds of cells. It may contain *sieve tubes, companion cells, parenchyma,* and *fibers* (sclerenchyma). Sieve tubes and companion cells are types of parenchyma. Sieve tubes are elongated cells attached end to end, thus forming tubes. The end walls are perforated like a sieve, through which the cytoplasm of one cell is continuous with that of another. They have no nuclei at maturity. Sieve tubes conduct manufactured food and organic substances through the plant. Adjacent to sieve tubes are the companion cells, which are also elongated but have small diameters. They have large nuclei and dense protoplasm and they are thought to have some controlling influence over the functioning of the cytoplasm in the adjacent sieve tubes. Fibers as well as parenchymatous cells are also present in the phloem.

Xylem (Fig. 4.16 F), the other complex tissue, is in everyday language simply wood. The bulk of it (secondary xylem of woody plants) is composed of *tracheae* (vessels) and fibers. In addition it may contain a little parenchyma, some radially arranged into rays. Parenchymatous cells, if present, are storage cells, or they may line tubes and secrete materials like latex or resins. Furthermore, the rays are routes of lateral transport. The tracheae are derived from columns of cells whose end walls have disappeared and thus left tubes. They are sclerenchyma. Cross sections of tracheae, called pores, are clearly visible with the naked eye in transverse sections of wood. Fibers are spindle-shaped with small diameters. Both tracheae and fibers are dead at maturity. The movement of large amounts of water and minerals throughout the plant occurs in the xylem. In addition, mechanical support is provided by the thick-walled fibrous cells.

ORGANS AND SYSTEMS ARE THE MOST COMPLEX LEVELS OF ORGANIZATION

Tissues working together as a functional unit make up an *organ.* Human examples are brain, heart, kidney, stomach, and liver. Organs in plants are usually distinguishable externally as differentiated parts of a higher plant such as roots, stems, or leaves.

Organs may act together as a functional unit known as a *system*. Examples of systems in higher animals are muscular, circulatory, nervous, skeletal, integumentary, digestive, respiratory, excretory, endocrine, and reproductive. It is difficult to recognize any equivalent kind of higher organization in plants.

MOST ORGANISMS SHOW TYPICAL BODY SYMMETRY

Organisms, almost without exception, are formed into symmetrically shaped bodies. If the body is irregularly shaped such as the *Amoeba*, it is *asymmetrical* (Fig. 4.17 A). If it has a definite shape, it is symmetrical in one style or other. The most common symmetrical shapes are *spherical, radial*, and *bilateral*. Spherical symmetry (Fig. 4.17 B) is exemplified by the ball shape of the Protistan organism *Volvox* (although this is actually a colony of cells, not a true multicellular organism). Radial symmetry (Fig. 4.17 C) is the shape of a plate, light bulb, or jar. The objects are circular in one dimension. Actually, if

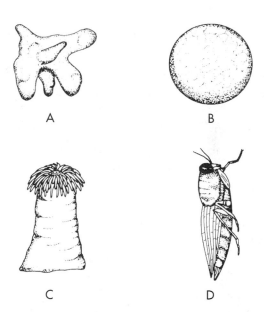

FIGURE 4.17.   *Asymmetry and symmetry in bodies. A: asymmetry of Amoeba. B: spherical symmetry of* Volvox. *C: radial symmetry of sea anemone. D: bilateral symmetry of insect.*

one draws a line or cuts through any radius from the margin through the center to the opposite side, the object will be divided into two similar halves. Living examples are jellyfish, sea urchins, and morning glory flowers. Bilateral symmetry (Fig. 4.17 D) is a shape in which there is only one plane through which an imaginary line or division can be made to produce two halves (left and right) nearly alike. The human body can be divided into a top and bottom or a front and back that are not alike. There is only one plane that can divide it into almost identical right and left sides. Additional examples of bilateral symmetry are most animals (all that are not radially symmetrical) and orchid flowers.

Other terms are useful to describe orientation with respect to direction of movement or gravity. The part that is in front as the animal moves is *anterior* and the back is *posterior*. In most animals, the anterior portion includes the head with its sensory apparatus. This arrangement is valuable in that the portion usually moving first into a new and potentially hazardous environment is also the portion best equipped to sense those hazards. The part toward the earth is *ventral*; the part away from the earth is *dorsal*. It is probably best to be consistent and use these terms in a comparative rather than a literal sense; for example, the chest and abdomen of an upright mammal like the human is ventral, not anterior. A similar situation is encountered in some mollusks such as squids and octopuses. Their heads are actually ventral.

STUDY QUESTIONS

1. What is the relationship between the resolving power of any microscope and the wavelength of the illuminating medium used with the microscope? In this context, what is the value of an electron microscope?

2. Why are viruses not considered to be cellular?

3. List at least five places in a eukaryotic cell that include membranes, and describe the role(s) of the membrane at each location.

4. What is the difference between smooth and rough endoplasmic reticulum?

5. What is the relationship (physical and functional) between the Golgi complex and lysosomes?

6. Of what use are contractile vacuoles to some Protozoa? What groups of Protozoa do not have them? Why are they not needed by those groups?

7. Describe the components of a typical plasma membrane, and indicate why the term *fluid mosaic* is appropriate for it.

8. Discuss the relation of cell size to function.

9. What do mitochondria and chloroplasts have in common? How are they different?

10. There are three general types of cytoskeletal materials. Name and describe each.

11. What are some cells that become so specialized that they die? Do they serve a purpose though dead?

12. What is meant by the term *cellular differentiation?*

13. During what phase of mitosis do the chromosomes align themselves in the equatorial region of the spindle?

14. In what ways does mitosis of plants and animals differ?

15. As clearly as possible, state the primary functions of the similar processes of mitosis and meiosis. In what important ways do they differ? How are those differences related to the functions of mitosis and meiosis?

16. Of the two divisions occurring during meiosis, which is reductional? Why is it so called?

17. There are two things wrong with the sentence that follows; indicate these areas, and correct them. "Cytokinesis, which occurs during the division of somatic cells but not during meiosis, is the building of a cell wall."

18. What is the functional difference between a gamete and a spore?

19. What is the relationship between cancer and mitosis?

20. What is the experimental evidence that there are mitosis inhibitors located in the cytoplasm of nondividing cells? What does this have to do with question 18?

21. What are the problems that must be overcome if an organism is to successfully regenerate a lost body part?

22. Define these terms: *tissue, organ, system.* Why is the order of these terms in the previous sentence appropriate?

23. What are the mature tissues of higher animals? Which of them may be ciliated? Which may have a considerable amount of intercellular substance?

24. Distinguish between neurons and nerves. Name the three kinds of neurons.

25. In what way does cardiac muscle resemble skeletal muscle? In what ways does it differ?

26. What are the tissues found in higher plants? Which are complex in the sense that they are composed of two or more simple tissues?

27. What is the difference between radial symmetry and bilateral symmetry? Which describes the shape of a cow? Of a starfish?

5
ENERGY SOURCES
FOR ORGANISMS

What do we know about:
—*the physical laws governing energy?*
—*how some organisms are equipped to capture the Sun's energy?*
—*the molecules that store energy for organisms?*
—*limiting environmental factors for photosynthesis?*
—*how nonphotosynthesizers get energy-rich molecules?*
—*the nature of predator–prey and symbiotic relationships?*
—*why decomposition of dead organisms is an important activity?*
These and other intriguing topics will be pursued in this chapter.

ENERGY CAN EXIST IN TWO FORMS

Not all the relationships of energy to life are understood; however, it is well known that life cannot be sustained without its availability. Hardly any cellular activity, however simple, is carried on without it. Although physicists have devised much more precise (mathematical) definitions for energy, we can use a very simple description—*energy is the capacity to do work upon some physical object.* In an organism, the work to be done includes building or tearing down the complex structures described in the previous two chapters (molecules, cells, tissues, etc.), moving objects (chromosomes, cilia, muscles, etc.), warming or cooling the body, and a few other activities.

Energy plays a major part in the complex story of food synthesis and use. For convenience, *foods* may be described as molecules or (combinations of molecules) that store energy for future release. If energy is stored in foods or any other materials, it is said to be in its *potential* form. That is, it is not currently performing work, but it is present and may be available to do work at a later time. When that time arrives, if the stored energy can be released and actually perform an act of work, it is said to have been converted to *kinetic* energy. These two forms of energy are interconvertible, a fact that has been formalized by physicists within a general statement often called the *first law of thermodynamics.* That law also includes the concept that all of the energy of the universe is in a constant amount. So, if one observes that a quantity of energy has appeared where it was not formerly present, it may be reliably said that this energy did not suddenly come into existence, but that it moved there from somewhere else or became more obvious for measurement by being converted from one form to the other.

Another law of physics has importance for biologists—*the second law of thermodynamics.* Simply put, it states that energy conversions (from kinetic to potential, or vice versa) are never totally efficient. That is, they result in less than 100 percent of the original energy being available. The lost energy does not disappear, but it is converted into heat and is no longer available for doing useful work. Another aspect of the second law is that the universe has a long-term tendency toward becoming less ordered, more random in its structure. As we will see in this chapter and the next, living organisms have the ability to use energy for building complex molecules out of simpler ones, but both their construction and their ensuing maintenance require input of energy. Whenever energy for maintenance becomes unavailable (as when the organism dies), those complex structures show the effect of the second law of thermodynamics as they decompose to simpler materials.

Except for a few types of bacteria, all organisms obtain the bulk of their stored energy by breaking the bonds of organic molecules—notably carbohydrates, fats, and proteins. (Those few exceptional bacteria break bonds of *inorganic* molecules to obtain energy.) All chemical bonds are formed by the use of energy; the atoms have to be pushed into close proximity in order for bonding interactions to happen. Consequently, if any bond can be induced to become unstable and break, the result will include the release of nearly the same amount of energy that was stored in it. Potential energy will have been converted to kinetic energy. The key to gaining energy by breaking bonds of food molecules is to make them unstable by adding a smaller amount of energy than will be released when they break. This is not unlike what happens when a person applies a match (providing energy) to the surface of a piece of charcoal. Once some bonds have become destabilized by the added energy, much more energy is released from the burning of the charcoal than was provided by the match.

The foods of green plants, some Protista, and some bacteria originate through the process of photosynthesis, a major topic in this chapter. Animals and other organisms unable to perform the chemistry of photosynthesis must consume foods originally produced by photosynthesizers. Most foods are complex and have to be simplified (digested) into substances that can pass through cell membranes. They are then used directly for energy or stored in various forms.

PHOTOSYNTHESIS PROVIDES ENERGY-RICH MOLECULES FOR ORGANISMS

All organic substances of organisms, whether or not they are storing energy for later use, can be traced back to a vastly important set of chemical reactions collectively called *photosynthesis*. Much of this chapter's emphasis is on producing and using the molecules that store energy, but the broader role of photosynthesis should not be forgotten.

As the prefix *photo-* suggests, light is a necessary ingredient of photosynthesis. To begin to understand the process of converting light (kinetic) energy into the potential energy of chemical bonds, it will be necessary to examine the respective roles of light, chlorophyll, water, carbon dioxide, and enzymes, since all of them are necessary for photosynthesis.

Only Some Forms of Light Energy Can Be Captured

The relationship of light to plants has been known for a long time. Even Aristotle noticed that it was necessary for the greening of plants. It is now recognized that the green pigment is chlorophyll and that it is able to harvest the energy of light for use in chemical transformations. The main source of light on Earth is the Sun. Sunlight reaches chlorophyll in different qualities and quantities, dependent upon many influences of the environment.

When sunlight is passed through a prism it separates into its component wavelengths. Some of those wavelengths are smaller or larger than those that our eyes can perceive (e.g., ultraviolet or infrared). Since they do not drive photosynthesis, they will not be discussed here. The wavelengths that we perceive, collectively called visible light, are the regions of a prism-generated spectrum having the colors of the rainbow. Some colors are more important to plants than others. The very fact that leaves look green means that green wavelengths are not used as much as other wavelengths, for their reflection to the eye gives the color. Of all the colors, red is best for photosynthesis; blue is almost as good. Whenever light passes through some medium, its quality (proportions of various wavelengths) is altered because of absorption and reflection. Thus, light passing through clouds, a forest canopy, or water will be altered to some extent. Green plants growing at a depth of 20 meters in water get chiefly blue-green rays. Even the distance through which light travels in the atmosphere affects its quality. This means that mountain plants are exposed to a different quality from those growing at sea level.

Plants grow in situations where there are wide variations in light intensities. Tremendous variations arise from the position of the earth's surface with respect to the direction of the sun. Sunlight hitting the surface head-on in the low latitudes is much more intense than light striking it a glancing blow in the high latitudes (Fig. 5.1). Daily variation in a particular spot is due also to the angle of exposure, just as are seasonal variations. Even at the same latitude at the same time of day, variations in the earth's surface may produce striking differences. A mountain side rising at a 45-degree angle does not get the same amount of light as a level surface. Similarly, slopes exposed in different directions (as those facing north, contrasted with those facing south) vary so much that they may have distinctly different groups of plants. Shading by clouds, fog, smog, buildings, or other plants may also produce significant effects.

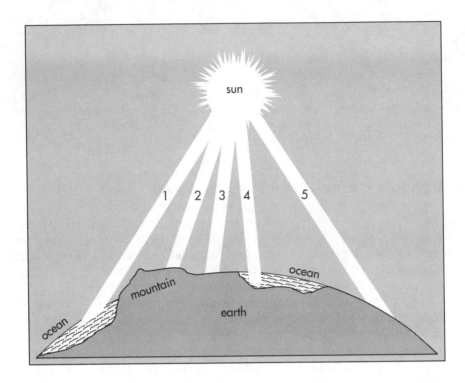

FIGURE 5.1.   *Effects of equal quantities of light striking surfaces at different angles. Compare beam 3 striking the surface almost at a right angle near the equator with beam 5 striking the surface at an oblique angle near the poles. Beam 2 strikes a mountain slope at a right angle in middle (and high) latitudes. Beam 4 striking ocean at almost a right angle penetrates water much deeper than beam 1 striking ocean at an oblique angle near the poles.*

In forest areas there is considerable variation in the amount of light received by trees of the forest canopy and the plants growing beneath them. Some of the smaller plants such as bloodroot compensate for their short stature by appearing and maturing before the leaves emerge on the trees above them; others have broader but thinner leaves that expose a larger light-absorbing surface. Leaves on trees growing in the shade commonly have twice as much surface area as leaves on trees of the same species growing in the open sunshine. Of course everybody recognizes that some plants simply cannot grow in the shade. In any dense forest many of the smaller trees die, it is believed, primarily because they do not get enough light. For the same reason, some pines cannot get started in a close hardwood forest. The lower branches of trees die and fall as a result of too little light. Green plants living in water cannot possibly grow deeper than light of sufficient quantity and quality for photosynthesis can penetrate, which under most conditions is less than 20 meters.

When water contains many impurities, light penetration is limited because most of it is reflected or absorbed almost at the surface.

Not all light falling upon leaves is used in photosynthesis. Transeau found that only 1.6 percent of radiant energy falling upon a field was used by a corn crop. Of the light that strikes a leaf, between 60 and 80 percent is absorbed. Some of the remainder is reflected from its surface and some is transmitted through the leaf. The relative percentages of absorbed, reflected, and transmitted light depend on the leaf surface, internal structure, and thickness of the leaf. Of the light that is absorbed by the plant, only a small percentage is actually absorbed by chlorophyll (the first molecule that can react to capture light—see below). It is estimated that only about 4 percent is actually absorbed by chlorophyll. Photosynthesis seems to be a most inefficient process until one realizes that about 90 percent of the light that does get absorbed by chlorophyll is eventually converted into chemical energy.

Any source of visible light may be used for photosynthetic activity. If the sun is the source, photosynthesis obviously ceases at night (except in bright moonlight when the rate is low). If artificial light is present as it is under street lights, vegetation may be distinctly larger there than it is in adjacent places less favorably exposed.

As our discussion approaches the actual mechanisms by which light energy is captured, it is necessary to add a complication. Light can be considered a wave-like entity, as above, but it also sometimes acts as if it were a stream of tiny particles called *photons*. For some, it is easiest to envision the conversion of light's energy into chemical bonds by imagining the impact of these very fast particles upon atoms within a plant's leaf. Just as a ball would be set into motion by the impact of another ball, so do some components of an atom gain extra motion if hit by photons. In the next section, we begin to see how photons do this and how it sets into action a series of motions until finally atoms are pushed together to make bonds.

Chlorophyll Is the First Molecule Reacting Usefully to Light's Impact

Although chlorophyll was discovered earlier, its essential role in photosynthesis was not proved until Dutrochet's work in 1837. Now we know that all chlorophyll is not of one type, but exists in several slightly different forms. Furthermore, it is apparently combined with other substances in the chloroplast in a chemical complex much different from extracted chlorophyll. Of the several kinds of chlorophyll known, only two of them—*chlorophyll a* and *b*—are found in higher plants. Both forms are very complex (each containing 55 carbon atoms), not fitting any of the major categories of organic molecules that were presented in Chapter 3. Chlorophyll a is the only form that initiates the first step in photosynthesis. However, chlorophyll b, which absorbs a slightly different range of wavelengths than does chlorophyll a, can pass its captured energy to chlorophyll a. Thus, the combination of both forms in a leaf allows the utilization of more of the light that strikes there.

Chlorophyll is located in the thylakoid membranes of distinct organelles, the chloroplasts (Fig. 4.3). However, the bacteria that also perform photosynthesis do not have chloroplasts. Instead, they have the same sort of membranes scattered through their interior. In chloroplasts, the patches of compacted membranes are the grana shown in Figure 4.3.

The Kinetic Energy of Photons Is Converted to Bond Energy of Molecules

It is appropriate to look first at an overview of what happens when chlorophyll is struck by light. In general, what happens is that light causes an electron of a chlorophyll molecule to become more energized. Such chlorophyll is described as *excited* or *activated* and is able to initiate important chemical changes. The energized electrons quickly leave chlorophyll and are captured for a moment by a nearby molecule before being transferred through a chain of molecules in two complicated processes known as *cyclic* and *noncyclic photophosphorylation*. The end products of this complex series of reactions are ATP (adenosine triphosphate), NADPH (reduced nicotinamide adenine dinucleotide phosphate), and O_2 (molecular oxygen). The first two of these products carry some of the energy that was originally provided by the impact of the light (photons). They can be used by the plant cell for a variety of energy-requiring activities, including the manufacture of glucose. Now, let us examine these activities in more detail.

Cyclic and Noncyclic Photosynthesis Produce Energy-storing Molecules

Existing in many copies throughout the thylakoid membranes of a chloroplast are complex sets of molecules. These *photosystems* include sets of chlorophyll molecules surrounding one special chlorophyll at a place called the *reaction center* (Fig. 5.2). Light hitting any chlorophyll initiates the passing of its excited electrons toward this reaction center. When the chlorophyll at the reaction center receives a pair of electrons, it quickly passes them on to a complex molecule called the *primary electron acceptor.*

There are two slightly different photosystems, called *photosystem I* and *photosystem II*. Each handles light in the same general way, but the electrons passed to the primary electron acceptor are eventually used in quite different ways. Look first at Photosystem I as it initiates the process of cyclic photophosphorylation.

Cyclic Photophosphorylation Produces ATP

As Figure 5.3 shows, when photosystem I generates a pair of energy-rich electrons they are passed to a row of molecules collectively called an *electron transport chain* (or *electron transport system*). Embedded in the thylakoid membrane very close to the photosystem, these molecules (mostly proteins)

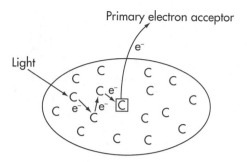

FIGURE 5.2.   *How a photosystem converts energy of photons into energy of electrons. Each C represents a chlorophyll molecule and the boxed C is the chlorophyll at the reaction center.*

pass electrons from one to the next. As this occurs, some of the electrons' energy is used to create a chemical bond. The reaction that occurs involves a modified nucleotide, adenosine diphosphate, to which is added a third phosphate group (symbolized as P in Figure 5.3) to create one molecule of adenosine triphosphate (Fig. 5.4). The covalent bond that links the added phosphate group to the ADP portion holds energy that can be traced all the way back to the kinetic energy imparted by the impact of light. ATP can be stored in a cell for only

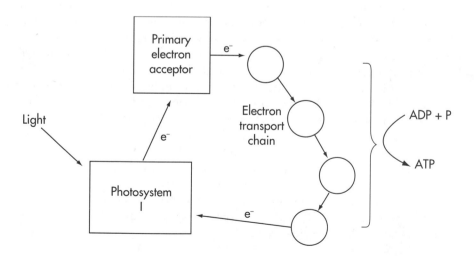

FIGURE 5.3.   *Cyclic photophosphorylation converts the kinetic energy of electrons to potential energy of a bond in ATP. One ATP is made for every two electrons moved through these pathways.*

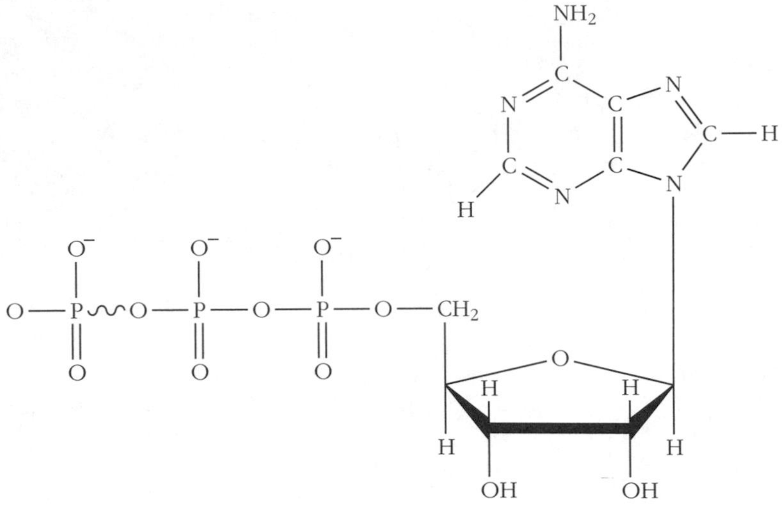

FIGURE 5.4.   *A molecule of adenosine triphosphate (ATP). The wavy line represents the bond between the terminal phosphate group of ADP and the phosphate group added to create ATP.*

a short time, after which it breaks that same bond and releases the stored energy. It is ATP that is most often the chemical source of energy for a cell when it must build, tear down, move materials, and so on. Indeed, the success of almost any activity in giving a cell some energy can be measured in the number of ATP's that are built. The precise manner in which the energy of electron flow is converted to the building of a bond of ATP is now fairly well known, and will be described in a section below (see *chemiosmosis*).

Why is this entire process called cyclic photophosphorylation? The first word refers to the final destination of the electrons that have moved through the electron transport system. They complete a circle by returning to the chlorophylls from which they were removed by the original impact of photons (Fig. 5.3). This is valuable, since these chlorophyll molecules could not repeat their valuable response to light unless replenished with electrons. The second descriptive word, photophosphorylation, is a compound word. The first portion reminds us that photons drive the process and the second portion indicates that something (ADP) has been phosphorylated. That is, a phosphate group has been added to yield ATP.

Noncyclic Photophosphorylation Produces Both ATP and NADPH

Noncyclic photophosphorylation also involves photosystem I, but we will first examine the impact of light upon chlorophyll in photosystem II (Fig. 5.5).

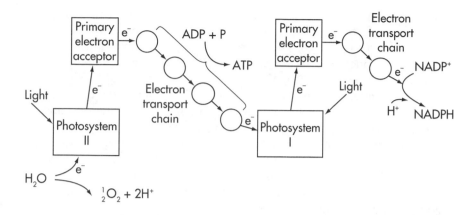

FIGURE 5.5. *Noncyclic photophosphorylation converts the kinetic energy of light into the energy of bonds within both ATP and NADPH. Note that both photosystems are involved in this set of reactions. Photolysis is shown just below the box marked Photosystem II.*

It also causes light's energy to be converted to bond energy, but two different kinds of energy-storing molecules are made: ATP (as in cyclic photophosphorylation) and NADPH.

When photons strike photosystem II, the same sort of electron movement occurs as is seen in cyclic photophosphorylation. The initial results are nearly identical to those of cyclic photophosphorylation: some of the energy of flowing electrons is used to make ATP. This pathway is more efficient than the cyclic method, as two molecules of ATP are made for every pair of electrons transported.

Rather than returning to the chlorophyll of the photosystem that released them, these electrons drop into chlorophyll of photosystem I. Why does the chlorophyll of photosystem I have room for these electrons? It is because this chlorophyll has also been stimulated by light to lose some electrons for running a quite different energy-capturing pathway. The energy-rich electrons released from photosystem I move to a different electron transport system than that used in cyclic photophosphorylation. This system passes two electrons (negatively charged, of course) to a molecule of NADP, which is then able to capture a positively charged atom of hydrogen (H^+) from the surrounding fluid. The resulting molecule, NADPH, carries both a hydrogen atom and some energy that can be traced back to light energy that had arrived at photosystem I. As will be seen below, both the hydrogen and the energy will eventually be used to build a more permanent energy-storing molecule.

Since the electrons leaving photosystem II are used to replenish the chlorophyll of photosystem I, there must be some way to replace them so that photosystem II can repeat its activity. This is accomplished by a reaction called *pho-*

tolysis. As shown in Figure 5.5, a molecule of water breaks down, yielding three materials: two electrons, two positively charged hydrogens (equivalent to protons), and one oxygen atom. The electrons are given to chlorophyll in photosystem II. After photolysis has occurred twice, the resulting oxygens combine to form molecular oxygen (O_2). Although oxygen is not useful during any step of photosynthesis and much of it is passed out of the plant into the atmosphere, it has enormous value in releasing energy within mitochondria of plants and other organisms (see Chapter 6). The hydrogen ions released by photolysis are dissolved in the fluid of the chloroplast; some of them ultimately become attached to NADP as described above. Hydrogen ions are also used to power the building of ATP as described under the topic of chemiosmosis, below.

The ability of a plant to produce chlorophyll for photosynthesis is influenced by hereditary conditions, general health of the plant, and availability of elements composing its molecule. The yellowing of crop plants due to the absence of chlorophyll often is traceable to a deficiency of nitrogen, a component of the molecule.

Chemiosmosis Is the Process of Making ATP at a Membrane

It is appropriate now to examine the production of ATP in more detail. Peter Mitchell was awarded the Nobel Prize in 1978 for his contribution to understanding the mechanism usually called *chemiosmosis*. As shown in Figure 5.6, molecules embedded in a membrane are the major players in this ac-

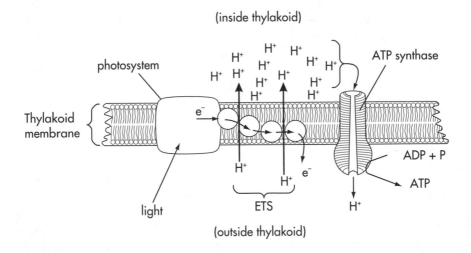

FIGURE 5.6.   *Chemiosmosis to produce ATP. Protons are pushed to the inside of a thylakoid, then flow passively back to the stroma side of the thylakoid membrane. ETS: electron transport system. H^+: hydrogen ion (proton). e^-: electron.*

tivity. When ATP is built during photosynthesis, the location is the thylakoid membrane of a chloroplast.

The process begins with the flow of electrons along molecules of an electron transport system, as described earlier. Some of the energy of the electrons is used to push protons (H^+) across the membrane. Protons accumulate on one side of the membrane, much as water might accumulate if pumped into a reservoir behind a dam. Just as the water holds potential energy that could be used to push a piece of machinery if it were allowed to flow down over the dam, the protons held on one side of the membrane also have much potential energy.

Another molecule also embedded in the thylakoid membrane is an enzyme called *ATP synthase* (Fig. 5.6). The shape of this enzyme is such that protons can passively flow across the membrane by passing through the synthase. The enzyme is also able to capture ADP and phosphate and to hold them close together to enhance the likelihood of their bonding. As the protons flow through the ATP synthase, the force of their movement provides the energy to create the bond uniting the ADP and phosphate to become ATP (that is, to phosphorylate the ADP). The kinetic energy of fast-moving electrons has been converted to (a) potential energy of protons held outside the membrane, then to (b) the kinetic energy of these protons flowing through ATP synthase, and finally to (c) the potential energy of a bond in ATP. All of this energy conversion is, of course, consistent with the first law of thermodynamics. It is also expected that not all of the energy that begins the trip is successfully converted to bond energy (remember the second law of thermodynamics), and that is the case.

It is important to note that chemiosmosis is also the mechanism for making ATP in processes other than photosynthesis, as will be shown in Chapter 6.

Light-independent Photosynthesis Reactions Produce Stable Molecules

Both cyclic and noncyclic photophosphorylation require continuous arrival of light. Therefore, they are often said to comprise the *light-dependent* portion of photosynthesis, or the light reactions. But there is another set of chemical reactions that is *light-independent*, the so-called dark reactions. Melvin Calvin, working with others in the 1940s, discovered many of the steps involved in these reactions, so they are often called the *Calvin Cycle*. A more descriptive title for them is *carbon fixation*, since their major feature is the fixing (i.e., making more avaliable for use) the element carbon. In summary, the carbon fixation portion of photosynthesis begins with the gas carbon dioxide (CO_2) and ends with a much more complex molecule, a carbohydrate. Figure 5.7 summarizes these reactions, but does not show all intermediate steps.

One at a time, six molecules of CO_2 are linked to form two 3-carbon molecules. To these are added hydrogen and oxygen atoms from cellular H_2O, additional hydrogen and electrons carried via NADP from the light-dependent reactions, and energy to link all of these together. The energy comes from

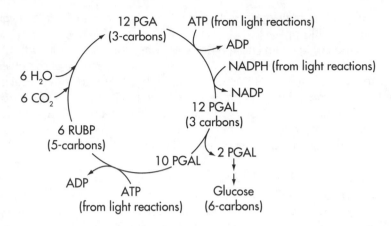

FIGURE 5.7.   *The light-independent reactions of photosynthesis (Calvin Cycle, carbon fixation). Beginning materials include H_2O and CO_2 from the environment and H^+, electrons, and energy from light-dependent reactions. The end product is a carbohydrate, glyceraldehyde-3-phosphate (PGAL). Intermediate materials include phosphoglycerate (PGA) and ribulose bisphosphate (RUBP).*

bonds of 18 molecules of ATP and 12 molecules of NADPH, which were made in the light-dependent reactions. So carbon fixation really does depend upon light; it is said to be light-independent only because it does not require light to be shining at the moment it occurs.

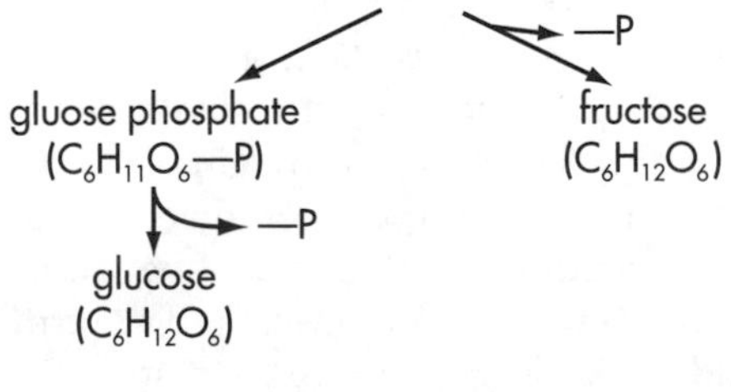

PGAL
↓
fructose diphospate
$(P—C_6H_{10}O_6—P)$

P— ↙↖↓

fructose phosphate
$(C_6H_{11}O_6—P)$
↙   ↘ —P

gluose phosphate   fructose
$(C_6H_{11}O_6—P)$ $(C_6H_{12}O_6)$
↘ —P

glucose
$(C_6H_{12}O_6)$

FIGURE 5.8.   *Synthesis of the monosaccharides glucose and fructose from PGAL, the product of light-independent reactions.*

The direct products of carbon fixation are the 3-carbon sugars known as glyceraldehyde-3-phosphate (PGAL in Fig. 5.7). PGAL does not accumulate in cells. It is a food that may be used directly or converted into something more stable. The more stable product is commonly glucose. Two molecules of PGAL unite to form fructose diphosphate (Fig. 5.8). Fructose diphosphate may lose its phosphate group and become the sugar fructose, or it may be converted into glucose phosphate. Glucose phosphate loses the phosphate group to become glucose.

THE ENVIRONMENT PROVIDES RAW MATERIALS FOR PHOTOSYNTHESIS

We turn our focus now to the sources of the molecules that are needed for photosynthesis. As will be demonstrated, the environment around plants is a critical factor in determining whether they will thrive.

Availability of Water Is a Critical Factor

At some time or other, water is a limiting factor in almost all places except in oceans and permanent lakes or swamps. It is especially limiting in dry deserts where rain seldom falls and in frozen deserts where the water is solid and not absorbable. People used to irrigate crops only in the very dry areas of the earth. Now farmers recognize that irrigation can more than pay for itself in regions where rainfall is fairly high. They know that there is seldom a continuous optimum supply of water but more frequently intermittent wet and dry spells. Yields, therefore, are increased—some crops even saved—by irrigating during dry periods. This practice has proved to be especially profitable to fruit and vegetable farmers. In some areas of the world rainfall comes during the dormant season, making it necessary to irrigate economically useful plants during the growing season.

Plants get their water in various ways. Submerged plants absorb directly from surfaces exposed to the water. A few terrestrial plants absorb water from the atmosphere, especially while it is raining. Some plants, such as mosses growing on nonsoil surfaces, must be able to resist periods of extreme drying and grow only when water is available. Some plants, such as Spanish moss or orchids that are attached to trees, grow in places where the humidity is high. The majority of plants absorb water via roots and conduct it to the leaves, where it is used or evaporated. Were it not for the fact that much of the water is lost by evaporation, water would not be as limiting as it is. A single corn plant growing in Kansas is said to require about 200 liters of water. More than 190 liters of that amount are lost by evaporation from the leaves.

The Atmosphere Provides Carbon Dioxide for Photosynthesis

Of the absolutely necessary materials, carbon dioxide is probably the least fluctuating in nature. The atmosphere contains about 0.03 percent by volume.

It is higher in some local areas—for instance, near the ground in some woods where there is considerable respiration by decay organisms or around factories where there is much combustion. The major source of atmospheric carbon dioxide is from the chemical breakdown of organic molecules by living organisms, although burning of materials by human societies is becoming more important as our numbers and industrial activities increase.

It has been shown experimentally that if all other factors are in abundant supply, the rate of photosynthesis can be speeded by increasing the concentration of carbon dioxide. For most plants the increase continues until the concentration is from 20 to 30 times the average amount in the atmosphere.

The proportion of carbon dioxide and oxygen in nature seems to be in a rather delicate quantitative balance. While photosynthesis and respiration (see Chapter 6) are opposites in the consumption and discharge of carbon dioxide and oxygen, certain nonliving processes like combustion and mineral oxidations also figure in the relative proportions of these gases in the atmosphere. The ocean contains an enormous reservoir of carbon dioxide that can be yielded to the atmosphere if the quantity should drop; or conversely, it can absorb large quantities of carbon dioxide from the atmosphere if the quantity should increase. There are great numbers of marine organisms that become a part of the oxygen-carbon dioxide balance through their activity in photosynthesis, respiration, or both.

It is unlikely that the proportion of carbon dioxide and oxygen has remained constant throughout geological history. The lush growth of vegetation that is now locked in coal beds must originally have grown in an atmosphere considerably richer in carbon dioxide. If the hypothesis about the most ancient atmosphere is correct, there was little or no free oxygen then. This means that the 21 percent oxygen content of the present atmosphere must have come primarily from photosynthesis. It is known that the unlocking of carbon (changing to carbon dioxide) by the burning of fossil fuels in the current era is increasing the carbon dioxide content of the air. The increase has been about 30 percent over the last 200 years. However, not all increase of carbon dioxide can be attributed to recent human industrial activity. Analysis of the air trapped in polar ice (as bubbles) indicates that carbon dioxide has been gradually increasing for at least the last 16,000 years. Whatever its cause, the phenomenon should be increasing the rate of photosynthesis by plants in natural settings since it is known to do so in controlled experiments.

Many scientists believe that an increased amount of carbon dioxide (along with other gases) in the atmosphere is forming a sort of blanket that will hold atmospheric heat and result in a warming of the earth. If this proves to be true, the effect upon many aspects of the environment may be profound.

Environmental Temperature Influences Photosynthesis

Just what temperature is optimum for photosynthesis for each kind of plant under any set of conditions is unknown, but it is known that extremes in temperature retard or stop the process, either directly or indirectly. Perennial seed plants adjust to low temperatures by shedding leaves and becoming dormant during the cold season; or if they retain their leaves, they have protective adaptations that prevent cold-weather damage while at the same time allowing photosynthesis to take place slowly. Annual seed plants do not have these characteristics; instead, they produce seeds containing a dormant embryo protected from the elements and supplied with a store of food sufficiently large to take care of the embryo until it can do its own photosynthesis.

SYNTHESIS OF OTHER MOLECULES IS LINKED TO THE PRODUCTS OF PHOTOSYNTHESIS

Glucose and fructose, which are monosaccharides, are not the best kinds of storage products. They and other simple sugars are quickly converted into more complex carbohydrates like disaccharides, starches, or cellulose. Starch is by far the best storage material because it takes up less space and is relatively insoluble in water. Being insoluble, it cannot affect the osmotic properties of the cell. Of course, starch and some other polysaccharides are made by linking together many individual glucose molecules. When energy is needed by the cell, the first step in getting it from polysaccharides is to break away individual glucose molecules.

Fats may be synthesized by either plants or animals. The basic building block in plants is PGAL available from photosynthesis. Glycerol is made from it by accepting hydrogen and fatty acids are made indirectly from the same material by the removal of oxygen. Glycerol and fatty acids then combine to form a fat and water (Fig. 3.6).

Proteins are made from amino acids that are manufactured by green plants and some bacteria and fungi. Amino acids of plants have their origin in PGAL, but the route of synthesis is complex. One important step in their synthesis is the addition of the amino group ($-NH_2$). Amino acids are put together in varying numbers and positions to make proteins, as described in Chapters 3 and 12.

PLANTS TYPICALLY PRODUCE MORE FOOD THAN NEEDED IMMEDIATELY

Seed plants generally produce a surplus of foods that goes into increasing their own bulk and into storage structures, particularly reproductive parts like seeds and fruits or dormant underground parts like fleshy stems and roots. It is estimated that 85 percent of food produced in photosynthesis is stored. An abundance of stored food in seeds and fruits serves the embryonic plant well in getting it off to a good start. Many plants, like the Irish potato, rarely produce seeds but do the same job by storing food in the tuber, which serves the reproductive function. Many perennials, like the day lily and iris, die to the ground and get a fresh start each year from underground parts that contain stored food. Of course, many animals including humans take advantage of these plant adaptations that concentrate foods, obtaining the vast majority of their vegetable foods (grains, fruits, nuts, potatoes, and so on) from reproductive structures.

Aquatic algae provide (directly or indirectly) the major part of the food consumed by fish and other aquatic animals. Because algae have no special storage organs like seeds, fruits, stems, roots, or leaves, the entire plant is usually consumed. If ponds are provided with abundant inorganic nutrients that promote rapid growth of algae, these Protistans serve as food for invertebrates such as worms, crustaceans, and insect larvae. Those animals in turn are used as food by fish.

FERTILIZERS CAN BE ADDED TO ENHANCE PLANT GROWTH

The synthesis of foods and other important organic molecules from PGAL usually involves a complex chain of events. It may involve the addition of atoms to the food molecule or to enzymes that catalyze the reactions. Then if any product synthesized directly or indirectly from PGAL contains an element other than carbon, hydrogen, and oxygen, the element must come from some source other than carbon dioxide and water, since these are the only compounds used in photosynthesis. Proteins have nitrogen and sometimes sulfur in their molecules and require these elements from some other source than carbon dioxide and water. Nucleic acids require both nitrogen and phosphorus. Chlorophyll molecules have an additional requirement of nitrogen and magnesium. These essentials come from the soil, usually in the form of ions.

Whenever essential minerals are deficient, they can be artificially supplied by adding fertilizers to the soil. They sometimes produce startling results, almost as if a half-starved plant is suddenly supplied with an abundance of food. As a consequence, it is a common practice to refer to fertilizers as plant foods. This practice is not only misleading but erroneous, for foods are organic compounds that release energy when broken down. Fertilizers do not release energy. It is better to say that they are chemical supplements.

Quite a few elements are found in cells, but not all are useful. Some are rather universal in distribution and are seldom deficient. The three most likely to be deficient—nitrogen, phosphorus, and potassium—are supplied in quantity in fertilizers along with smaller amounts of other elements. Fertilizers have formulae written 8-10-10, 8-8-8, and the like. The three figures are percentages of usable elements. The first stands for nitrogen; the second, phosphorus; the third, potassium. The following is a list of some elements and their uses in plants:

NITROGEN: A part of many organic molecules, including proteins, chlorophyll, DNA, and RNA.

PHOSPHORUS: A part of many molecules including ADP, ATP, DNA, RNA, NAD and NADP. Very important in energy transfers.

POTASSIUM: Used in carbohydrate metabolism.

MAGNESIUM: A part of the chlorophyll molecule.

MANGANESE: Important in breaking down carbohydrates to yield energy.

IRON: Necessary for chlorophyll formation but not a part of its molecule.

CALCIUM: Part of the cell wall substance calcium pectate.

SULFUR: A part of protein molecules.

ZINC: A constituent of certain enzymes and hormones.

COPPER: A constituent of certain enzymes.

NONPHOTOSYNTHETIC ORGANISMS MUST OBTAIN ENERGY FROM OUTSIDE THEMSELVES

Since green plants, some Protista, and a few bacterial species are the only truly independent organisms, all other forms of life are dependent on them for food. The closeness of dependence varies. Some utilize these photosynthesizers directly, others utilize dead bodies, and still others live upon or within another living organism from which they absorb or withdraw food. Whether the organism is a cow grazing in the meadow, a person eating at the table, a cat devouring a mouse, mold growing on a piece of bread, or tuberculosis bacteria living in the lungs of a person, they all exist at the expense of some other living organism. Two useful terms to describe the organisms of the world with respect to their energy sources are *autotroph* and *heterotroph*. The former is an organism independently capable of capturing energy and storing it in molecules of food; the latter is one incapable of this and therefore dependent upon autotrophs to provide energy in the form of food.

Many Heterotrophs Are Predators

Animals and other heterotrophs may be classified according to the kind of foods they consume as *herbivores* (autotroph-eaters), *carnivores* (heterotroph-eaters), or *omnivores* (mixed-diet-eaters). All of them make great or even fatal demands on other organisms.

Organisms that capture and eat others are *predators*. They are often larger than their victims, or at least superior enough to conquer them. As in the case of a bird eating an insect, it is obvious that the predator does considerably more damage to the victim than is nominally done by any of its parasites. Although predators are most often thought of as being carnivores, hervbivores are also technically in this category.

Symbiosis Is a Long-term Relation Between Species

Many dissimilar organisms live together in a close relationship known as *symbiosis*. Such a relationship need not be involved with transfer of food, but many cases do involve this. The degree of association may vary from the intimate contact of algae and fungi in a lichen thallus to a rather loose proximity between some ants and aphids. Symbiosis may be *mutualism, commensalism,* or *parasitism.* In mutualism, both associates benefit; in commensalism, one benefits and the other is not damaged; in parasitism, one benefits and the other is harmed.

Mutualism Is Often Easy to Identify

The symbiotic association whereby both members are benefited is mutualism. Indeed, some people use symbiosis in a narrow sense to be synonymous with mutualism. Several well known examples of mutualism are familiar to most people. There is a case where a sea anemone attaches itself to the back of a crab. In doing so, it camouflages the crab from its enemies and salvages bits of wasted food from the crab's meals. Another case involves the yucca plant and the pronuba moth. The plant is absolutely dependent on the moth for pollination; the moth is equally dependent on the plant for nourishing its larvae. The lichens illustrate an extremely close relationship between algae and fungi. The former provides food and the latter protection. An interesting mutualistic combination is the Protozoa and bacteria in the intestines of termites. The insects furnish a home and wood, while the Protozoa and bacteria digest the wood efficiently enough to provide glucose for both their host and themselves.

Of particular interest to farmers is the mutualistic relationship between certain bacteria and the roots of pea plants (vetch, alfalfa, clovers, and others). The bacteria penetrate cells and stimulate the growth of nodules. They (as do some bacteria and fungi living free in the soil) convert the pure nitrogen of the air, which plants cannot use, into a usable form. Nitrogen is thus made available to the larger plant, which in turn furnishes food and a place to live for the bacteria. Some legumes, like alfalfa and clovers, do not grow well where the bacteria are absent. Farmers have learned to inoculate seeds with the proper types of bacteria at the time of planting. The process involves mixing the seeds with a commercial preparation containing the bacteria.

Commensalism Can Be Difficult to Identify Correctly

It is possible for two kinds of organisms to live together so that one is benefited while the other is unaffected. This relationship is commensalism. A marine worm, *Chaetopterus*, lives in a parchment-like, U-shaped tube, usually with a small male and female crab. The openings of the worm tube are too small for either the worm or crabs to escape. (The crabs get into the tube when they are small larvae and grow to maturity there.) Apparently, worm and crabs simply live together without doing harm to one another. The crabs benefit to some extent by the protection of the tube. However, as with all cases of apparent commensalism, it is possible that the seemingly neutral situation for the worm is simply a case of our ignorance about some subtle benefit that it obtains from the sharing of its residence. There are a number of cases where relations labeled commensalism have had to be recategorized when closer examination revealed that both species either gained or lost during the association.

Parasitism Involves Gain by One of the Participants at the Expense of the Other

Parasites are organisms that have among their characteristics the dependence on other living organisms for food. They could be called herbivores or carnivores, depending upon the nature of their victims. They are predators, but differ from most in maintaining a long-term relationship with a prey victim. The relationship differs from the other two types of symbiosis in that the parasite causes harm to the host, by damaging it or robbing it of food. Parasites are generally smaller than their host, living upon or within the host. They belong mainly to the lower phyla or divisions, the vast majority being Protozoa, bacteria, Fungi, viruses, worms, insects, and arachnids (mites, ticks, and chiggers). Only occasionally do parasites (hagfish, lamprey eel, cancer-root, and dodder) belong to the higher phyla or divisions. Unlike the predaceous carnivore, which may consume many individuals, the parasite steals primarily from only one host.

It is likely that parasites are most abundant in the tropics, where warm, moist conditions favor their survival for a longer period during intervals when they are without a host. Regardless, it seems that most organisms throughout the world are parasitized to some extent. It is surprising for some to learn that small animals like the Betsy beetle, found in decaying logs, usually have mites. Even one-celled organisms like the diatoms (of Kingdom Protista) and bacteria are not exempt from parasites. Furthermore, parasites may also have parasites.

The geographic range of a specific kind of parasite coincides roughly with the range of the host or hosts upon which it depends. Where two or more hosts are necessary during the life cycle, it is limited to approximately the area where the host ranges overlap. For instance, malaria (*Plasmodium*) parasitizes humans only when they live within a region inhabited by a certain species of mosquito, since a portion of this Protozoan host lives in the mosquito. The ranges of parasites having complex cycles are much more limited than the ranges of those with simple cycles because meeting all the conditions for survival is more difficult. A complex cycle is also probably more hazardous to the parasite because destruction of any host in the cycle will destroy the parasite itself.

The future of a parasite, if artificial controls are excluded, depends on a successful relationship with the host. Its success cannot be measured by the maximum amount it can extract from the host, for a parasite could damage enough to kill the host, thus killing itself. The best adjustment would be to require so little that the host would suffer little from its presence.

Parasites are accustomed to more limited environments than free-living creatures, and they cannot tolerate conditions for which they are not adapted. Long dependence on a host results in adjustments to a definite set of conditions that are seldom found elsewhere. For that reason, any particular stage of a parasite usually attacks a definite host or a closely related species. This is obvious in the well-known cases of disease parasites. Humans do not contract oak tree wilt from oak trees nor dog distemper from dogs but can get polio or blood flukes from some closely related primates.

Complex life histories may involve several stages and several hosts. Predictably, each stage is host-specific. In the case of a human liver fluke (a flatworm, phylum Platyhelminthes), for instance, the parasite passes through stages in the snail, then a fish, and finally back to a human, where it becomes sexually mature. The final stage parasitizing humans cannot live in the snail or fish; neither can the snail or fish stages live in humans.

The damage suffered by a host may not be noticeable or may produce effects that range from mild symptoms to serious disease and death. Some parasites use a part of the host's food supply (intestinal worms), some damage tissues (hookworm, damping-off, or smut), some block passages (chestnut blight, intestinal worms, or filariae), some produce poisons (bacteria that cause diseases), some irritate (lice or athlete's foot), and some may do damage in other ways. A considerable number of human illnesses can be traced to microorganisms and a few larger parasites that damage in one or more of the above ways.

Certain viruses, the *bacteriophages*, parasitize bacteria. The damage caused by them results in eventual death and breakdown of the bacterial cell. Bacteriophages are of varying sizes, but almost all are too small to be seen with the light microscope. A bacteriophage consists of a core of nucleic acid surrounded by an envelope of proteins. Certain of the proteins can adhere to the bacterial cell wall. Shortly thereafter, the nucleic acid is injected into the bacterium. The outer, protein layer is left behind. Upon entering, the nucleic acid of the core replicates and dictates the construction of appropriate viral proteins. The replicated nucleic acid and the synthesized protein units assemble into complete new viruses. Shortly thereafter the victimized bacterium dissolves and releases the new viruses. A single bacteriophage is known to generate 100 to 300 offspring within a single bacterium of the species *Escherichia coli*. The complete process of invasion and multiplication takes from about 15 to 45 minutes in several bacteriophages that have been studied.

SOME ORGANISMS DECOMPOSE DEAD BODIES

Fungi on dead wood (Fig. 5.9) and some bacteria illustrate cases where dependent organisms extract their foods from dead bodies. The process is *decomposition* and the organisms who perform it are *decomposers*. They are also known as *saprophytes*. The suffix *-phyte* means plant, but modern classification categories make the use of this term archaic. The preferred synonym for decomposers is now *saprobes*.

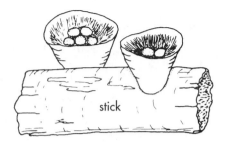

FIGURE 5.9.   *Bird's-nest fungus on a decaying stick.*

Decomposers, while feeding themselves, perform the crucial function of freeing chemicals locked up in dead bodies. Were this not possible, it would be only a short time until essential elements would be exhausted. Decomposers chiefly responsible for decay are bacteria and filamentous fungi. Decay of complex organic substances to such end products as carbon dioxide, water, or ammonia often requires the sequential work of numerous kinds of organisms in many steps, each accomplished with the aid of enzymes. Decomposers secrete enzymes outside their cells and digest organic matter into simpler substances. The decomposers then absorb whatever food materials can go through their membranes. Absorbed foods are then broken down further, with a consequent release of energy for use by the decomposers. Thus decay involves not only digestion outside the body but also further processing after molecules have been brought inside. Both actions cause the breakdown of complex molecules and release of the resulting simple units that other organisms can utilize. Decay of proteins is known as putrefaction and may produce foul-smelling compounds, especially where free oxygen is deficient.

It is customary to refer to the decay of organic substances that we call food as spoiling. We control food decay by refrigeration to kill or stop the growth of decomposers (which are ubiquitous in the environment); canning to destroy those present and exclude others; pasteurizing to reduce them to small numbers; drying to create an arid material unfavorable to their growth; salting and sugaring to dehydrate the material or any organism that may try to grow upon it; and pickling to poison them.

STUDY QUESTIONS

1. What are the two forms that can be taken by energy and what is their relationship to each other?

2. How does the second law of thermodynamics have impact upon the lives of organisms?

3. What colors of light are most used in photosynthesis? What environmental conditions alter the quality of light available to plants?

4. Of the sunlight absorbed by a plant, approximately what proportion is absorbed by chlorophyll? What percentage of that is used in photosynthesis?

5. Identify these: photosystems, reaction center, primary electron acceptor.

6. What is the location of thylakoid membranes? What kingdoms include organisms that have thylakoid membranes?

7. Trace the path of electrons through cyclic photophosphorylation.

8. What is the importance of photolysis for (a) noncyclic photophosphorylation and (b) the light-independent reactions of photosynthesis?

9. Name two valuable products of noncyclic photophosphorylation.

10. What structural category of molecule is ATP?

11. What happens during carbon fixation, the second phase of photosynthesis? What is regarded as the immediate end product of photosynthesis?

12. Describe how the energy of electrons is captured in ATP, by the process of chemiosmosis.

13. What are the roles of water (from the environment) in making photosynthesis possible? (Be very specific.)

14. Make a clear distinction between foods and fertilizers.

15. Can predators be parasites? Can parasites be herbivores?

16. What are three forms of symbiosis? Give an example of each.

17. Discuss specificity of the parasitic relationship.

18. What are bacteriophages? How do they reproduce? Why are they discussed in a chapter dealing with the gain of energy by organisms?

19. Discuss the role of decomposers with relation to the other organisms of the world.

6
RELEASE AND UTILIZATION OF ENERGY

What do we know about:
—the central role of glucose in energy storage?
—why we do not simply burn glucose to obtain energy?
—the role of enzymes in helping reactions?
—the specific steps of glycolysis and the Krebs cycle?
—how chemiosmosis converts electron energy to ATP energy?
—exactly why we must breathe O_2 and release CO_2?
—the role of fermentation in organisms and in industry?
—the remarkable efficiency of cellular respiration?
—how vital elements cycle through an ecosystem?
These and other intriguing topics will be pursued in this chapter.

ORGANISMS USE THE ENERGY OF LIGHT

The major original source of energy for sustaining life is light. As was described in the previous chapter, light energy activates chlorophyll which is then used in photosynthesis. During the process some of the energy is stored in a bond of the molecule ATP and some is used to make a 3-carbon molecule, PGAL. The latter becomes part of a sugar, glucose. Or, to say all of this differently, much of the energy of light is transformed into chemical bonds of molecules, where, in a potential form, it is held in reserve.

ORGANISMS USE STORED ENERGY IN MANY WAYS

Certainly not all the energy transfers in organisms are known. Here, however, are several obvious uses of energy that has been stored in molecules:

movement of cilia and flagella
changing shapes of cells
muscular contraction
transmission of nerve messages
synthesis of needed molecules and organelles; destruction of those no longer
     needed
active transport of ions and molecules across plasma membranes
cell division
bioluminescence (production of visible light)

THE TERM RESPIRATION
IS SOMETIMES MISUSED

Since the word *respiration* is sometimes used when breathing or gas exchange would be the more accurate terms, it is necessary to understand clearly
the distinction among them. Respiration is a chemical process of nearly all living cells whereas breathing is a mechanical function of many multicellular animals. Breathing is a renewal of air (terrestrial animals) or water (aquatic animals) on a surface through which an exchange of oxygen and carbon dioxide
(i.e., gas exchange) can take place. Actually, gas exchange is necessary for all
organisms that are performing respiration most efficiently, since they all must
take in oxygen and release carbon dioxide in order to do so.

The most accurate meaning of respiration, and the topic of this chapter, is
a set of chemical processes within living cells that releases energy from foods.
Naturally, the more active a cell is, the more energy it requires. Thus, an active cell like a contracting muscle fiber will respire faster than relatively inactive cells of dormant seeds. The activity of dormant seeds is so low that it is
barely measurable. The low water content in seeds or spores limits any digestion of stored food, since water is necessary in that process. The first step in
germination, which sets digestion in motion, is the absorption of water.

MANY MOLECULES PROVIDE ENERGY,
BUT GLUCOSE IS CENTRAL

It will be recalled from the previous chapter that glucose is the most useful product made after photosynthesis has led to the capture of energy in the
bonds of molecules. Later in the current chapter, it will be shown that other
food molecules (amino acids, polysaccharides, fats) can sometimes be used to
provide energy to a cell, but all of these fit into the central scheme of glucose
utilization. Therefore, we should begin the story of energy release by following glucose as it enters into the reactions of respiration.

Look back at Figure 3.5, to remind yourself that glucose is a monosaccharide with six atoms of carbon. It includes 23 bonds between atoms, all of which hold potential energy that could be used to power activities of the cell. As will be shown, not all of these bonds are broken in respiration, but many are.

THE BONDS OF GLUCOSE MUST BE BROKEN INDIVIDUALLY, NOT SIMULTANEOUSLY

The breaking of chemical bonds in respiration is a slow, regulated process. Each step is controlled by enzymes, and much of the energy is transferred to other chemical bonds. An alternative way to release glucose's bond energy would be to burn it (in the process called *combustion*). This would cause so much energy to be released in a very short time that the cell would (a) not be able to capture it for particular needs and (b) be damaged by the large amount of heat energy generated. Actually, respiration has sometimes been called slow combustion, or at least it has been compared to combustion because of some similarities. If respiration uses free oxygen (O_2), both respiration and glucose combustion are alike in doing so and in making the end products carbon dioxide and water. However, a major distinguishing characteristic between the two processes is that combustion requires a significant input of energy (usually as heat) in order to get underway, while cellular respiration needs much less.

ENZYMES ARE CATALYSTS THAT ALLOW LOW-TEMPERATURE BREAKING OF BONDS

The key to understanding how cells control the release of bond energy and manage to do it without a large initial input of energy is to see how enzymes catalyze reactions. Any *catalyst*, enzymatic or otherwise, enhances the likelihood of a chemical reaction because it holds the reactant(s) in proper position so that a minimum of additional energy is necessary to cause a change in bonding. Consider the reaction involving hypothetical molecule A, in which a bond breaks and product B is made. Since a bond has been broken, with the release of energy, the potential energy within the products will be less than that of the starting material, molecule A. (This is a model for some of the reactions that occur in respiration.) Now, examine the graph that is Figure 6.1a. It shows that a certain amount of energy (called *activation energy*) must be applied to molecule A before the bond can be broken. Figure 6.1b plots the same reaction helped by a catalyst. Molecule A is held in such a way that the

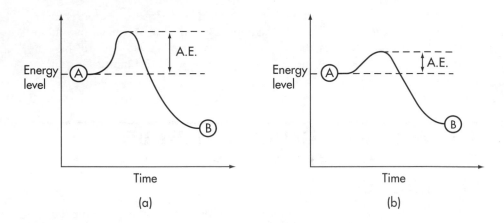

FIGURE 6.1.   *(a) An energy yielding reaction occurring without the help of a catalyst.
(b) The same reaction, aided by a catalyst (such as an enzyme). The required activation energy is lower. In both, A.E. = Activation Energy.*

important bond is made more likely to break, even if provided with less activation energy. Thus, a catalyst makes a reaction more likely to happen, with
less use of energy to start it. Another way of viewing this is that the reaction
is more likely to happen at a lower environmental temperature. This is important for a cell, of course, since high temperature would be apt to damage delicate structures.

The catalysts of biological systems are called *enzymes*. They are almost always proteins (a few are RNA, distinguished by the name *ribozymes*.) A review of Chapter 3 will remind the reader that proteins are very complex polypeptides, which fold into highly specific three-dimensional shapes. The rather
small portion of an enzyme that actually participates in holding a reactant is
called the *active site*. Any material that is to be changed during the catalyzed
reaction is called a *substrate*. The active site must be of a proper shape to hold
one or more substrates in an appropriate position to enhance the reaction.
Even the slightest change in an enzyme's amino acid sequence altering the
shape of the active site can render the enzyme disabled. Because of the precise substrate-to-active site match-up that must occur if catalysis is to happen,
each enzyme is likely to catalyze only one or a few related kinds of chemical
reaction. Since literally thousands of different chemical reactions must occur
in an organism, it follows that thousands of different enzymes (or genetic instructions for their manufacture) must be present. Although the emphasis in
this chapter will be on enzymatically catalyzed reactions that break down large
molecules into smaller units, it should not be forgotten that reactions to synthesize larger molecules also require enzymatic help if they are to occur use-
fully fast in a cell.

THE FIRST STEPS IN GLUCOSE BREAKDOWN ARE CALLED GLYCOLYSIS

Figure 6.2 shows a series of reactions collectively known as *glycolysis*. Free glucose is the molecule that is acted upon to cause a transfer of some of its bond energy (directly or indirectly) into the bond energy of ATP. In the process, each 6-carbon glucose is eventually converted into two 3-carbon molecules called pyruvic acid.

Glycolysis Begins with a Loss of ATP

Just as the energy of a lighted match must be added to a lump of charcoal before the charcoal can be expected to release energy by burning, a molecule of glucose must be made unstable by the addition of energy before its bonds can begin to yield their energy. As with very many reactions in cells, this is done by breaking down some ATP. The first two reactions of glycolysis involve the release of a phosphate group from each of two ATPs and transfer of the phosphates (with attendant energy) to the sugar molecule. Each phosphate replaces a hydrogen atom of the original glucose molecule. Along the way, the glucose 5-carbon ring is modified to a 4-carbon ring. The new sugar is *fructose* and the complete molecule with the added phosphates is called *fructose 1, 6-bisphosphate.* Since it has additional bond energy provided by the breakdown of two ATP molecules, it is less stable than glucose and is thus primed to break down into simpler compounds.

If glycolysis were stopped after the first two steps, the cell would have lost ATP rather than gained. But, of course, glycolysis continues.

The 6-carbon Sugar then Breaks into Two 3-carbon Fragments

Fructose 1, 6-bisphosphate soon undergoes the breakage of two bonds in its backbone structure, yielding two nearly identical 3-carbon molecules. Their names and exact structures are not important in this description, but it is important to note that each contains one of the phosphate groups provided by ATP in the initial steps of glycolysis. It is also important to understand that this reaction does not yield usable energy for the cell, even though bonds have been broken and energy released. The cell is not equipped to capture the energy of every such event.

Immediately after the two 3-carbon molecules are made, one of them is enzymatically converted to an exact copy of the other. From this point onward, each of these two undergoes identical reactions. This is taken into account in Figure 6.2 and is important to remember when a final tally of new ATPs is made. Shortly thereafter, each molecule picks up an additional phosphate group, giving it a total of two. These phosphates had been dissolved in the cytosol of the cell, and their addition does not confer significant new energy

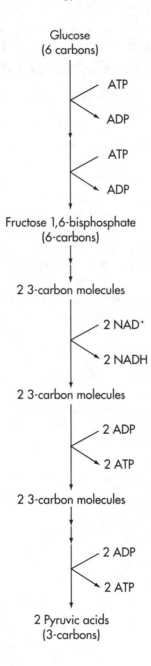

FIGURE 6.2.   *The steps of glycolysis.*

to the molecules that we are following. Addition of phosphates at this point does, however, allow important events in the following steps.

The Next Glycolysis Step Is Removal of Hydrogen and Energy

Each of the newly produced 3-carbon molecules includes five atoms of hydrogen, each of them attached to the rest of the molecule by an energy-containing bond. The cell can use this energy to build ATP if given the opportunity. The first time for this to happen is at the next step of glycolysis. Each 3-carbon molecule transfers one hydrogen (and its energy) to a molecule abbreviated NAD⁺ in Figure 6.2. The last word of its full name, nicotinamide adenine dinucleotide, identifies the general structure of this important molecule. NAD⁺ is a molecule used to transfer hydrogen and its energy from one place to another in a cell. It can be used repeatedly in this action, not being altered by temporarily carrying hydrogen. When it has hydrogen bonded to it, the correct designation for it is NADH (or reduced NAD, since the addition of hydrogen is a chemical event called reduction). Since NADH is not of further importance in glycolysis itself, we will not follow it to its destination at this time. However, it will be of great importance in later events.

The Final Steps of Glycolysis Transfer Energy for Building New ATP

After NADH has been formed, each of the two 3-carbon molecules undergoes a series of four reactions that re-arrange its atoms and lead to loss of its phosphate groups. It is not important to examine each of these reactions. The important fact is that two of the reactions involve a loss of a phosphate group, which transfers to a molecule of ADP, producing ATP. The energy of the newly formed bond in ATP is traceable to energy that had been holding together atoms of glucose before glycolysis began. The number of ATPs made at these steps is seen to be a total of four, when it is remembered that two 3-carbon molecules are simultaneously undergoing the reactions and that each yields two phosphate groups. It is only after the last ATP-building reaction that the cell has gained more ATP than was used to start glycolysis. The net gain is two ATP for every glucose that begins the trip.

The two identical carbon-containing materials that remain after glycolysis has been run are the 3-carbon molecule called pyruvic acid. Each still has a rather large quantity of energy that could be harvested. Glycolysis is (under certain conditions) only the first of several activities that can be used to wring out glucose's energy, as will be seen below. It is, however, the necessary first activity if a cell is using glucose as its starting material for energy utilization.

Glycolysis Occurs in a Cell's Cytosol

It is important to remember that every step in glycolysis is a chemical reaction made faster by the help of a specific enzyme. If one wants to determine where glycolysis happens in cells, one must find the location of these enzymes. Careful studies have shown that they are all dissolved in the cytosol. If pyruvic acid and NADH are to yield more energy for the cell, they both must be transferred from the cytosol where they were made to the interior of mitochondria. We will follow these energy-rich molecules to that place.

THE PRESENCE OF OXYGEN ENABLES FURTHER ENERGY UTILIZATION VIA THE KREBS CYCLE

Two products of glycolysis, pyruvic acid and NADH, have much energy that could be utilized by making more ATP. We will begin by following pyruvic acid. At the outset, it must be understood that none of the reactions to be described below can happen unless a cell is supplied with molecular oxygen (O_2). That is, the cell must be operating under *aerobic* conditions (opposite: *anaerobic*). The reason for oxygen requirement must be deferred, but will become clear by the end of this section.

A general idea of what follows is appropriate at this point. Pyruvic acid, an end product of glycolysis, will be seen to feed into another major series of steps, usually called the *Krebs cycle*. Those steps will result in the complete separation of all adjoining carbons (remember that the starting material was glucose, with six carbons linked together). During the Krebs cycle, some ATP will be made directly, but the major energy transfer will be in the form of several hydrogen atoms (with their attendant energy) being stripped from the remains of glucose and being moved toward important ATP-building reactions. Thus, although the Krebs cycle does not itself yield much new ATP, it enables the cell to make much more *after* it has been run.

Glycolysis Is Linked to the Krebs Cycle by an Energy-transferring Step

Each of the two molecules of pyruvic acid originating from the glycolysis of glucose performs identical steps as shown in Figure 6.3. The initial step is geographical: each pyruvic acid must diffuse through both membranes of a mitochondrion. It is within the liquid at the center of mitochondria that the various Krebs cycle enzymes are located, as well as those performing the steps described in the current paragraph. Several chemical things then happen to each pyruvic acid to link glycolysis to the Krebs cycle. One of the carbons that was once part of glucose is split off, along with some oxygen. The resulting CO_2 molecule has no further importance for the cell and eventually leaves as a waste product. Almost simultaneously, a hydrogen atom is transferred to a molecule of NAD^+ to make NADH. Since two pyruvic acids are doing this, the cell gains two NADH's that can be used for ATP production later (see below).

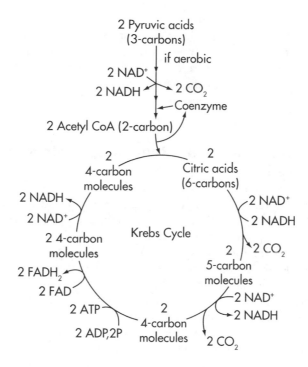

FIGURE 6.3. *The Krebs cycle and reactions feeding into it.*

The material that remains after these changes is a simple 2-carbon formation called an *acetyl group*. It is immediately attached to a complex organic molecule called *Coenzyme A*, which acts to carry it to the first step of the Krebs cycle. When coenzyme A is carrying an acetyl group, the combination of the two is abbreviated *acetyl CoA* (Fig. 6.3). The steps described here are important not only because they link glycolysis to the Krebs cycle, but also because other energy-carrying molecules such as amino acids and fatty acids can be converted to acetyl groups and be fed into the Krebs cycle by linking to coenzyme A in the same fashion. In other words, these steps act to funnel a number of quite different molecules into the Krebs cycle to yield energy for the cell. A summary of the various molecules that converge toward the Krebs cycle will be provided later in this chapter.

However, as we move toward discussing the Krebs cycle itself, we will follow acetyl groups whose origins were from glycolysis.

The Krebs Cycle Requires the Presence of Existing 4-Carbon Molecules

The first step of the Krebs cycle involves the transfer of an acetyl group from its carrier, CoA, to a 4-carbon molecule that had been made earlier. The resulting 6-carbon molecule, citric acid, has two carbons (with their accompanying hydrogens and oxygens) that have just been added, but the other four carbons can be traced back to previous identical inputs from glycolysis. That is, each turn of the Krebs cycle adds a 2-carbon fragment that was once part of glucose to others that had the same origin. An alternative name for the Krebs cycle is *citric acid cycle* because of the central role played by this 6-carbon molecule.

The Krebs Cycle Causes All Remaining Hydrogens to Be Removed

At four distinct points along the series of reactions of the Krebs cycle, hydrogen atoms and accompanying energy are removed and carried off for further processing. Three of these activities are familiar from our earlier descriptions: hydrogen is captured and carried by NAD^+. Since each glucose provided two acetyl groups for the Krebs cycle, a total of six NADHs are produced. Again, we will defer discussion of the fate of NADH until later, but the reader is reminded that each has the potential to cause ATP to be formed.

At one step of the Krebs cycle two atoms of hydrogen are removed to a slightly different carrier molecule, flavine adenine dinucleotide, usually called *FAD*. Although the resulting molecule, $FADH_2$, is handled in a slightly different way, the result of its being built during the Krebs cycle is nearly identical to what happens when NADH is formed: ATP will eventually be made with the help of the transported energy.

In summary, as the Krebs cycle operates, some of the energy from one glucose molecule is converted to bonds of six NADH molecules and two $FADH_2$ molecules.

Some ATP Is Made Directly as the Krebs Cycle Operates

There is a direct transfer of energy into a bond of ATP during the Krebs cycle just as happened during glycolysis. At one point, a free phosphate group is joined to an existing ADP, with the energy coming from one of the intermediate molecules that is moving through the cycle. This is called *substrate-level phosphorylation*, to distinguish it from the building of ATP from the energy provided by NADH or $FADH_2$.

Carbon Dioxide Is Made During the Krebs Cycle

At two points along the cycle of reactions, carbon is broken away in a molecule of CO_2. As occurred when the first CO_2 was removed (just before coenzyme A entered the picture), these are no longer able to provide energy and are excreted from the cell. The net result is that a 2-carbon unit (an acetyl group) has entered the Krebs cycle and two 1-carbon molecules have left it during one turn of the cycle. They are not the identical carbon atoms that entered at the beginning of a single turn, but, as the cycle continues to recur, each entering carbon eventually leaves as part of CO_2.

When the entire set of reactions (glycolysis through the Krebs cycle) is considered, every 6-carbon glucose that begins the trip is eventually released so that six CO_2s are made. During the same complete run-through, the following additional materials have been made: 4 ATPs (net gain), 10 NADHs, and 2 $FADH_2$s.

OXIDATIVE PHOSPHORYLATION MAKES MANY MORE ATPS

It is time now to follow NADH and $FADH_2$ to the point where they pass on their energy in activities leading to the production of much more ATP. Since NADH is the more prevalent molecule, this will be the first one traced. It travels to the inner membrane of a mitochondrion (see Chapter 4, especially Fig. 4.1 and 4.2) where it finds an assembly of proteins and other molecules embedded in the membrane. The actions that follow are collectively called *oxidative phosphorylation*, to produce ATP by phosphorylating ADP in the presence of oxygen.

As Figure 6.4 shows, each respiratory assembly is quite similar to the arrangement of molecules operating in a chloroplast's thylakoid membranes to put together ATP. Indeed, both systems operate by the mechanism called

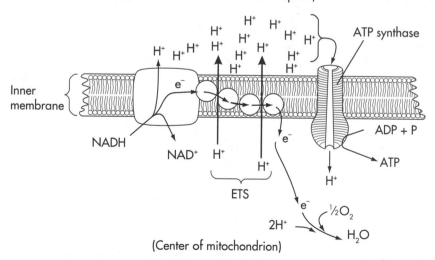

FIGURE 6.4. *Oxidative phosphorylation and chemiosmosis at the inner membrane of a mitochondrion. ETS: electron transport system. H^+: hydrogen (proton). e^-: electron. Number and shapes of molecules comprising the ETS are not meant to be accurate.*

chemiosmosis that was described in the previous chapter (Fig. 5.6 and the topic of photophosphorylation). There are two major differences. First, the source of electrons (and their energy) for driving the H^+ pump is NADH, not chlorophyll. Second, the final resting place for these electrons after they have transferred some of their energy to a bond of ATP is in a newly made molecule of water, not chlorophyll or NADPH.

As NADH approaches the first molecule of the mitochondrial electron transport system, it releases its hydrogen into the surrounding medium. But it also releases an electron that had been associated with the transported hydrogen nucleus. This electron, which carries much energy that can be traced all the way back to bond energy of glucose, is picked up by the electron transport system and moved from one molecule to another. Along the way, some of its energy is used to power the pumping of hydrogen ions across the membrane. Just as in the story of phosphorylation during photosynthesis, these hydrogen ions are then allowed to flow through the protein called ATP synthase, and the power of this flow is used to put together a bond to make ATP. Actually, each arrival of NADH at the mitochondrial membrane results in the production of three ATP molecules (on average).

Examination of Figure 6.4 reveals the all-important function of O_2 in this series of events. When two electrons have done the job described in the previous paragraph, they must become part of some molecule. That molecule is water, which is formed from two hydrogen ions (available from the watery medium within the mitochondrion), one atom of oxygen (formed by splitting a molecule of O_2), and the two electrons. If oxygen is not available in this part of the cell (i.e., it is anaerobic), the entire process of oxidative phosphorylation comes to a halt. This, in essence, is why organisms try to provide oxygen to their cells as much as possible.

The carrier FAD can bring its pair of acquired hydrogens to the same mitochondrial system. However, it is not as efficient in giving the electron transport system the energy of its carried electrons, and does not produce as much ATP, yielding only two ATPs per transfer.

Since 12 hydrogen atoms and 12 pairs of electrons are brought to oxidative phosphorylation as a result of one molecule of glucose having been processed by glycolysis and the Krebs cycle, the total number of H_2O molecules made during oxidative phosphorylation is six. The overall equation describing the fate of a molecule of glucose is:

$$C_6H_{12}O_6 + 6\ O_2 \rightarrow 6\ CO_2 + 6\ H_2O$$

THIRTY-EIGHT (OR THIRTY-SIX) ATPS ARE GAINED PER GLUCOSE MOLECULE

A valuable exercise at this point is to tally up the number of ATP molecules gained for a cell if glycolysis, the Krebs cycle, and oxidative phosphorylation have cooperated to break down one molecule of glucose into six molecules of CO_2 and six molecules of H_2O.

Gained directly during glycolysis	2 ATP
Gained because of NADH made before Krebs	12 ATP
Gained because of NADH and $FADH_2$ made in Krebs	22 ATP
Gained directly during Krebs	2 ATP
Total gained if O_2 available for oxidative phosphorylation	**38 ATP**

Sometimes the total number of ATPs gained is listed as 36 rather than 38. This is probably more accurate, for the following reason. Two of the ten molecules of NADH are made during glycolysis; therefore, they must travel from the cytosol into the mitochondrion before they can participate in ATP production. Actually, they cannot pass through the inner mitochondrial membranes and must pass their energy-rich load to a special shuttle system. In some animal tissues, this activity requires the energy of two previously made ATP molecules. So, the net gain should be lowered by two, to 36, at least for some tissues.

Whether the accurate number is 36 or 38, either of these totals is far higher than the net gain of two ATP made during glycolysis. Remember that this difference depends entirely upon the presence (~38 ATP) or absence (2 ATP) of the molecule O_2. Although most cells can continue existence at a low level of activity if they are anaerobic, they are far more capable of normal life when given the aerobic opportunity to harvest the maximum number of ATPs from each glucose. An important exception should be noted—brain cells of the higher vertebrate animals, including humans, are so active that they cannot continue living under anaerobic conditions for more than a few minutes. This is why the human brain is so richly supplied with blood (carrying glucose and O_2).

Under aerobic conditions, the efficiency of transferring energy from the bonds of glucose to those of ATPs is quite remarkable. Approximately 38 percent of the energy of glucose is captured. Although this may not seem like a high figure, it holds up very well when compared with the efficiency of a typical automobile in converting the energy of gasoline into movement: only about 25 percent. Furthermore, much of the un-captured energy from glucose is in the form of heat, which some organisms (such as mammals and birds) retain to help their chemical reactions run faster.

FOODS OTHER THAN GLUCOSE CAN ENTER THE KREBS CYCLE

As was briefly mentioned above, glucose is not the only energy-carrying molecule that can be modified to enter the Krebs cycle. Of course, any polysaccharide containing glucose (such as glycogen or starch) can be broken down if the proper enzyme is present to yield monosaccharides for feeding into the beginning of glycolysis (Fig. 6.5).

Other important food molecules are amino acid. Remember that each amino acid has a backbone of two carbons and one nitrogen. If the nitrogen and its two attendant hydrogens are removed, the remaining 2-carbon portion can be modified to become an organic acid, such as the acetyl group that is carried by coenzyme A into the Krebs cycle (Fig. 6.5). The removal of the —NH$_2$ (amino) group creates a problem for the cell: it quickly becomes NH$_3$ (ammonia), a toxic material. The removal of ammonia and related materials is a topic to be pursued in Chapter 9, under the topic nitrogenous waste removal.

Another rich source of energy is fat. Here, both the glycerol backbone (containing three carbons—see Fig. 3.6) and the attached long carbon chains of fatty acids can be modified to enter energy-yielding reactions. As Figure 6.5 indicates, fatty acids can be enzymatically chopped into 2-carbon fragments which enter the Krebs cycle with the help of Coenzyme A. The glycerol becomes modified to be identical with one of the 3-carbon molecules moving through glycolysis. Stored fat is probably the most-used energy source when a person wakes in the morning, before a meal is eaten. Then, after breakfast has provided glucose, this becomes the favored energy source.

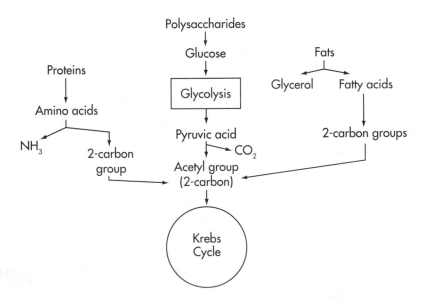

FIGURE 6.5.   *How various energy-storing molecules feed into glycolysis and/or the Krebs Cycle.*

ANAEROBIC CELLS MUST PERFORM FERMENTATION AFTER GLYCOLYSIS

If a cell lacks molecular oxygen, it can still perform glycolysis. However, there is a limited supply of NAD$^+$, which is needed during glycolysis. When all available molecules of this carrier have been converted to NADH, glycolysis will no longer continue. Of course, the way out of this problem is to provide an alternative molecule that will accept hydrogen from NAD, freeing NAD to go back to glycolysis and repeat its activity. Such an alternative must be not involve O$_2$.

Figure 6.6 shows that two end products of glycolysis, pyruvic acid and NADH, interact in the process called *fermentation*. The figure also shows that two alternative versions of fermentation exist, depending upon the organism in which the reactions are happening. In both cases, the following features occur. First, NAD yields its hydrogen and becomes free to help glycolysis happen. Second, the pyruvic acid accepts this hydrogen. Third, the final products of fermentation are molecules that the cell can easily remove. This is important, since a buildup of end products could lead to a clogging of the system, slowing the entire glycolysis process that precedes fermentation. Fourth, no new ATP forms as a direct result of the actual fermentation steps. However, because it allows the continuation of ATP production in glycolysis, fermentation indirectly participates in this all-important activity of making energy available for the cell.

Sometimes, an organism whose cells have had to use lactic acid fermentation can later gain more energy by modifying the end product and sending it back toward energy-releasing reactions. For instance, in the person who has

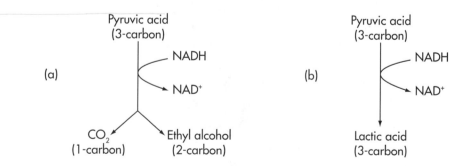

FIGURE 6.6. *Fermentation follows glycolysis if a cell is anaerobic. A: alcohol fermentation, the version occurring in yeasts and some bacteria. B: lactic acid fermentation, the version occurring in most other organisms, including humans.*

been working so hard that muscle cells temporarily became anaerobic, the resulting molecules of lactic acid are transported (via the bloodstream) to the liver. There, they can be manipulated and combined to make glucose. This requires some use of ATP (providing energy to make new bonds), but the value of providing new glucose makes the expenditure worthwhile. Much of what is commonly referred to as *oxygen debt* is the need for extra oxygen (O_2) that is used to aerobically provide the ATP for combining lactic acids in the liver. This is part of the reason why a person who has just done rapid exercise must breathe heavily for awhile.

Organisms can be classified according to whether or not they regularly perform anaerobic respiration in their cells. Yeasts can get along with or without oxygen and are termed *facultative anaerobes*. Anaerobes (such as some bacteria) that cannot use oxygen are termed *obligate anaerobes*. For them, oxygen is actually a poison.

Some end products of anaerobic respiration are economically significant in several respects. This is particularly true of alcoholic fermentation to produce ethyl alcohol and carbon dioxide. Under the most favorable conditions alcohol accumulates in quantities up to about 16 percent before it becomes toxic enough to kill off the microorganisms and stops the process. One of our most important lipid solvents is alcohol, which is used in many industrial processes and medicines. Of course, alcohol made by fermentation is also used in some beverages. Carbon dioxide from alcoholic fermentation bubbles away (Fig. 6.7). Carbon dioxide emission by the same process fluffs up dough in its prep-

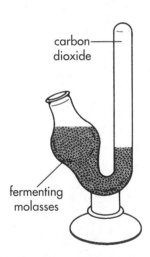

carbon—
dioxide

fermenting
molasses

FIGURE 6.7.   *Fermentation tube containing yeast in molasses. The closed end of tube was filled with the solution at the beginning of experiment but quickly accumulated carbon dioxide from anaerobic respiration.*

aration for baking. Lactic acid bacteria produce lactic acid in sauerkraut, sour milk, and ensilage.

A few anaerobic bacteria are serious infestors of wounds. Tetanus organisms, *Clostridium tetani*, produce a very poisonous end product that affects nervous tissue. Gas gangrene, which is especially known because it is associated with war wounds, is caused by several species of bacteria. End products produced by them kill and disintegrate local tissue.

Sometimes in alcoholic fermentation, alcohol and carbon dioxide are not the end products. Acetic acid bacteria produce another 2-carbon molecule, acetic acid (vinegar), from the alcohol. The souring of wine or cider results from the action of these bacteria. Thus, pure apple vinegar is made from hard (alcoholic) cider by acetic acid bacteria. Dregs in the bottom of a vinegar container are a concentration of the organisms commonly called mother-of-vinegar. Because apple vinegar is more expensive, it is commonly replaced by artificially colored or clear acetic acid obtained by a synthetic process.

ENERGY YIELDS FROM FOOD MOLECULES CAN BE MEASURED

Unless there were some yardstick by which measurements could be made, it would be impossible to compare the potential energy of fuel substances—in the biological sense, foods—or to determine how much energy is used under any particular set of circumstances. Since potential energy sources can be completely broken down by burning and since the bulk of energy released is in the form of heat, a measurement of the amount of heat liberated can be a useful standard for comparison. Of course, the simplest calculations are those involving the complete burning of specific amounts of a food outside the body and measuring the amount of heat generated. The measurement of energy released by aerobic respiration in living organisms is much more complicated. It requires calculations based on oxygen consumption and carbon dioxide elimination from a living organism using a mixture of foods with different energy values.

Heat is measured in units called *calories.* One thousand calories (or 1 kilocalorie) is the amount of heat required to raise the temperature of one liter of water one degree Celsius. Relative energy (heat) values of foods are obtained by direct calorimetry. Foods are burned in a device called the calorimeter. One type is a bomb calorimeter, a closed metal cylinder containing a certain amount of food to be burned and oxygen to burn it. It is submerged in a measured quantity of water and the fuel ignited by an electrical current. Heat from combustion raises the temperature of water, which is measured and calculated in kilocalories. It must be obvious that foods are as variable in their energy potentials as they are variable in their chemical differences. The average yield for the classes of foods in a bomb calorimeter is about 4.1 kcal per gram for carbohydrates, about 9.5 kcal per gram for fats, and about 5.6 kcal per gram for proteins. The figures show that fats are ideal as storage foods, for roughly twice as much energy is packed into a given quantity by weight. Because of this fact, plants reap an advantage when they store fats in reproductive structures like cotton seeds or peanuts. Being small in size, the reproductive structures are easier to distribute, and having large quantities of energy stored in proportion to size, seedlings have a better chance of getting off to a good start. Animals profit in the same way by concentrating excess potential energy in this space- and weight-saving form.

A measurement of food input is not a satisfactory measure of energy utilization in a living organism for the simple reason that one cannot know what percentage of food may actually be converted to the energy of ATP in respiration. It is possible, of course, to measure directly the heat production of an entire organism by putting it in a closed, insulated room for a time. The heat given off by the body can be absorbed by water circulating through tubes and then measured. This method is so complicated and expensive that an indirect method based on the consumption of oxygen (and liberation of carbon dioxide) is a more practical approach.

An instrument used to measure respiration by the indirect method is a *respirometer*. Or if used on humans, it is commonly referred to as the *basal metabolism apparatus*. The amount of oxygen consumed is measured, as is also the amount of carbon dioxide expired. The ratio of oxygen consumed to carbon dioxide expired is quite important in revealing what kind of food is being used. Then by knowing the kind of food and the amount of oxygen used, it is possible to calculate the quantity of food actually moved through the respiration process. One must keep in mind though, that respiration will ordinarily consume a mixture of classes of foods simultaneously.

The rate of respiration is by no means uniform in all organisms, nor in all tissues in the same organism. It is especially high in microorganisms and active animals. The rate is considerably influenced by environmental temperatures in plants and in animals unable to maintain a constant internal temperature (cold-blooded). Within limits, increasing temperatures usually mean a faster rate of respiration. Animals able to regulate their temperature (warm-blooded), however, are not so directly affected. But they do have a problem in heat regulation. Opposite of what happens to cold-blooded animals, they have to speed up respiration as the temperature falls in order to maintain their operating temperatures. This is accomplished by voluntary muscular activity or involuntary shivering, as well as certain hormonal activities. When it is too hot, animals have the problem of dissipating enough heat to keep the level normal. Heat is lost through the body surface, a phenomenon meaning that small animals have a considerable advantage over larger ones. A large animal like an elephant would soon die if it had the metabolic rate of a shrew because it could not dissipate enough heat from its surface (see Chapter 4 discussion on surface-area/volume ratios). In fact, size and rate of metabolism are enough related so that it is almost correct to say that the smaller the warm-blooded animal, the higher its rate of metabolism. Within a complex organism it is obvious that a more active tissue like muscle uses much more oxygen than less active tissues.

Since heat is a by-product of an organism's energy conversions, its production can be taken as evidence that respiration is in progress. The generation of heat in hot beds for planting seeds, in manure heaps, and in green hay is primarily the result of respiration by bacteria. Sometimes enough heat is generated by green hay stored in barns to set them on fire spontaneously. Everybody knows that increased activity in the human requires increased energy and results in faster breathing and a higher temperature. Consequently, one can keep warm on cold days by keeping active.

PHOTOSYNTHESIS AND RESPIRATION CAN OCCUR SIMULTANEOUSLY IN PLANTS

One point often misunderstood is the fact that green plants respire at all times. The misconception probably arises from the statement that plants give off oxygen in the daytime (as a by-product of photosynthesis) and carbon dioxide at night (as a by-product of respiration). The truth is that both photosynthesis and respiration go on in daylight, but photosynthesis is usually faster. This means that during the day carbon dioxide from respiration, plus additional quantities taken from the air, is used immediately in photosynthesis. The only gas apparently eliminated is oxygen. With photosynthesis stopped at night and respiration continuing, carbon dioxide is given off. As far as one can tell, the ratio of photosynthesis to respiration on the earth is close to constant, as indicated by their relative proportion from air analyses. Some may fear that human-caused rapid destruction of vegetation on the earth will upset the balance. It should be remembered that there are large areas of the world that are relatively untouched and that where vegetation is destroyed other plants usually come in quickly to heal the scar. Of more significance is the addition of carbon dioxide from combustion.

DECAY ALLOWS CONTINUATION OF ELEMENT CYCLING IN AN ECOSYSTEM

Regardless of what dead organic material it may be, organisms known as decomposers will inevitably attack it. As described in the previous chapter, decay is extremely important, otherwise many elements necessary for life would soon be tied up in the flesh of dead bodies or in other organic materials such a chitin, hair, rubber, petroleum, or horn. In and near decomposing bacteria and fungi, there are two processes involved in decay—digestion and respiration. If plenty of oxygen is present, the end products are primarily carbon dioxide and water; if oxygen is deficient, breakdown is incomplete and results in such substances as organic acids, methane gas, gaseous hydrogen, and alcohols.

When simple products are released from dead bodies by decay, they are made available to living organisms in their environments. They are variously used and again returned to the environment for reuse. The use, release, and reuse of the same materials are cyclic. The most common cycles of elements related to life are the carbon, oxygen, hydrogen, and nitrogen cycles. Sometimes one speaks of a water cycle, which, of course, is a part of the hydrogen and oxygen cycles. In fact, all of the cycles are interrelated. For simplicity, the carbon, oxygen, and nitrogen cycles will be considered separately.

Proteins, containing nitrogen, undergo changes resulting in ammonia, nitrites, and nitrates, which can be used by plants, or in the freeing of gaseous nitrogen into the atmosphere. Other elements remain as solid residue.

THE CARBON CYCLE INVOLVES INTERCONVERSION FROM AND TO CARBON DIOXIDE

One of the ingredients of organic molecules synthesized by green plants is inorganic carbon dioxide, which comes from air, soil, or water. Organic molecules are incorporated into the substance of the green plant until some of them are used in respiration (their carbons released as CO_2), or until the plant is de-

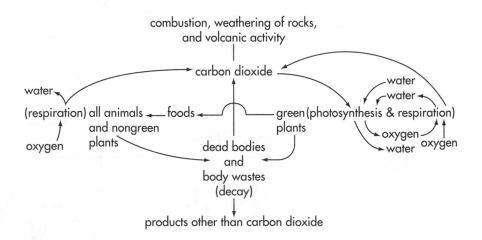

FIGURE 6.8. *Simplified carbon cycle.*

stroyed. The destruction of the plant may result from its being consumed as a food or from its decaying after death. Should it be consumed as food, organic substances will be respired or used to build structures. In any event, the destiny of organic materials is destruction by respiration or decay. Occasionally, a third activity releases carbon: burning of the organism, as in forest fires. By whatever means, organic molecules are thereby broken into inorganic molecules, including carbon dioxide. Other sources of carbon dioxide than the destruction of organic molecules are volcanic activity and weathering of rocks (Fig. 6.8).

THE NITROGEN CYCLE DEPENDS ON NITROGEN FIXING BY BACTERIA

Nitrogen comprises 78 percent of the atmosphere. The source of atmospheric nitrogen is denitrification (breaking NO_3^- into gaseous nitrogen and oxygen) by bacteria in the soil, combustion, or volcanic activity (Fig. 6.9). Despite the large quantity of nitrogen in the air, it is not usable by plants in the gaseous form. It is, however, usable by green plants as the ammonium ion (NH_4^+) nitrites (NO_2^-), or nitrates (NO_3^-). These ions are absorbed from the soil or water and are used by green plants to synthesize amino acids, proteins, and nucleotides. It should not be inferred that these syntheses are part of the photosynthetic process, for they do not require light or chlorophyll. Plant proteins are used by heterotrophs for food. Proteins in any organisms are even-

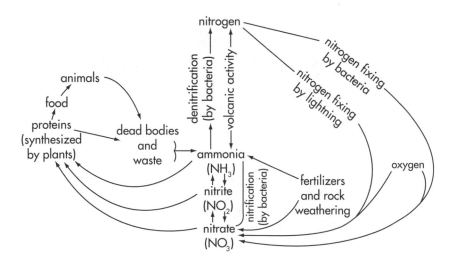

FIGURE 6.9. *Simplified nitrogen cycle.*

tually broken down by respiration, decay (respiration in part), or combustion to various ions or even gaseous nitrogen.

Nitrogen combinations that enter the green plants come mostly from nitrogen fixation and decay. A much smaller source is volcanic activity (ammonia). *Nitrogen fixation* simply means that atmospheric nitrogen is converted to a form that can be used by green plants. Fixation is accomplished by lightning, human-generated electrical processes, and bacterial action. The bacteria, known as nitrogen-fixing bacteria, may live free in the soil or in cells within the nodules on the roots of some plants (legumes and others). One of the processes occurring in decay is nitrification, which is also a bacterial action. If it proceeds through the entire chain of events, ammonia (which is produced during breakdown of amino acids) is changed to nitrites, and nitrites to nitrates.

THE OXYGEN CYCLE IS INTIMATELY CONNECTED WITH PHOTOSYNTHESIS

Molecular oxygen (O_2) of the atmosphere, soil, and water comes from photosynthesis. It comprises 20 percent of the atmosphere and smaller amounts of soil and water. It is available for respiration, combustion, and decay (respiration in part). Two end products of either of these three reactions are carbon dioxide and water, both of which contain the element oxygen in their molecules. Both carbon dioxide and water enter the photosynthetic process with the end result that organic compounds, water, and molecular oxygen are produced. Oxygen is still chemically bound in the first two and may not be re-

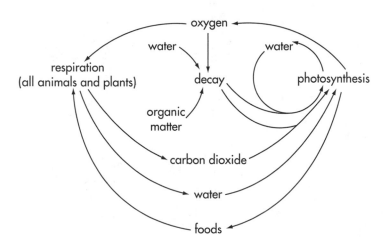

FIGURE 6.10. *Simplified oxygen cycle.*

leased as molecular oxygen before it has gone through an indefinite number of transfers until finally freed by photosynthesis (Fig. 6.10).

STUDY QUESTIONS

1. What is the relationship between energy storage and the bonds of food molecules?

2. List a few uses of energy in your body.

3. Distinguish between breathing and respiration. Where does respiration occur?

4. Why is it both inefficient and dangerous to break many bonds of glucose molecule simultaneously in a living cell?

5. Any enzyme speeds a reaction by lowering its _____ energy.

6. Explain why glycolysis must begin by using up some already-existing ATP.

7. Follow the route of NADH from glycolysis if (a) the cell lacks O_2, and (b) the cell contains O_2. Which is more productive?

8. What is the role of acetyl CoA in the energy-producing activity of a cell?

9. What is the ultimate fate of the carbons of glucose, if both glycolysis and the Krebs cycle are able to operate?

10. Why is the Krebs cycle also known as the citric acid cycle?

11. Compare oxidative phosphorylation in mitochondria with photophosphorylation in chloroplasts.

12. At what point do some amino acids enter the respiratory process? At what points do glycerol and fatty acids enter the process?

13. Describe the highlights of the Krebs cycle.

14. What unit of measurement is used to express the energy values of foods? What are the relative values of carbohydrates, fats, and proteins?

15. How efficient is the respiratory process in transferring energy from glucose to ATP? What happens to the remainder of the energy?

16. What is meant by such terms as *carbon cycle* and *nitrogen cycle*? Through what are these elements being cycled?

17. Describe nitrogen fixation. What importance is it for humans?

7
OBTAINING AND DIGESTING FOOD

What do we know about:
—*how organisms are equipped to maintain optimal internal conditions?*
—*the many ways animals use to bring food into themselves?*
—*how fungi are similar to animals in their need for food?*
—*why some animals do not need to digest their food?*
—*how food is moved through a digestive tract, and how it is processed?*
—*the paradox of some plants having the ability to eat animals?*
—*the roles of vitamins?*
—*the specific steps of food processing in humans?*
—*how food and its energy are passed through complex ecosystems?*
These and other intriguing topics will be pursued in this chapter.

PHYSIOLOGY RELATES AN ORGANISM'S ANATOMY TO ITS NECESSARY FUNCTIONS

This is the first of several chapters that will explore the physiology of organisms, so it is appropriate to begin by discussing the nature of physiology in general before looking closely at any single topic. The word is used at many levels, from subcellular through organismal, but it always relates to the study of how various necessary activities get done and how their accomplishment is linked to anatomy. The first paragraph of Chapter 4 contained the famous "form follows function" statement attributed to the architect Louis Sullivan; this encapsulates much of the emphasis of physiology.

For the most part, our level of focus will be on tissues, organs, and systems, but we will sometimes have to return to cellular and biochemical phenomena to explain what is happening. Two major emphases will be observed as we examine particular activities. First, we will try to understand how an organism maintains the optimum internal environment within which to operate. This is the idea of *homeostasis*. Second, we will look at how the body of an organism interacts with its environment. However, this will not usually be on the levels of behavioral or ecosystem interaction, but rather on how individual organisms exchange materials (food, gases, wastes, and so on) with their surroundings and how they react to environmental stresses such as heat or cold. Obviously, these two major themes of physiology are really closely related; e.g., interacting with the environment to obtain food is a way of keeping the best internal concentration of glucose.

It is worthwhile to examine the mechanisms by which homeostasis occurs. The goals of any homeostatic system are to sense change in some feature of the organism's internal environment, then to perform the proper action to bring it back to the best condition for the organism. For instance, if a person's internal temperature drops below normal, certain brain cells sense this and set into motion events leading to increased heat production and decreased heat loss. In order for this to be a fine-tuning mechanism, it is usually necessary for a *negative feedback* system to be in place. As shown in Figure 7.1, such a system operates by the end product of some positive activity serving as an inhibitor of its own further production when enough has been made. The units of such a system usually include (a) a *sensor* that can measure some material or action and compare it to some built-in standard, (b) a control center to receive a message from the sensor and send it to (c) the appropriate *effector* that produces a (d) *response* (release of the material or action), and (e) some means of communication between sensor and effector. Furthermore, the sensor must be capable of sending a positive message to the control and effector when there is not enough response measured, but also be inhibited from sending the positive message by a high amount of sensed response. If all of these components are in place, the result is a fine homeostatic ability, with some material or activity remaining within a narrow optimal range for the organism. We will see several examples of this happening as we examine physiology. It also occurs at the level of gene and enzyme control (Chapter 12).

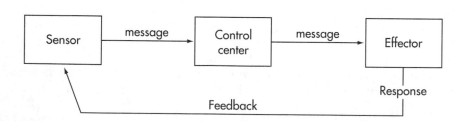

FIGURE 7.1.   *A generalized homeostatic system relying on negative feedback. The control center is a portion of the central nervous system in an aninal, but absent in plants and when feedback is occurring entirely within a single cell.*

Another thing that will become obvious as we explore various physiological actions is that most organisms share a common set of challenges, even if the organisms are very different from each other in their general shapes and origins. For instance, nearly all organisms must exchange the gases oxygen and carbon dioxide with their environments, even if they are as different as plants are from animals. We will see that some diverse organisms use strikingly similar methods to solve problems that they share. But we will also find examples of very different solutions being found. That is, there are sometimes several good ways to arrive at the same result. To help emphasize the common problems faced by organisms, most chapters on physiology will combine animal and plant studies, as well as the occasional example from the other kingdoms.

With this general introduction behind us, let us now examine the topics of food procurement and processing as our first specific topics of physiology.

HETEROTROPHS NEED TO CAPTURE ENERGY BY INGESTING FOOD

Chapter 5 introduced the terms autotroph and heterotroph to describe organisms with respect to their sources of energy-containing foods. The major emphasis of the current chapter is on how the heterotrophs of the world do two things: ingest food and prepare it for energy-releasing reactions. Then we will examine how organisms eliminate any ingested materials that proved to be unusable as food.

Foods are not the only requirements that must come from outside sources. Heterotrophs must acquire most of their vitamins, which like the foods are also synthesized by autotrophic plants. In addition, minerals like calcium, iron, potassium, iodine, and magnesium are ingredients of essential processes, and the compound water has many uses. Ordinarily all of these essentials, with the possible exception of water, are obtained in sufficient quantities along with the foods. All of these materials are considered *nutrients.*

HETEROTROPHS USE VARIOUS METHODS OF EATING

Since the specific foods of animals vary tremendously, it should not be surprising that there are many methods of capturing those foods. They fall into several categories. Some animals living in water are *filter feeders*, setting up some sort of fine-meshed net device to capture very small organisms or organic fragments. This can be a productive activity since many bodies of water are rich in *plankton*—organisms at or near microscopically small size. It is surprising to realize the range of animals doing filter feeding. A sampler includes sponges, clams and oysters, sea-dwelling Annelid worms, and even the world's largest animals, certain whales. In each case, the key is some part of the anatomy that has extremely small holes in a sheet-like structure. Usually this filtering region is covered with a thin coat of sticky mucus to hold whatever objects are caught. It is necessary to provide a near-constant flow of water through the net, to enhance the likelihood of enough tiny prey organisms being brought to it. Some filter-feeders place themselves in areas of high current flow (fanworms); others make their own current by pulling water into themselves (clams, sponges) or swimming swiftly through water and then spitting the gathered water back out through the net (baleen whales).

Another category is the *fluid feeders*. These animals take advantage of the already-processed rich nutrients within the body fluids of other organisms. Good examples are mosquitoes, leeches, and fish (lampreys) that temporarily attach to larger animals and tap into their blood supply. Aphids and some nematode worms do the same to plants, removing sugar-rich fluids from their vascular tubes.

A third category is the *substrate feeders*, which eat the materials upon (or within) which they spend much or all of their lives. In this group are earthworms eating the soil in which they burrow, snails scraping algae and bacteria off surfaces, and caterpillars eating their way through fruits or leaves of plants.

It should not be forgotten that some nonanimals are also heterotrophic. The members of kingdom Fungi could be classified as substrate feeders, since they obtain their food from the organisms or remains upon which they rest. The major difference between them and the animals that are substrate feeders is that the former do not ingest whole food. Rather, they secrete digestive enzymes that break down complex structures to simple food molecules and then ingest those by simple absorption. The same procedure is employed by those few plants that are carnivorous as well as being photosynthetic. They are described in detail later in this chapter.

The animals we are most familiar with are those that eat all or a large part of their prey, using a mouth. These are the animals (including ourselves) equipped to capture, hold, tear, and grind food before beginning the digestion process. Much of the remainder of this chapter will concern them.

PROCESSING OF FOOD BEGINS WITH INGESTION

Ingestion, the act of eating, is usually through a mouth. Capturing and holding the prey is the first task to be attempted here. This is accomplished with rather obvious tools, such as the lips of mammals, beak of birds, and specialized appendages of insects.

Most mammals and many other vertebrate animals use teeth to capture and hold prey and to perform mechanical tearing and/or grinding. If a biologist were presented with only the skull of some mammal (such as a fossil fragment), he or she could make a fairly accurate hypothesis as to the diet of that animal when it was alive just by examining the types and relative numbers of teeth in the mouth. Mammalian teeth fall into the categories of *incisors* (for cutting and holding), *canines* (also for cutting and holding), and *molars* (for grinding). A large number of prominent molars in the mouth (such as in a cow) indicates a herbivorous lifestyle.

The only ingestion-like activity of land plants occurs at the roots, where water and dissolved minerals enter. This is better called *absorption*, since it involves taking in materials that are at or below the molecule level. Over a very extensive surface area created by branching of roots, water enters by simple diffusion. Ions (K^+, Na^+, NO_3^-, etc.) are initially caught on root cell wall surfaces because of their electrical charges. They are then transported internally by active transport pumps, with the expenditure of energy on the part of root cells.

INGESTED FOOD IS THEN DIGESTED

Most food enters the body as complex tissues, much too large to be moved across cell membranes. Even when individual cells like the leukocytes or independent Protozoa ingest solid particles, they are initially contained within vacuoles and are not incorporated into the cytoplasm until changed into a material simple enough to pass through vacuolar membranes. The breakdown of food materials into smaller molecules is *digestion*. The molecules must be of a certain size (and configuration) before they can be absorbed and distributed.

Intracellular Digestion Occurs in Plants and Primitive Heterotrophs

Plants naturally digest substances within their cells. Simple foods constructed by photosynthesis (such as glucose) are changed into complex foods (such as starch) and stored. Before the complex foods can be *translocated* (transferred) to other places of use or storage, they must be broken again into simpler molecules that can pass through membranes. Plants store, digest, and translocate in all directions for a variety of purposes. For instance, glucose that is newly manufactured may be synthesized into starch and temporarily stored in the leaves of the potato plant. The starch is digested back into individual glucose molecules, which are then translocated to the tubers, where starch is again synthesized and stored (Fig. 4.8 D). As another example, germinating seeds digest concentrated storage materials like fats and starch upon which they depend until they can develop photosynthesizing parts.

Intracellular digestion is typical of animal cells as well as of plant cells. Foods stored as glycogen or fat are drawn upon as needed. When they are di-

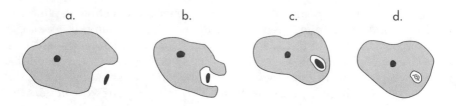

FIGURE 7.2.   *Phagocytosis and intracellular digestion by an* Amoeba *or leukocyte.*

gested, the simpler materials move to places of use, either within the cells that stored them or elsewhere via blood flow.

All of the above examples are digestion of molecules that had been produced within organisms that made them from simpler molecules. The type of intracellular digestion that is more to the point of the current chapter is that which follows the capture of food from outside the organism. Intracellular digestion of a sort is common in Protozoa and several invertebrate animal groups. Bits of food are taken inside each individual cell in a vacuole by a process called *phagocytosis* (Fig. 7.2). Lysosomes are attracted (by unknown means) to such a *food vacuole* and fuse their membranes with it. This causes contact of the lysosomes' digestive enzymes with the captured material , which is then broken down to simple molecules. They are slowly absorbed from the vacuole.

The kind of digestion just described is the method used by Protozoa, but some of the more primitive multicellular animals also rely on it wholly or in part. Intracellular digestion is the only method available to members of phylum Porifera (sponges). It is a supplemental method in Cnidaria (such as sea anemones) and Platyhelminthes (such as planaria). The Cnidaria and Platyhelminthes have a cavity (digestive tract) where some extracellular digestion occurs, but the cells lining the cavity also perform intracellular digestion. Any animal relying on intracellular digestion in whole or in part is limited to working only on very small food particles. Also, if it uses this kind of digestion as its major method of obtaining nutrition, it must have a large surface at which the digesting cells are located, since many cells must participate. Many animals of the phyla named above devote large portions of their body to the digestive tract.

Extracellular Digestion Is the More Efficient Method

The better way to accomplish digestion is to hold food in a sac or tube where digestive enzymes can break it down to small molecules. The major animal phyla not named above do this *extracellular digestion* as their sole method. This allows relatively few cells forming the tract lining to contribute digestive fluids to a large open space where the enzymes can surround and work on large pieces of food over an extended time.

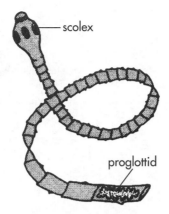

scolex

proglottid

FIGURE 7.3. *Tapeworm with flattened body and no digestive system. The scolex (head) is equipped with hooks for attachment, but lacks a mouth. Proglottids are reproductive segments, filled with embryos when mature.*

Some Internal Parasites Do Not Need to Digest

Some parasites (e.g., flukes) actively eat cells of their hosts but many that live surrounded by intestinal fluids or blood have no need to digest since they can absorb what has already been digested by their host. The tapeworm (Fig. 7.3), for instance, does this. Its flattened body eliminates the need for a distributing system, since no cells are far from the outside.

Complete Digestive Tracts Are More Efficient than Incomplete Tracts

There are two anatomical styles of digestive tract seen in animals—*incomplete* tracts and *complete* tracts (Fig. 7.4). The incomplete digestive system is less efficient. It has a single opening through which food must enter and waste must leave. Such systems are characteristic of Cnidaria and Platyhelminthes, the same phyla whose animals carry on both intracellular and extracellular digestion. In such an animal, digestive enzymes are secreted into the cavity to digest the food there. The resulting individual molecules are absorbed by cells lining the cavity. Some larger food pieces are chemically and mechanically broken down to small enough size to be intracellularly digested at the same time that extracellular digestion is occurring. In these animals, all body cells are quite close to the digestive cavity, either because the body wall is thin as in most Cnidaria or because the tract branches into the tissues as in many Platyhelminthes. As a consequence these animals do not need and do not have a distributing (circulatory) system.

The complete digestive tract of other animals has both an entrance (*mouth*) and an exit (*anus*). As food moves through the tube, it is processed and ab-

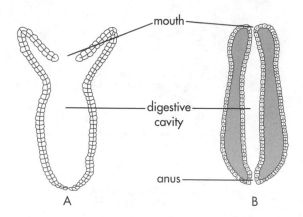

FIGURE 7.4.   *Two variations in digestive systems. A: the incomplete system. B: the complete system. The incomplete variety has only one opening, which is used both as an entrance and an exit.*

sorbed. Some of the ingested material is likely to be indigestible and is eliminated at the end of the system. The tract is modified along its length to perform specific functions. The ability to move food in a single direction throughout the tract, with sequential actions performed upon it, is what gives a complete tract the advantage over the other variety.

Complete Digestive Tracts Exist in Many Varieties

The specific set of activities along a complete tract varies considerably in relation to the type of food used and the particular manner in which an animal handles it. Animals that feed intermittently usually have a place such as a stomach or crop where foods can be stored temporarily. As a general rule animals that use bulky foods with a large amount of hard-to-digest material have a longer tract. This is true of herbivores, which consume much cellulose and lignin. Carnivores, however, have shorter tubes. In either case, typical digestive tubes are so long that they have to bend back and coil in order to be accommodated in a short body. The parasitic nematode *Ascaris*, which consumes predigested foods within the tract of a host organism, has a digestive tube exactly the same length as the body, whereas the human, which consumes bulky unprocessed foods, has a tube about five times as long as the body is tall.

Fragmentation of bulky materials is an important adjunct to digestion. Of course, chewing animals pulverize their food in the mouth, but nonchewing animals (such as earthworms and birds) swallow it whole to be pulverized farther along the tract. Both earthworms and birds accomplish this in a pouch called a *gizzard* that grinds food to bits. The cow has a multichambered stomach in which food may be temporarily stored until there is a more opportune time for chewing. Then it is regurgitated and chewed more thoroughly (an activity called chewing the cud). While it sits in two of the four chambers, food is partially digested by symbiotic bacteria and Protozoa that provide enzymes for breaking open tough plant cell walls.

Peristalsis Provides Efficient One-way Movement of Food

Most animals with a complete digestive tract are equipped with circular muscles along most of this route. When they contract, they squeeze the interior (*lumen*) of the tract and push food along. This squeezing, which moves in a series of wave-like motions from the mouth end to the anus end, is called *peristalsis*. The effect is to keep food moving in the proper direction as it is processed. Of course, peristalsis also helps digestion by mechanically churning food chunks, helping them to break into smaller fragments.

Movement of food through the digestive tract must proceed at a suitable speed for ideal efficiency. If the action is too fast, digestion and absorption are not completed. If the action is too slow, there is too much reabsorption of water resulting in constipation. Valves are often present that regulate the speed and direction of movement.

Chemical Digestion Breaks Down Food into Individual Small Molecules

Chewing and other mechanical actions cannot yield separated molecules small enough to be removed from the intestine for distribution. This requires the activity of a number of chemical reactions in the lumen of the tract. Most of these reactions are catalyzed by specific enzymes secreted into the vicinity of the food. They come from intestinal cells and sometimes also from auxiliary organs (generally called *digestive glands*) with ducts emptying into the intestine. The result of complete digestion is the accumulation of amino acids, simple sugars and small polysaccharides, fatty acids, and other small organic molecules. These are the materials finally removed from the tract.

Absorption of Digested Molecules Requires Contact with Cell Membranes

The next logical step in the whole process is efficient removal of these small molecules from the intestinal lumen for distribution throughout the body. This is accomplished by a variety of movements, all of which involve passage through the membranes of individual intestinal cells. It should be obvious that *absorption*, as it is called, is enhanced if there is a very large membrane surface area in contact with the digested molecules.

Typically, the absorbing portion of the tract is quite long, to provide more membrane surface. To further increase it, many animal species provide some device within the intestine that modifies it from a simple circular wall. Some such modifications are shown in Figure 7.5. They increase wall surface in proportion to the space through which food can move. One of the simplest adaptations is a flat intestine as in the nematode worm *Ascaris*. Since it is an

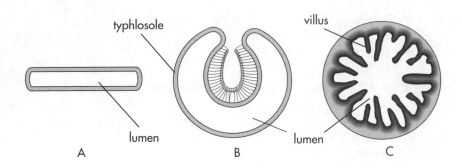

FIGURE 7.5. *Adaptations of intestines that increase the absorbing surface in proportion to the space through which food moves. A: flattened intestine as in* Ascaris. *B: tubular intestine with a dorsal fold (typhlosole) as in an earthworm. C: tubular intestine with fingerlike projections (villi) as in a human. All views are cross-sections.*

intestinal parasite, its own tract is probably more important as an organ for absorption than for digestion. The earthworm increases intestinal surface by having a fold, the *typhlosole*, which projects down into the lumen from the dorsal side all along the tract. Many vertebrate animals, including humans, have a huge number of tiny fingerlike projections, *villi*, each of which provides the same sort of surface area enhancement as a typhlosole. Each villus may have a myriad of its own tiny projections into the lumen, aptly called *microvilli*. These increase the total surface area but are also equipped with contractile microfilaments so that they can wave about, thus helping to move the liquid of the intestinal contents and enhance the likelihood of absorption. The total effect of the presence of villi and microvilli is to provide a tremendous absorptive surface in a small portion of the body. It is estimated that they increase the internal surface area of the human intestine by a factor of at least 500.

The Last Processing Step Is Elimination of Wastes

For most animals, not all that was eaten can be broken down to absorbable molecules. What remains is *fecal matter* (or *feces*), which must be removed from the body. In animals with complete digestive tracts, this is rather easily accomplished by continuing peristalsis until the anus has been reached. If it is in a terrestrial animal (which must almost always be trying to retain water), the last portion of the intestinal tract should be active in *reabsorbing* some of the water that is still traveling with the wastes. The term *reabsorption* is descriptive, since much of the water taken back out of the lumen was previously added from tissues along the tract—carrying enzymes, and other materials that were secreted to help digestion. Fecal matter may also contain toxic chemicals that were secreted into the digestive tract as a means of eliminating them from the body. Finally, a considerable portion of fecal matter can be bacteria that have been benignly inhabiting certain portions of the tract.

Fungi and Carnivorous Plants also Perform Extracellular Digestion

Extracellular digestion is by no means an exclusive function of animals. All members of Kingdom Fungi break down substrate material before absorbing it into their cells. Perhaps the most surprising organisms to do such things are a few green plants capable of digesting captured animal bodies. These carnivorous plants (Fig. 7.6) are usually described as insectivorous because they capture so many insects. Their rarity makes them novelties to most people. The Venus's-flytrap, which grows only in the coastal plain of the Carolinas in the United States, has modified leaves that catch insects. At the end of each leaf are two lobes that can fold rather rapidly together and catch insects. Pitcher plants have tubular leaves into which small animals fall. Sundew plants have leaves covered with sticky hairs to which small animals adhere. The bladderwort has many submerged leaves about the size of a pin head and shaped like a fish basket (a fish trap). The leaf traps allow animals to enter, but block their exit. None of these carnivorous plants absolutely require animal food since they can photosynthesize to provide their organic needs and bring inorganic materials in via roots. The pitcher plant can be plugged with cotton to keep out insects and other animals without any apparent effect on the plant, showing that its carnivorous activity is only supplemental. However, such plants

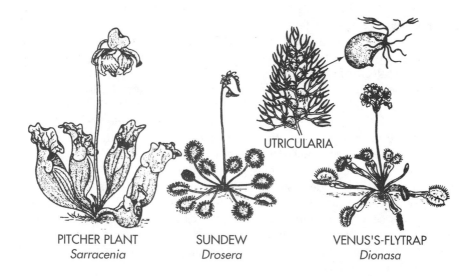

UTRICULARIA

| PITCHER PLANT | SUNDEW | VENUS'S-FLYTRAP |
| *Sarracenia* | *Drosera* | *Dionasa* |

FIGURE 7.6.   *Carnivorous plants. (Courtesy General Biological Supply House, Inc., Chicago.)*

tend to live in swampy areas where they might be expected to have some difficulty obtaining enough inorganic materials (particularly nitrogen), so their unusual activities are good adaptations.

DIETS OF ANIMALS VARY GREATLY

Although all animals digest for the same reason, the raw foods and the methods used to prepare them are highly variable. Foods comprising the usual diet of some animals are totally unusable to others. Along with many other items in the diet, humans eat some woody materials (cellulose and lignin), yet cannot digest them because they lack the genetic instructions to make appropriate enzymes to attack these tough materials. However, woody materials are valuable to us in providing roughage or bulk to prevent constipation. Other animals such as rabbits, cows, and larvae of insects use these same materials as one of their major food sources. Even such unlikely food material as hair is used by larvae of the clothes moth and feathers are used by bird lice.

Digestion is not always performed solely by the organism taking in the food, for a digestive tract frequently contains symbiotic organisms of one kind or another that act upon its contents. Bacteria in the digestive tract of vegetation-eaters (such as cows) can digest cellulose, thus making it available to the host as well as to themselves. They are able to do this because they manufacture and release the enzyme *cellulase*. Likewise, mutualistic bacteria and Protozoa in the intestines of termites digest the wood provided to them by the insects. Fungi are able to digest wood also, as is shown by the many species of bracket fungi found on fallen tree limbs in a forest.

Bacterial colonies in the digestive tracts of mammals are sometimes acting in the same mutualistic fashion. The aforementioned rabbits and cows that are such efficient users of plant materials succeed in breaking open plant cell walls only because of their intestinal bacteria. For such an enterprise to be successful, the bacteria must usually be located at the beginning of the tract so that there is sufficient length beyond them for absorbing the released molecules. Rabbits place their colony near the end of the tract, but eat their fecal pellets to give the digested plant material a chance for being absorbed on the second pass. The human intestine is teeming with bacteria, but almost all of them are in the large intestine, where significant absorption is not possible. Thus, we could not benefit from eating uncooked vegetation, even if our resident bacteria possess cellulase activity. Of course, we can overcome this shortcoming since we possess the knowledge to cook vegetable matter before eating it. This behavior leads to nonenzymatic (heat-driven) breaking of cell walls, with the same valuable effect.

One very useful action of our intestinal bacteria is the synthesis of some vitamins. It will be remembered from Chapter 2 that vitamins are organic molecules an organism needs in small quantity, but which it cannot synthesize.

Most vitamins for human health are readily obtained from plant material that we eat, but some of these are supplemented by the products of intestinal bacteria. Table 7.1 shows the sources and known activities of some common human vitamins.

Digestion Is Usually a Hydrolysis Reaction

In essence, the chemical activity that leads to breaking a complex food molecule into smaller units is the type of reaction called *hydrolysis*. In such a reaction, a molecule of water is split into its ionic components, H^+ and OH^-. A bond breaks between two units of a polymer and the two ions from water are placed where the bond had been. For example, if the food is a polypeptide, the following hydrolytic reaction takes place to split two of the amino acids from each other.

The reaction is said to be hydrolysis because a water molecule is split (hydro = water, lysis = split). With appropriate enzyme action, this can be repeated many times along the length of a polypeptide to yield individual amino acids, which are then absorbed from the intestinal lumen. Hydrolysis is also the means by which a polysaccharide such as starch is digested to single sugars that can be absorbed.

Fat degradation in the intestine begins with hydrolysis that detaches each of the three fatty acids that are linked to the three-carbon unit glycerol. This is a reversal of the synthesis of a fat shown in Figure 3.6, and is enzymatically mediated. The resulting free fatty acids, since they are hydrophobic like the lipid of cell membranes, easily move out of the lumen into epithelial cells of the intestinal wall. Once into these cells, individual fatty acids are re-joined to reconstruct fats. It is necessary to do this quickly, since accumulations of free fatty acids can act as detergents and destroy cellular membranes. However, this reconstruction creates a problem: whole fats cannot pass out of an intes-

TABLE 7.1. SOME COMMON VITAMINS FOR HUMANS.

Name	Solubility	Food Sources	Function	Deficiency Effect
Vitamin A	fat	green vegetables yellow foods liver	maintenance of epithelial cells of eyes, skin, digestive and respiratory tracts chemistry of vision bone formation	weakened resistance of epithelial tissue to disease night blindness abnormal growth
Vitamin B$_1$ (thiamine)	water	whole grain, especially the germ vegetables nuts yeast liver pork meat	carbohydrate metabolism	mild deficiency: loss of appetite, fatigue, etc. extreme deficiency: beriberi
Vitamin B$_2$ (riboflavin)	water	milk and its products yeast liver wheat germ meat eggs	coenzyme used in metabolism of glucose and amino acids and in some oxidative processes	retarded growth inflammation of eyes dermatitis cracking of corner of mouth
Panthothenic acid	water	liver meat eggs peanuts sweet potato	component of coenzyme A	convulsions, twitching, poor coordination
Niacin	water	yeast vegetables meat liver	important in formation of NAD and NADPH	pellagra, liver damage

Biotin	water	liver egg yolk meats vegetables	protein metabolism	dermatitis weakness
Folic acid	water	vegetables liver yeast bacterial synthesis in intestines	coenzyme in nucleic acid and amino acid metabolism	anemia leucopenia
Vitamin B$_{12}$	water	liver meat milk and its products	coenzyme in nucleic acid metabolism	anemia, nervous disorders
Vitamin C (ascorbic acid)	water	citrus fruits vegetables	maintenance of connective tissues antioxident	scurvy impaired immunity
Vitamin D (calciferol)	fat	liver oils milk products eggs manufactured by body	calcium and phosphorus metabolism	rickets brain damage
Vitamin E (tocopherol)	fat	wheat germ vegetable oils most other foods	antioxidant	none completely proven
Vitamin K	fat	most foods, especially vegetables bacterial synthesis in intestines	blood clotting component	hemorrhaging

tinal cell into a nearby capillary blood vessel for distribution to the rest of the body. Therefore, these fats are collected into small droplets, packaged within vesicles at the Golgi apparatus, and then dumped out of the intestinal cell into nearby lymphatic vessels called *lacteals*. Each villus of the intestine is supplied with a single lacteal, which joins others to become a set of lymphatic veins leading eventually to a connecting point into the general blood circulation system. Fatty acids are the only materials from digestion that are not transported directly to other parts of the body by blood vessels in the intestinal wall.

HUMANS USE A SOPHISTICATED SYSTEM FOR PROCESSING FOOD

Let us now pull together the entire story of food processing by examining it as it happens in one specific organism, the human. Here it occurs in a complete tract, and digestion is entirely extracellular. Digestion is aided significantly by the products of two auxiliary organs, the liver and the pancreas. Being omnivorous, humans consume a wide variety of foods. We are terrestrial animals, so there is an important component of water reabsorption.

During the chewing and lubrication of food in the mouth, enzymes are added to begin the digestion of polysaccharides. They are a part of the salivary secretions. Saliva also contains the antibacterial enzyme lysozyme, mineral ions, mucus, water, and some excretory products. Saliva is secreted via ducts from three pairs of glands; they are the parotids, submaxillaries, and sublinguals. The food-digesting enzymes contained therein are *amylase* and *maltase*. Amylase initiates the hydrolysis of polysaccharides such as starch and glycogen by changing them to disaccharides (Fig. 7.7). Some of the disaccharide maltose, in turn, is further changed to glucose with the help of maltase, which is produced in very small quantities. Maltose may originate from the digestion

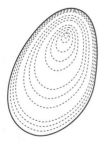  

FIGURE 7.7.   *Visual evidence of digestion of starch grains by amylase. (From* Plants: An Introduction to Modern Botany *by Greulach and Adams, John Wiley & Sons, Inc.)*

of starch or be consumed directly in foods. Oral digestion can proceed at pH values that are near the neutral point.

Food passes from the oral cavity (Fig. 7.8) to the stomach via the *esophagus*, which extends through the chest to the abdominal cavity. Food is pushed along in the tube by peristalsis. The esophagus serves as a passageway only and does not perform any direct digestive function.

The *stomach* is a large multiple-purpose organ. It is divided topographically into *fundus* (expanded part nearest the esophageal entrance), *body* (the central region), and *pylorus* (the part narrowing toward the small intestine). The stomach serves the triple purposes of temporary storing, churning, and digesting. Here protein and fat digestion begin. Carbohydrate digestion previously begun in the oral cavity continues until it is stopped by high acidity of the stomach interior.

The stomach is normally very acidic—as low as pH 2. Its churning action soon thoroughly mixes acid with the food. The acid secreted by the stomach is *hydrochloric acid*, which is not an enzyme but is needed to activate the en-

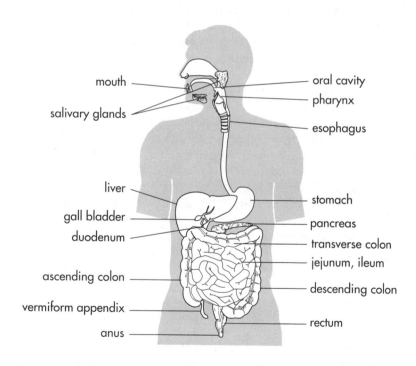

FIGURE 7.8.　*The human digestive tract with its associated organs. For illustrative purposes, the several organs of this system are somewhat separated from each other. In their normal positions, they are much more intimately associated. (Adapted from* Principles of Biology *by Buffaloe. Used with permission of Prentice-Hall, Inc., Englewood Cliffs, New Jersey, 1962.)*

zyme *pepsin*. In addition to activating pepsin, it helps macerate food and curdles milk. Hydrochloric acid is a strong irritant that would damage the stomach wall were it not for natural defenses, among which is the secretion of mucus that coats the inner surface. Whenever natural resistance breaks down, as when coarse foods cause abrasions, acids and pepsin contribute to the erosion of the wall that causes ulcers. (Ulceration can also be attributed to colonization by certain bacteria in the stomach, a condition treatable with specific antibiotics.) Oversecretion of acid is stimulated by some foods or drinks and by anxiety. Hyperacidity, sometimes called heartburn, is experienced by most people, who usually take alkaline substances to counteract it.

There are three enzymes secreted by the stomach. One is *rennin*, which curdles milk protein. Curdling of milk protein is a preparatory step preceding its digestion by pepsin. First, the protein is split into *casein* and *whey* (mostly water with small protein content). Casein then reacts with calcium to form a curd that separates from the liquid whey. Separation of the protein in the form of a solid has an advantage of retarding its movement through the system, thereby giving more time for digestion. Retardation of movement by curdling seems to be especially important in mammalian infants, who for a time are entirely dependent on milk. Rennin is less important in the digestive tract of adults, where hydrochloric acid curdles milk directly.

Another stomach enzyme is *pepsin*. It influences the fragmentation of large protein molecules into smaller molecules, even single amino acids. Ordinarily the time proteins are retained in the stomach is not sufficiently long for much of the protein digestion to go completely to amino acids. Rather, most protein molecules are partially digested into smaller polypeptides. Fortunately, pepsin is not secreted in an active form; otherwise there would be the problem of self-digestion. It is modified into its active form only after coming in contact with acid. Activated pepsin is separated from the stomach wall by an insulation of mucus.

Lipase is the third stomach enzyme. It is relatively insignificant in fat digestion by comparison to what occurs in the small intestine. Most fats are not yet emulsified (broken into small aggregates); thus, surface areas upon which lipase can work are limited. Fat digestion in the stomach is limited mainly to finely divided fats such as those in egg yolk, cheese, cream, or mayonnaise. Another unfavorable circumstance affecting fat digestion is the highly acid stomach environment.

The stomach is also involved in the regulation of food passage to the small intestine. Over a period of three or four hours after eating, it discharges food intermittently into the small intestine. A band of encircling muscles known as the *pyloric sphincter*, located between the stomach and the intestine, controls the discharge. A similar muscle between the esophagus and stomach, known as the *cardiac sphincter*, is ordinarily closed except when one swallows. It prevents reverse movement of food. When its proper functioning is upset by sickness or poisoning, the reverse movement of vomiting results.

By the time food enters the intestine, it has a consistency similar to cream and is called *chyme*. In the small intestine, digestion of all three classes of foods is carried to completion, and end products are absorbed. This takes from four to eight hours. The small intestine is by far the most important digestive segment of the entire tract. The inner surface has a velvety appearance, which is due to numerous fingerlike processes, the villi, that project inward. Each villus contains a network of capillaries and a central lymphatic vessel, a lacteal. As described before, its larger surface exposure and richness of vascular supply are important in increasing absorption and secretion. Intestinal surface area is further increased by circular folds that partially or completely encircle the inner surface. These are in every part of the intestine but are most prominent in the middle portion.

The intestine is a much-coiled tube about 2 centimeters in diameter and 7 to 10 meters long in an adult. It is divided into three regions—*duodenum, jejunum,* and *ileum.* The duodenum, which is the upper part joining the stomach, is about 20 centimeters long. It is the widest part of the small intestine. Both the liver, which is the largest organ of the body, and the pancreas pour their secretions into the duodenum, usually through a common opening. The jejunum is that part of the intestine usually found empty when autopsies are performed. There is no distinct line of separation between it and the other two parts, although there are some recognizable histological differences in the three parts. It is about 2.5 meters long. The ileum is the longest part of the small intestine, typically over 4 meters. It empties into the large intestine (colon) through the *ileocolic valve,* which is constructed from two folds of tissue. The valve is a safeguard against the back movement of fecal material into the small intestine.

Muscular contractions of the small intestine continue the mechanical action of mixing foods with enzymes secreted along the way. In addition, propulsion by peristalsis moves chyme through the tube.

The liver is active in hundreds of other ways, but its contribution to the digestive process is the secretion of *bile.* This product is stored in the *gallbladder* until the proper stimulus is received indicating that food is available for digestion. Then the gallbladder discharges bile into the duodenum. Being alkaline, bile contributes toward neutralizing the acid chyme coming from the stomach. Bile is not an enzyme, but it performs the important function of breaking fats into small droplets. This process is called *emulsification.* Emulsified fats have a much greater surface area in contact with the enzyme lipase than would otherwise be possible. Being water soluble, lipase does not mix with (penetrate) fats but must act at the contact surface.

The pancreas has the dual roles of secreting a hormone (insulin) into the blood stream and several enzymes into the intestine. Just before their secretions enter the duodenum, the pancreas and liver empty into a short tube they share. Pancreatic enzymes are *amylase, lipase, trypsin, chymotrypsin,* and *carboxypeptidase.* With this combination of enzymes, all three major categories of food—polysaccharides, polypeptides, and fats—can be digested.

In addition, the wall of the small intestine secretes a liquid containing enzymes. This fluid is alkaline and together with bile and pancreatic fluid makes the chyme slightly alkaline. Enzymes secreted by the intestinal wall are *amylase, lipase, disaccharidases* (lactase, maltase, and sucrase), *aminopeptidase,* and *dipeptidase.* Together with pancreatic enzymes, these finish the digestion job. Most of this enzymatic action occurs in the duodenal portion of the small intestine.

Digestible carbohydrates that enter the intestine are polysaccharides and disaccharides. Polysaccharides are digested to disaccharides with the help of intestinal and pancreatic amylase (Table 7.2). The resultant disaccharides are broken into monosaccharides. Of the disaccharides, sucrose is digested with the aid of sucrase, maltose with maltase, and lactose with lactase.

Fats are digested primarily in the small intestine (Table 7.3). Digestion is begun in the stomach and is continued in the small intestine, where bile emulsifies the fats and where intestinal and pancreatic lipase hydrolyzes them into fatty acids and glycerol.

Protein digestion (Table 7.4) begins in the stomach, where rennin curdles milk and gastric pepsin helps break some proteins into smaller polypeptides and—to a very limited extent—free amino acids. Any proteins not previously touched by stomach digestion are fragmented in the same manner with the help of trypsin and chymotrypsin (both pancreatic in origin) in the small intestine. The intermediate products are broken into dipeptides with the help of aminopeptidase, an intestinal secretion, and carboxypeptidase, a pancreatic secretion. The two enzymes just named are involved in the breaking of two different kinds of bonds. Dipeptides are hydrolyzed to amino acids with the help of dipeptidase, an intestinal secretion.

The three enzymes pepsin, trypsin, and chymotrypsin are produced in an inactive form, thus preventing destruction of the wall of the intestine by self-digestion. It has been seen that pepsin is activated by stomach acid. Trypsin is activated by *enterokinase,* an intestinal enzyme. Once activated, it in turn activates more trypsin and chymotrypsin. Rennin likewise is in an inactive state until activated by stomach acid.

Nearly all absorption of digested materials occurs in the small intestine, aided by the huge surface therein. It has been estimated that the adult human small intestine has a membranous surface of about 300 square meters! Also absorbed here is a great deal of the water that has been traveling with food. Some of this was ingested, but there was also much water added with saliva and stomach, liver, pancreas, and intestinal fluids. This *reabsorption* of water is very important, since it is estimated that as much as 9 liters of water could potentially be lost in a single day if not reabsorbed from the intestinal tract.

The *large intestine* (*colon*) is a tube of larger diameter than the small intestine and is nearly 2 meters long. It may be described as shaped like an inverted U and is puckered by longitudinal muscle bands. The primary functions of the large intestine is to continue the reabsorption of water and to absorb inorganic ions. In addition, some foods still being digested by enzymes from the small intestine and by bacteria are absorbed, but this is a minor activity. Also, as described before, the large intestine houses a very large colony of bacteria, some of which contribute vitamins that are absorbed there.

The materials eliminated by *defecation* are a combination of indigestible material that entered by the mouth, a huge number of living and dead bacte-

TABLE 7.2. SUMMARY OF CARBOHYDRATE DIGESTION.

Place	Food	Enzyme	Intermediate Product	End Product	Where Absorbed
oral cavity	polysaccharides	amylase	dextrin and disaccharides		
	disaccharide (maltose)	disaccharidase (maltase)		glucose	small intestine
stomach	continuation of digestion started in mouth				
small intestine	polysaccharides	amylase (intestinal & pancreatic)	dextrin and disaccharides		
	disaccharide (maltose)	disaccharidase (maltase)		glucose	small intestine
	disaccharide (sucrose)	disaccharidase (sucrase)		glucose and fructose	small intestine
	disaccharide (lactose)	disaccharidase (lactase)		glucose and galactose	small intestine
large intestine	continuation of digestion already underway			glucose, fructose, and galactose	large intestine

TABLE 7.3. SUMMARY OF FAT DIGESTION.

Place	Food	Enzyme	End Product	Where Absorbed
stomach	fats	lipase (gastric)	fatty acids & glycerol	small intestine
small intestine	fats*	lipase (intestinal & pancreatic)	fatty acids & glycerol	small intestine

*Fats emulsified by bile; some colloidal particles absorbed without digestion.

ria, some ions (particularly of heavy metals) that were moved into the large intestine lumen, and water. Additionally, there are also many intestinal cells that were rubbed off by food as it passed through. The intestinal epithelium must be in a state of constant cell division to replace those lost by wear and tear.

HUMANITY MAY HAVE TO SEARCH FOR NOVEL FOOD SOURCES

The current rapid increase in human population will eventually leave us with no alternative but to explore possibilities for increasing the food supply. Among some possible sources that seem promising for high yields are the algae and fungi. There are probably many of them capable of producing enormous yields very cheaply. It is said that some algae can produce nearly 50,000 kilograms of protein per hectare (or 44,000 pounds per acre) at a minimal cost for raw materials. By comparison, it is estimated that soy beans produce about 375 kilograms of edible protein per hectare at a much higher price.

CHAINS AND WEBS DESCRIBE THE PASSAGE OF ENERGY-STORING MOLECULES AMONG ORGANISMS

In the simplest relationships, the sequence of food passage (and therefore energy passage) among groups of organisms is referred to as a *food chain*. Significantly, a chain always begins with an autotrophic organism, which in this context is often called a *producer*. It may be eaten by a herbivore, referred to as a *primary consumer*. Primary consumers, if they are animals, are sometimes eaten by carnivores. The latter would, in the context of a food chain, be termed *secondary consumers*. It is even possible to have *tertiary consumers*,

TABLE 7.4. SUMMARY OF PROTEIN DIGESTION.

Place	Food	Enzyme	Intermediate Product	End Product	Where Absorbed
stomach	caseinogen (milk protein)	rennin	casein		
	{ proteins casein	pepsin	proteoses peptones peptides	amino acids	small intestine
small intestine	proteins	trypsin (pancreatic) chymotrypsin (pancreatic)	proteoses peptones polypeptides		
	proteoses peptones polypeptides	amino-peptidase (intestinal) carboxy-peptidase (pancreatic)	dipeptides		
	dipeptides	dipeptidase		amino acids	small intestine
large intestine	continuation of digestion already underway, plus some bacterial action				

animals eating secondary consumers. In the simplest set of such predator-prey relationships, the straight food chain would look like this:

producer → primary consumer → secondary consumer
(autotroph)   (heterotroph)   (heterotroph)

An example of a food chain might be found in fresh water ponds, where algae begin the series. They support insect larvae acting as primary consumers. The larvae serve as food for a small fish, which in turn are food for a species of larger fish. In this example the large fish is a tertiary consumer.

algae → larvae → small fish → large fish

Although such simple energy-passing relationships certainly do exist in real ecosystems, it is likely that much more complex interactions would happen. For example, the small fish species in the above illustration might very well be capable of eating both the insect larvae (acting as a secondary consumer) and the algae (acting as a primary consumer). The insect larvae might serve as prey for both species of fish. Individuals of all four species in this example would eventually die and their remains would be eaten by several species of decomposers. These would release simple chemicals from the dead bodies, which would become part of living algae and many other species of producers. All of these complex interactions, if outlined on a page with arrows connecting them, would make a diagram more reminiscent of a web than of a chain. Thus, typical energy-passing activities in most ecosystems are said to be *food webs*. Obviously, their elucidation requires much serious study by ecologists.

In view of the fact that organisms are so interdependent, reason dictates that prosperity of one group probably means prosperity for the group that depends upon it; or misfortune to one group will mean misfortune to the group that depends upon it. This interdependence of organisms can account for the surges and declines in populations. When any link in a food chain or web is altered, succeeding links will be affected correspondingly. Certain species in a food web, those upon whom several other species rely heavily, have greater than average impact upon the entire system if they suffer decline. They are called *keystone species* to indicate their importance to the health of the ecosystem. Weather conditions, pollution, and disease are among the factors that may upset the balance within an ecosystem.

STUDY QUESTIONS

1. Provide a simple definition of homeostasis. How is it related to feedback inhibition?

2. Since living membranes act as barriers to the free passage of many substances, what important process is necessary to alter the substances before absorption is possible?

3. What is intracellular digestion? In what organisms does it occur exclusively?

4. Give examples of animals having an incomplete digestive system. Is this type of system more efficient than a complete system?

5. Heterotrophs acquire food for its energy content, but what other materials must also be considered nutrients?

6. Describe the features that must be present to allow efficient filter feeding.

7. What is the difference between digestion and absorption?

8. Discuss modifications of the tube-type digestive system in relation to feeding habits and types of foods consumed.

9. What variations in digestive tracts of different animals increase their secreting and absorbing surfaces?

10. What is phagocytosis?

11. Why do some internal parasites require their own digestive tracts, while others do not?

12. Which would be likely to have the longer intestine relative to its body size—the herbivorous frog larva (tadpole) or the carnivorous adult frog? Why?

13. How are ingested materials moved through the human digestive tract?

14. What is the role of symbiotic microorganisms in a cow's stomach?

15. What is believed to be a reason some plants supplement photosynthesis with carnivorous activity?

16. What are the anatomical divisions of the human digestive tract, and what types of enzymes are secreted in each?

17. What is hydrolysis? What does it have to do with food processing in the body?

18. What prevents the stomach from being digested by pepsin, its own protein-digesting enzyme?

19. In what part of the tract does most digestion and absorption occur? Into which part do the pancreas and liver empty their secretions?

20. What is the role of bile in digestion? Where is it made? Where is it stored?

21. In humans, why is ingested fat distributed by a different route than other food materials, after digestion?

22. Do bacteria in the human intestine have any value for us?

23. Define and place these in the proper order within a food chain: secondary consumer, producer, primary consumer, tertiary consumer. Describe a real food chain involving these.

24. Why is food *web* a more accurate term than food *chain* for many real ecosystems?

25. What is a keystone species?

8
INTERNAL TRANSPORT AND ASSOCIATED SYSTEMS

What do we know about:
—*how ions and molecules get across membranes?*
—*the complex system of transporting tubes within a land plant?*
—*methods for bringing water and minerals into a plant's root system?*
—*how water can be delivered into the tallest portions of a tree?*
—*why humans could be considered to possess two hearts?*
—*the multiple roles of the lymphatic system?*
—*the advantage of carrying oxygen via hemoglobin?*
—*why land-dwelling animals tend to use lungs instead of gills?*
—*how two immune systems protect our bodies and how they can be compromised by AIDS?*
These and other intriguing topics will be pursued in this chapter.

INTERNAL TRANSPORT, GAS EXCHANGE, AND IMMUNITY ARE RELATED TOPICS

A fundamental problem to be solved by any organism is that of moving materials from place to place within itself. It can be as (seemingly) simple as getting single molecules through a cell membrane or as complex as efficiently moving cells and fluids throughout the body of a large multicellular organism. This chapter will explore the physiology of transportation. Certainly, a major topic must be that of blood—its properties and its movement through an animal's circulatory system. Since an important function of blood is picking up and carrying the two gases of respiration—oxygen and carbon dioxide—it is appropriate for this chapter to include the exchange of these molecules both with the environment and between blood and the internal tissues. And, since much of an animal's internal defense against invaders involves certain blood cells, the topic of immunity will also be explored in the chapter.

ATOMS AND MOLECULES MOVE
AT THE LEVEL OF CELLS

Regardless of the size of organisms, their living cells must acquire materials from the outside to sustain life processes and eliminate materials resulting from metabolic activities. In this section we will explore three major methods employed by cells to gain or lose materials that are molecules or smaller. It will be shown that only one of the three requires energy expenditure on the part of a cell.

Diffusion Is Predictable Movement Requiring No Cellular Energy Expenditure

The passage of some of these materials is accomplished by the intrinsic ability of atoms and molecules to move. Unless it is in an environment so lacking in heat energy (at absolute zero on the Kelvin temperature scale) that no movement is possible, each atom or molecule suspended in a liquid or gaseous milieu of others tends to travel in a straight line until it collides with one of its neighbors. The collision deflects both, sending each off on a new course until another collision occurs. The scattering of molecules caused by their continuous movement, collision, and deflection is called *diffusion.*

Since collision with other objects is more likely if an object is moving toward an area of high concentration than if it is moving toward a more dispersed area, the *net* movement of the objects is predictably *away* from the higher concentration area and *toward* the region of lower concentration (Fig. 8.1 A). Net movement should be expected to continued until all of the objects are equally dispersed. At that point, the objects continue to move individually, but further change in concentration is not expected.

Simple diffusion can also be accomplished through any membrane that does not present itself as an impenetrable barricade (Fig. 8.1 B). Artificial membranes, such as those used in kidney dialysis machines, have pores of sufficient size to allow passage of small molecules. Membranes of living cells (plasma membranes or those forming internal organelles) do not usually have pores as such, but do allow some molecules and smaller chemical entities to pass through them almost as quickly as if the membranes were not present. (An exception is the pair of membranes collectively called the nuclear envelope, with rather large pores. See Chapter 4.)

For either artificial or cellular membranes, the term *semipermeable* is descriptive of the fact that only certain materials can pass through. Recall that a cell membrane is a complex structure, with phospholipids being its dominant chemical entities (Fig. 4.5). It should not be surprising that the materials able to pass through such a barrier are those that are soluble in lipids. They tend to be hydrophobic (as are lipids), carrying little or no electrical charge (nonionic), and small. On a list of such items are the following important mole-

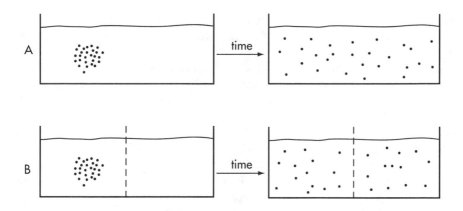

FIGURE 8.1.   *Diffusion illustrated. A: Sugar molecules move by diffusion through a tank of water. In time, they become equally dispersed among the water molecules and remain so. B: Diffusion can occur in the same setting with a membrane added. If the membrane has pores that are sufficiently wide, equal dispersion of sugar molecules is again the final result.*

cules: oxygen (O_2), carbon dioxide, ethyl alcohol, ammonia, urea, and fatty acids. Each of these melts through the cell's various membranes, not needing any open space such as a pore. Although water is polar, its small size allows it to pass through a membrane also.

Osmosis Is Diffusion of Water through a Membrane

The movement of water through membranes is so important that it has been given a particular name: *osmosis.* Osmosis is simple diffusion. The direction of osmotic movement is predictable under the same rules as for any diffusion. If the concentration of water is different on two sides of a membrane, it will flow through the membrane toward the area of its lower concentration. If, for instance, the cytosol of a cell has more ions, sugars, and so on than does the liquid outside the cell, this means that any given volume of the cytosol has a lower concentration of water than does the same volume outside the cell. In this case, it would be expected that the cell would gain water. If this difference is large, and if it cannot be changed by the cell's dissolved molecules leaving through the membrane, the inflow of water might put the integrity of the cell at risk—it might burst under the pressure of its contents. A cell that has taken in so much water that it is rather rigid is said to be *turgid.*

One might ask why water would not be forced back out of the cell in such a situation of increasingly higher internal pressure. This does happen, but not until the cell has gained more water than one would have expected. It can be explained by the fact that water (with its slightly polar nature) is prone to clinging to charged molecules in the cytosol and is therefore likely to become literally stuck inside the cell after it has entered by osmosis.

Some relative terms are useful when considering the effect of osmosis upon a cell. If the concentration of solutes (dissolved materials) inside the cell is equal to their concentration outside the cell, that cell is said to be *isotonic* to its environment. Such a cell has, of course, the same concentration of water as its environment and would not be expected to gain or lose water by diffusion. A cell with a higher concentration of solutes (and therefore a lower concentration of water) than its environment is likely to gain water. It is said to be *hypertonic* to its environment. The opposite situation, where the cell is expected to lose water, is one in which the cell is *hypotonic* to the environment. In Chapter 9 we will examine situations in which cells (and whole organisms) are either hypertonic or hypotonic and how they avoid gaining or losing too much water by osmosis (see the topic Osmoregulation).

Facilitated Diffusion Is Made Possible by Membrane Proteins

Many biologically important materials cannot pass quickly through a cell membrane unless special help is provided. On this list are ions of all sorts (including large molecules with multiple charges on their surfaces such as proteins and nucleic acids) and uncharged molecules about the size of glucose or greater. Of course, many materials of biological importance fit these descriptions and must somehow pass the barrier that a membrane presents. Diffusion can occur through specific proteins embedded in the phospholipid bilayer of a membrane. Such a method is still diffusion (still involving movement from high concentration side to low concentration side, with no cellular energy required), but it is given the name *facilitated diffusion* to distinguish it from movement directly through the phospholipids.

Two varieties of facilitated diffusion are recognized. In one, a protein is shaped to form a water-filled *channel* right through the phospholipid bilayer (Fig. 8.2 A). The channel is just large enough to allow a particular material through. Such channels specialize in allowing the passage of inorganic ions, such as Na^+ and Cl^-. Even though it is thought that a channel requires the passage of ions in single file, as many as one million can pass through every second. It is important to note that channel proteins are equipped with gates capable of closing when movement through them is not appropriate. The other sort of protein for facilitated diffusion is called a *carrier* protein (Fig. 8.2 B). Here, the protein must recognize (by shape) the molecule that is to be allowed through. It then alters its own conformation to allow the molecule to move across the phospholipid bilayer. This too is a highly selective mechanism; only the molecule with the recognizable shape will activate the shape

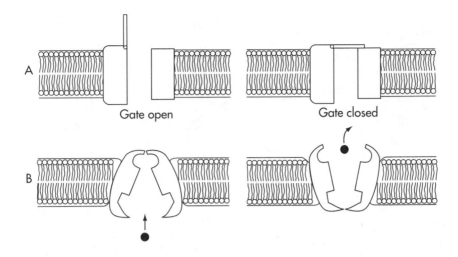

A

Gate open   Gate closed

B

FIGURE 8.2. *Two varieties of facilitated diffusion. A: Channel-mediated, where a protein provides a water-filled passageway. B: Carrier-mediated, where a passageway opens only after the proper molecule contacts the protein and induces it to change shape.*

change. Facilitated diffusion via carrier proteins is much slower than the variety involving channel proteins.

Active Transport Opposes Diffusion and Requires Cellular Energy

Sometimes a cell must move materials across a membrane in the direction opposite that expected by passive diffusion. For instance, the disaccharide sucrose should be moved from the lumen of the intestine into the epithelial cells of the lining, even if those cells already have a higher concentration than is in the lumen. Otherwise, some of these valuable energy-carrying food molecules will be lost by elimination from the body. Facilitated diffusion moves sucrose in the useful direction only until the cells have the same concentration as the lumen, and no longer. However, if the cell membranes are equipped with the appropriate protein, it can help them carry on *active transport* (Fig. 8.3). This protein is a carrier protein not unlike that described above in connection with facilitated diffusion. However, it is equipped to move a particular molecule across the membrane in the direction *opposite* that expected for diffusion. This is a work activity and therefore requires the expenditure of energy. Most active transport mechanisms find this energy by breaking the bond of one or more ATP molecules. Obviously, only a living cell with the ability to provide energy can be expected to do active transport.

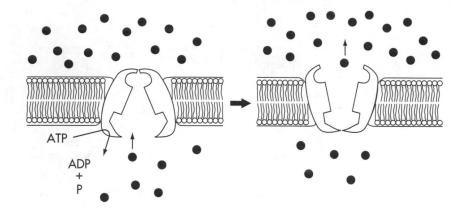

FIGURE 8.3.   *Active transport across a membrane. Movement from the low-concentration side is accomplished with expenditure of cellular energy (breakdown of ATP in this case).*

We have encountered active transport earlier, although it was not labeled as such. Refer to Chapters 5 and 6, where chemiosmosis is described. In both photophosphorylation and oxidative phosphorylation, the movement of hydrogen ions across the membrane of a thylakoid or a mitochondrion is accomplished by active transport (e.g., Fig. 5.6). In both of those cases, active transport is driven by energy from a non-ATP source—the kinetic energy of fast-moving electrons.

LARGER OBJECTS CAN CROSS
A PLASMA MEMBRANE BY ENDOCYTOSIS
OR EXOCYTOSIS

Bulky objects, such as congregations of molecules or even whole cells, can be taken into a cell by a maneuver called *endocytosis*. One type of endocytosis has already been described in Chapters 4 and 7. In phagocytosis, the cell changes its shape to surround an object (Fig. 7.2). When complete, the process has made a vesicle containing the captured material. This is the method used to initiate intracellular digestion along the lining of the intestine and to destroy bacteria and other potentially harmful living cells (see the activity of the cellular immune system later in this chapter).

A second variety of endocytosis, for the uptake of fluid (with dissolved or suspended materials) is *pinocytosis*. This is sometimes called cellular drinking. Here, tiny vesicles are produced by infolding of plasma membrane.

A more selective method of bringing small objects into a cell is called *receptor-mediated endocytosis*. It differs from pinocytosis in that the portion of the plasma membrane that forms a vesicle is lined with molecules acting to capture specific types of molecules. Then, when this portion of the membrane folds inward to become a vesicle, these clinging molecules are brought into the cell with the vesicle.

Exocytosis is the reverse of endocytosis, exporting objects to the exterior of a cell. As described in Chapter 4, the Golgi apparatus makes vesicles loaded with molecules. If these chemicals are for export, the vesicles migrate to the plasma membrane, merge with it, and thereby dump its contents to the exterior. This is an effective way to release many copies of a large molecule without having to provide individual transport through an intact membrane. For instance, digestive enzymes are sent into an intestinal lumen in this fashion. Exocytosis is also the method for removing the contents of food vacuoles if there is any waste left after intracellular digestion.

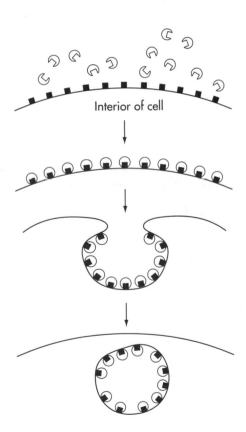

FIGURE 8.4.   *Receptor-mediated endocytosis.*

MULTICELLULAR ORGANISMS MUST MOVE MATERIALS FROM CELL TO CELL

Of course, it is necessary to transport materials throughout a body, not just into and out of individual cells. Coming to mind immediately in this context is the need to move digested food, O_2, CO_2, H_2O, toxic wastes, and hormones from place to place. As we will see, some (but not all) multicellular organisms have developed transport systems for the extracellular delivery of these and other materials. We will find that plants and animals use some methods in common, but are also quite different.

NONVASCULAR PLANTS ARE LIMITED BY THEIR LACK OF EFFICIENT TRANSPORT

The more primitive terrestrial plant groups, including mosses and their relatives in division Bryophyta, lack any tubular system for internal transport. The general descriptive word for such a system is *vascular*, so these plants are said to be *nonvascular*. The inability to efficiently distribute nutrients is an important factor in the small size seen among such plants. Their inability to move water via tubes limits them to a moist environment where all parts of the plant can be provided with water directly from outside.

VASCULAR PLANTS POSSESS AN EXTENSIVE TUBE SYSTEM FOR TRANSPORT

The bodies of higher plants generally have distinct roots, stems, and leaves. Any other parts, such as flowers or thorns, are often interpreted as being modifications of one of these. Roots and leaves are the primary organs of absorption from soil. All three organs have conducting tissue forming tubes, through which materials are transported from place to place.

Roots Absorb Water and Inorganic Materials from Soil

One of the primary functions performed by root systems is the absorption of water and other inorganic substances essential to the plant. Most absorption occurs through the root tips, since just a short distance behind them the roots are covered with tough cork. Beginning at the tip and going backward, recognizable zones are *root cap, growth zone, absorption zone* (area of root hairs), and *suberized* (waterproofed) *zone* (Fig. 8.5).

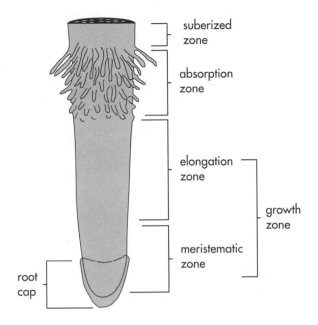

FIGURE 8.5.   *Root tip.*

The root cap functions in shielding the tip of the root as it grows through the soil. Although abrasion destroys some of the cells, the cap is maintained by the addition of cells from the meristematic zone behind it.

The growth zone behind the cap has dividing cells near the tip, comprising an *apical meristematic zone*, and enlarging cells farther back, comprising the *elongation zone*. After cells stop dividing, they grow (especially elongating in a direction that increases the length of the root).

As the cells get older, they become specialized. Specialized tissues derived from the apical meristem are primary ones (see Chapter 4). Differentiation of primary tissues is represented in the absorption zone immediately behind the growth zone. The outside layer, the epidermis, has many cells with elongated, delicate extensions known as *root* hairs. Each root hair cell, including its body and root hair extension, is a single cell (Fig. 4.8 B). Root epidermis, unlike epidermis of young stems or leaves, is not coated with the lipid cutin that would interfere with absorption. The bulk of absorption of water and soil solutes occurs in this zone. The region of cells internal to the epidermis is *cortex* (Fig. 8.6), which is composed of parenchymatous cells (Fig. 4.16 B). The internal layer of the cortex is a rather distinct set of cells known as the *endodermis*. This layer is much more characteristic of roots than it is of stems. The remaining tissues in the center comprise the *vascular cylinder* (or *stele*). Its outer layer is the *pericycle.* In the roots of dicot plants (such as oak trees or sunflowers), the center is occupied by xylem, which characteristically radiates toward the outside. When the radiations are four in number, the cross section of xylem will look somewhat like a plus (+) sign (Fig. 8.6). The phloem is located in the angles of the radiating arms of xylem. All other cells of the vascular bundle are thin-walled cells, some of which will divide and increase the diameter of the root.

As growth continues, secondary tissues are added to the primary ones. The consequence of these changes can be seen in the suberized zone behind the absorption zone, where meristematic tissue called *cambium* adds layers of

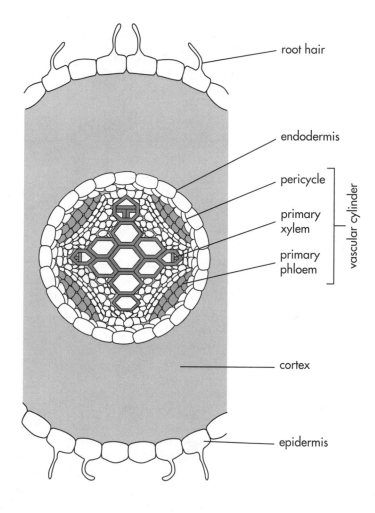

FIGURE 8.6.   *Primary tissues in the cross section through the absorption zone in a root of a dicot plant.*

secondary tissues. One kind of cambium, *vascular cambium*, between the primary xylem and primary phloem, produces secondary xylem and phloem. Xylem is added in layers, ordinarily one each year, which are the rings of growth (tree rings of roots and stems). Since xylem is a rigid tissue that keeps expanding by growth, it occupies space formerly occupied by weaker tissues outside it. As a consequence, cells of the bark are crushed and disrupted, and the bark becomes cracked and crumbled. Phloem is continually crushed and replaced by new (secondary) phloem produced by the vascular cambium. In other words, vascular cambium produces secondary xylem toward the inside and secondary phloem toward the outside. Another kind of cambium, *cork cambium*, which usually originates in the pericycle, produces the cork that replaces epidermis.

The commonly used terms *wood* and *bark* should be related to the more technical terms already used. Wood is the same as xylem; bark includes all tissues outside of the cambium (Fig. 8.7).

Branch roots originate from the pericycle and grow outward. They do not come from root hairs. They may originate in the absorption zone or a little farther back. As they grow, they themselves have all the zones previously described.

Monocot roots (such as those of grasses) do not ordinarily have cambium; therefore they do not have secondary tissues. They have pith in the center, as do the stems.

The bulk of absorption from the soil occurs through the root hair cells and other epidermal cells of the same zone. Root hairs greatly increase the surface area through which absorption can occur. Absorbed water and solutes move from cell to cell across the cortex until they reach the conducting units of the xylem (Fig. 8.8). However, it should not be assumed that water movement to-

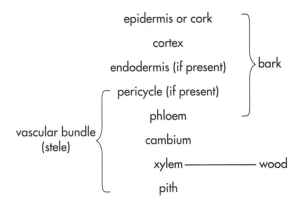

FIGURE 8.7.   *Relationship of tissues or regions in roots and stems. Pith is absent in some roots.*

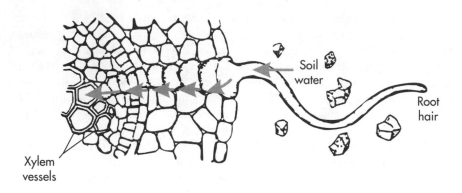

Xylem
vessels

FIGURE 8.8.   *Path of water from root hair to xylem. Most of the trip is between cells, not through them.*

ward xylem is *through* individual cells. Most of it is *between* cells, along the cell walls, until the endodermis is reached. The walls of the endodermis cells are coated with a line of wax, called the *Casparian strip*. This effectively blocks further intercellular movement of water, forcing it to enter the vascular cylinder by crossing through cell membranes of the endodermis. Such a path allows much more control of the amount and quality of water eventually entering the vascular system and moving throughout the plant. Upon entering the vascular cylinder, water soon moves into the xylem tubes and thence upward from the root.

By contrast, minerals of the soil take a different route to the xylem. Most travel directly through cells of the root, rather than along cell walls. The first activity is an electrostatic attraction between soil ions and the cell walls of roots. After adhering to the cell wall, an ion might enter the cell by simple diffusion through the plasma membrane if the concentration gradient is proper for this direction of movement. However, there is likely to be a higher concentration of ions inside the root cells than in the nearby soil. In such cases mineral ions are still brought into the cells, by active transport across membranes. As with all active transport, the drawing in of soil minerals is quite selective.

Foods that have been synthesized elsewhere come into the roots by way of the tubes of the phloem. They move from cell to cell to the tip and laterally to the epidermis. In many plants surplus food is stored in cells of the root, especially the cortex.

Roots often have other uses than absorption, conduction, and storage, which have already been mentioned. They are also used for anchorage in most plants. English ivy has roots that are used for climbing. Dahlias and sweet potatoes have roots that are capable of reproducing vegetatively. Corn has roots from the lower nodes of the stem that help prop up the plant.

Stems Provide Support and Transportation

Primary tissues of stems are derived from apical meristem of the buds just as they are derived from apical meristem in roots. A short distance behind the growing tip the conducting tissues xylem and phloem (as well as the protective tissue epidermis) begin to differentiate. A cross section through that region reveals the xylem to be in bundles arranged in a circle (Fig. 8.9 B). Each bundle is separated from other bundles by bands of parenchymatous cells that radiate from similar parenchyma located in the center and known as *pith*. Such bands are called *pith rays*. Primary phloem occurs in patches immediately outside each bundle of xylem. This arrangement is unlike that in roots where bundles of primary phloem alternate with bundles (or arms) of primary xylem. Also unlike roots, there is no distinct endodermis or pericycle. The cortex is outside of the phloem, and epidermis is on the outside of that. Leaf and bud primordia (earliest stages) grow from the peripheral cells in what would be epidermis or outer cortex if they were distinctly differentiated.

As development of a stem proceeds, secondary tissues are added. The changes that occur can be seen by cutting through the stem behind the zone of primary tissues. Here a vascular cambium forms in the bundle between the primary xylem and phloem and extends from bundle to bundle across the pith rays. If the stem is herbaceous (Fig. 8.9 A), it adds secondary xylem on its inner face and secondary phloem on its outer face in each bundle. It also adds parenchyma on both faces in the bands between the bundles. If the stem is

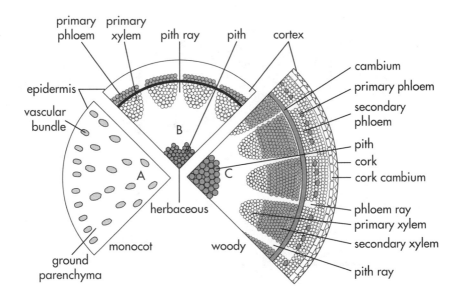

FIGURE 8.9. *Cross sections of stems. A: monocot (herbaceous). B: dicot (herbaceous). C: dicot (woody).*

woody (Fig. 8.9 C), the cambium produces secondary xylem and phloem in the bands between the bundles. In time the bundles are united into a more or less continuous ring. The only interruptions are narrow rays, which are called *pith rays* if they radiate from the pith and *wood rays* if they radiate from some point of origin within the wood. *Phloem ray* is the name given to that part of the pith or wood ray that extends through the phloem. As annular rings of rigid wood are added, weaker cells of the bark adjust by disrupting or collapsing. A *cork cambium* that produces *cork* usually originates in the cortex.

Wood has several characteristics that are of economic importance. It has become customary to designate the wood of cone-bearing trees (such as pines) as softwood and the wood of broad-leaved trees (such as maples) as hardwood. This distinction on the basis of softness or hardness is not necessarily true. For instance, a conifer could have harder wood than some nonconiferous hardwood tree. Hardwoods have *pores*, which are cross sections of vessels made of xylem. In some woods they are distinctly visible without any magnification. Woods are *ring-porous* or *diffuse-porous*, depending on the arrangement of pores (Fig. 8.10). Oak wood is a ring-porous wood because the pores are arranged in a rather distinct zone in the inner part of the growth ring. Maple wood is a diffuse-porous wood because the pores are more or less equally spaced throughout the growth ring. Softwoods are *nonporous* because

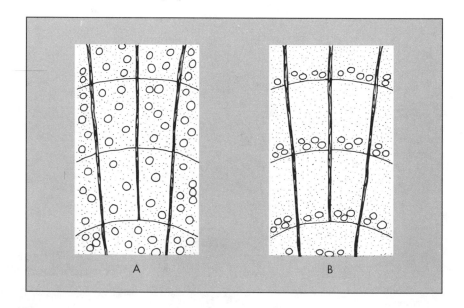

FIGURE 8.10.   *Wood types. A: diffuse-porous. B: ring-porous. Pores are cross sections of vessels; vertical lines are rays; horizontal curves are boundaries between annual rings of growth.*

they have no vessels. The older wood near the center of stems is often discolored with resins or other deposits. It is called *heartwood* as distinguished from the unaltered sapwood. The distinction between herbaceous and woody is an arbitrary one based on the proportion of wood to softer tissues. Of course, both types have wood for conduction, but herbs have much less wood in proportion to parenchyma.

Most conduction in stems is through xylem, phloem, and rays. Water and dissolved substances move up the stem through the xylem. Manufactured foods are transported down the stem (and occasionally back up into other stems) in phloem. Lateral transport occurs in the rays.

Leaves Import Water and Minerals and Export Sugars

The fundamental parts of a leaf are *blade, petiole,* and *stipules* (Fig. 8.11). The blades are of many shapes and sizes. They may be attached directly to the stem or have an interposing stalk known as the petiole. Some leaves, like those in roses, have a pair of basal appendages, the stipules. Some stipules are small and ephemeral while others are larger and permanent. They usually carry on the same processes as the blades.

Leaves are simple or compound, depending on whether they have one or more blades. The compound leaf, having two or more blades, may have a pinnate or palmate arrangement. If pinnate, the blades are in two rows along a central axis (*rachis*). If palmate, the blades are attached at one point, the tip of the petiole, and radiate in a fanlike pattern. There are frequently three to five blades in compound leaves. The blades of a compound leaf, together with their stalks if present, are called *leaflets*.

Leaves have a rather consistent relationship to lateral buds in being attached just beneath them. The particular position where they are attached is the *node* of the stem. They are *alternate* if only one is attached to a node, *opposite* if

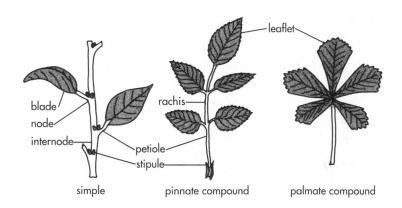

simple   pinnate compound   palmate compound

FIGURE 8.11. *Simple and compound leaves.*

two are attached on opposing sides of the node, and *whorled* if three or more are attached around the node.

Leaves are efficiently adapted to make exchanges of gases with the atmosphere. They are usually flattened, a shape that is ideal for increasing surface area in proportion to volume of tissue. Furthermore, leaves have porous surfaces and much air space among the cells, favoring easy circulation of gases. Gas exchange with the atmosphere is essential in photosynthesis and aerobic respiration. The arrangement that makes such an exchange possible is also ideal for evaporation of water, which in the case of plants is called *transpiration.*

A section through an ordinary leaf blade (Fig. 8.12) has an upper and lower epidermis between which there is a zone of chlorenchyma (parenchyma with chlorophyll) known as *mesophyll. Veins* in the mesophyll conduct materials to and from the leaves, being continuous with the vascular system of stems.

The epidermal layers of a leaf are covered with a cuticle to avoid excess water loss. They are often hairy or scaly. They have pores called *stomata* (singular: *stoma*). A typical leaf surface has thousands of stomata per square centimeter. Each stoma is surrounded by two or more modified epidermal cells that act as *guard cells* (Fig. 8.12, 8.13), which act as homeostatic devices for appropriate control of water loss. Guard cells change their shape to open or close the pores that they surround (Fig. 8.13). Generally, the pores are open during the

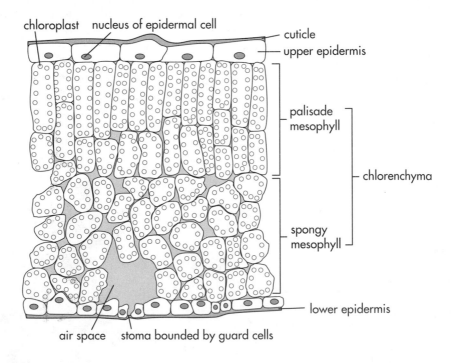

FIGURE 8.12. *Section of a leaf.*

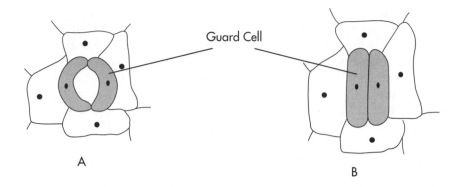

FIGURE 8.13. *The opening and closing of a stoma is determined by the shape of its two guard cells. A: Guard cells turgid, pore open. B: Guard cells flaccid, pore closed. This example is of a dicot plant; guard cells of monocots are somewhat different in shape.*

day, to facilitate the exchange of oxygen and carbon dioxide with the surrounding atmosphere and to cool the leaf by allowing evaporation of water. However, in times of drought the need to save water overrides the need to photosynthesize and cool, and the stomata close during the day. They are normally closed during night hours, to conserve water when cooling is not necessary.

A pair of guard cells that are turgid with extra water from surrounding epidermal cells take shapes that create an opening into the mesophyll region of the leaf (Fig. 8.13 A). Passage of water into guard cells is by osmosis and is driven by the movement of potassium ions (K^+) into them by active transport. This energy-requiring activity is triggered by the impact of light upon guard cells. In a fashion not thoroughly understood, blue light is absorbed by pigments in the guard cells, which somehow activate the K^+ pump. However, two other conditions can also cause the pump to operate. One is a low concentration of carbon dioxide (which will be alleviated by opening the leaf to the atmosphere). The other factor is a little-understood internal clock that causes stomata to open at a certain time in the morning even if light does not strike the leaf. In late afternoon or early evening the K^+ pump ceases operation, which leads to diffusion of K^+ from the guard cells and consequent shift of water causing the guard cells to become flaccid and the stoma to close (Fig. 8.13 B). Thus, stomata are normally open only when it is appropriate to gain carbon dioxide from the atmosphere and to cool the leaf by water evaporation. Leaves that are normally displayed in a more or less horizontal position generally have more stomata in the lower epidermis, while those that are more vertical tend to have somewhat the same number in both surfaces.

The mesophyll zone is several layers thick. Horizontal leaves have the upper layer or layers of mesophyll modified into column-shaped cells called *palisade mesophyll*. Since a horizontal leaf gets most of its light through the upper epidermis, these are the cells receiving the greatest exposure. While it may seem paradoxical to say that a leaf needs light for photosynthesis and at the same time suggest that a plant can get too much of it, the fact remains that maximum exposure is harmful. The orientation of palisade cells with their ends toward the sun is probably important in protection against overexposure. Their contents keep circulating, thereby moving chloroplasts out of the position of maximum exposure. Other mesophyll cells in horizontal leaves tend to be spherical and surrounded by more air space, hence their description as *spongy mesophyll*. Vertically exposed leaves have a rather spongy mesophyll throughout.

Water Movement in Xylem Is Driven by Multiple Forces

Unlike animals, plants have no muscles and thus have no pumping organ like the heart. Furthermore, the movement of water and dissolved materials is one-way, rather than in a circulatory fashion. Nearly 99 percent of the water entering at roots exits the plant at leaves and stems. The movement of water and minerals through the plant is called *translocation*. It may entail movement over more than 100 meters against the force of gravity in tall trees. What is the transporting mechanism strong enough to account for such movement? It appears to be a combination of several things.

As mentioned earlier, water initially enters a terrestrial plant at the roots, by osmosis. This may be called *root pressure* and is sufficient to account for much of the movement of water through a small plant. It is a push from outside, unlike the forces about to be discussed.

A second factor is that of *cohesion*. It will be remembered that water is a polar molecule, leading to its ability to create hydrogen bonds with neighboring water molecules and other charged molecules. If water molecules are pulled from above, they transmit this pull to molecules below them. An entire column of water in xylem tubes will show this cohesiveness and move upward if there is force applied from above.

The third factor is the evaporation at leaf surfaces, providing the needed pulling force from above. Called *transpirational pull*, it is driven by the difference in water concentration between the leaf interior and the atmosphere exposed at the stomata. As individual water molecules follow this gradient and diffuse into the air, each pulls along the molecule to which it was adhering by hydrogen bonding. That one, in turn, pulls another water molecule, and so forth down the entire column of water in xylem. This sucking of water upward extends down to roots and even into the soil around the roots. Its measured force is sufficient to move water against gravity for a distance of at least 150 meters, which is greater than the height of any tree.

Phloem Tubes Conduct Manufactured Foods from Leaves

Since foods (sugars) are first made in the green parts of plants, they have to be transported to cells that cannot synthesize them. Foods are conducted by phloem tubes, which begin in individual leaves and lead to stems and roots.

The primary food passing into phloem is the disaccharide sucrose, which may be in as high as a 30 percent concentration in the *sap* that flows through phloem. This is a much higher concentration than the sucrose in mesophyll of a leaf, so an active transport pump mechanism operates to push sucrose against its gradient into the phloem.

Sap moves through phloem by bulk flow. This is driven from above because of the osmotic entry of water. Water is attracted into phloem by the high solute (sucrose) concentration. It continually flows inward, pushing sap ahead of it. When sap reaches nonphotosynthetic cells of stems and roots, the sucrose exits. In some plants, this is by active transport, but it is often by simple diffusion into neighboring cells that are constantly metabolizing sucrose to gain energy.

Phloem also carries hormones, amino acids, and minerals from place to place in a plant, by the same bulk flow mechanism.

MOST MULTICELLULAR ANIMALS ACCOMPLISH INTERNAL TRANSPORT BY BLOOD CIRCULATION

All but the smallest (or thinnest) animals have the same need as most plants: to transport water, gases, food, hormones, and the like throughout the body by some method more speedy than simple diffusion. Animals accomplish this with a system of tubes called the *circulatory system.* As the name indicates, it involves a somewhat circular movement of materials, rather than the one-way motion seen in a plant's xylem and phloem tubes.

Another difference between plants and animals is that the latter accomplish most of their internal transport by an active pumping of liquid (blood). In most animals this is accomplished by one specialized portion (or rarely several portions) of the circulatory system, the *heart.* This organ, really an enlarged muscular component of the tube system, operates by the periodic contraction of its muscular walls to force blood into the non-contractile remainder of the system. A few primitive animals (worms of the phylum Nemertea) move blood by a peristaltic contraction of all of their blood vessels, rather than giving the duty of pumping solely to a heart. The earthworm and other members of phylum Annelida have a set of several specialized portions of their blood vessels that have thickened walls and contract to push blood. These may be considered multiple hearts or simply specialized portions of certain blood vessels. Most animals of other phyla with circulatory systems have a single large heart that is distinctly different from the vessels.

Two General Types of Circulation Exist

Not all animals have the *closed* circulatory system with which we are familiar in our bodies. A closed system carries blood through the entire circuit within vessels: from heart to arteries, to capillaries, to veins, and back to the heart. The less efficient method is to use an *open* system, where a vessel might pour its contents into a large chamber within which the blood moves more or less randomly before being brought again into a vessel. This results in lower blood pressure and a longer time to complete a circuit. Members of phylum Arthropoda (insects, crustacea, etc.) and most members of phylum Mollusca (clams, snails, etc.) have open systems. Annelid worms sometimes have rather small open spaces, but their circulation is usually classified as closed.

Most Circulatory Systems Require Capillary Vessels

No animal could tolerate a truly closed system, since it would not allow interchange of transported materials with tissues. To allow transfer of food molecules, gases, hormones, and wastes, *capillary beds* run through tissues. As arteries branch and rebranch to become smaller and more numerous within an organ, the smallest arterioles give way to capillary vessels. It is easy to detect the point at which this happens. Even the smallest artery has a tough wall with muscle and elastic connective tissue making it incapable of releasing or receiving materials from neighboring tissues. By contrast, a capillary's wall is composed only of a single layer of epithelial cells, which are not tightly joined to each other. Although red blood cells cannot pass through this wall, the liquid portion of blood (*plasma*) readily leaves the capillary to bathe nearby cells before returning to circulation. Any molecules dissolved in plasma make this trip, too. Capillaries leaving a region regather to become a vein, which is not as thick-walled as an artery of the same diameter but is just as impermeable to plasma.

Capillary Beds Create a Mechanical Problem

As described earlier, it is absolutely necessary to include capillary beds in a closed circulatory system. However, their presence creates a serious problem—each bed is an area where much friction is generated between vessel walls and moving blood. If an artery supplying a bed were to be compared with the many capillaries in that bed, it would be found that there is far more total surface area of internal walls in the capillaries of the bed than there is in the artery. Of course, this enhancement of surface area greatly increases the ability to exchange materials between blood and surrounding cells, a good thing. But simultaneously it greatly increases the opportunity for blood to rub against vessel walls and to lose its kinetic energy—to slow down.

In general, there are two capillary beds that every portion of blood must traverse in every circuit around the body. After leaving the heart, it goes through a bed at the animal's gas exchange area—lungs or gills. As will be described later in the chapter, this is where an intricate mesh of capillaries lies just under the animal's surface to release carbon dioxide to the exterior and capture oxygen for transport to all other tissues. After blood leaves a lung or gill, it should carry its load of oxygen to some other capillary bed where it can reverse the gas exchange process. However, if it attempts to do this directly it will travel through two capillary beds before being given a boost from another heart beat. It will have lost much velocity and will return to the heart very slowly. Among vertebrate animals, only the fishes have this inefficient routing system.

The other vertebrate animals solve the problem of passing two capillary beds by bringing oxygenated blood back to the heart for a second boost before sending it to a bed in a tissue. For this to happen most efficiently, an animal must have a heart with four chambers: one pair to receive blood from the body and send it to a gas exchange area and the other pair to receive it from the gas exchange area and send it out to tissues that need oxygen. It is almost as if there were two hearts side by side. Because of this, such a system can be described as a *double circulation system* (Fig. 8.14). It is the system of mammals and birds. It is not coincidental that these are the types of animal that are best able to maintain a higher internal temperature than their environment, a feat that requires much energy and consequently a very efficient food and oxygen delivery system.

The other vertebrate classes, of the amphibians and the reptiles, also bring blood back to the heart after oxygenation. However, they possess only three

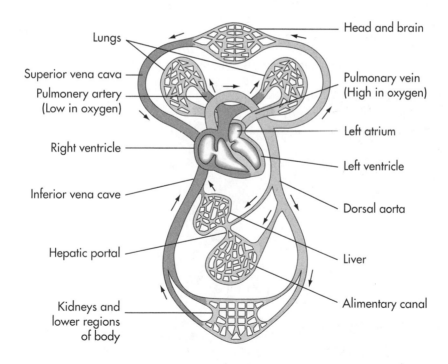

FIGURE 8.14.   *Distribution of blood in a complete double circulation. The relation of the capillaries to the arteries and veins cannot be shown in this diagram but is implied.*

chambers in their heart. This causes some mixing of oxygen-rich blood with oxygen-poor blood in a single chamber, meaning that the blood going out to oxygen-starved tissues will do so at high speed but with less than a full load of oxygen.

The Human Heart and Circulatory System Are Very Efficient

The human heart (Fig. 8.15) consists of two *atria* (singular: *atrium*) and two *ventricles*. An atrium's only job is to move blood into an adjacent ventricle, but a ventricle must provide sufficient force to move blood far out into the circulatory system. It is no surprise that the ventricles have much thicker walls to provide a stronger squeeze. Blood is pumped by a person's left ventricle through a large vessel known as the *dorsal aorta*, which branches into all parts of the body (Fig. 8.14). The larger branches of the aorta are arteries, which continue into smaller vessels called *arterioles*. They in turn continue into *capillaries*. Blood returns to the heart through the venous system. It enters *venules* from the capillaries, and from the venules it enters *veins*. Veins from most of the body converge into two large vessels, the *superior vena cava* (also known as the *precava*) and the *inferior vena cava* (or the *postcava*). The superior vena cava drains the head, neck, and arms, while the inferior vena cava drains the rest of the body. Both of these veins converge and their oxygen-depleted blood enters the heart's *right atrium*.

Blood contained in the right atrium flows to the right ventricle through an opening guarded by a *tricuspid valve*. (This and other valves in the heart and elsewhere are built so that they allow blood to flow in only the proper direction; if it attempts to return, such a valve closes.) The right ventricle then pumps the blood out via the *pulmonary artery*, which sends branches to the two lungs.

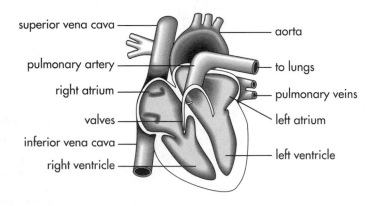

FIGURE 8.15.   *Diagram of the human heart.*

Oxygenated blood returns to the heart by way of *pulmonary veins,* which enter the *left atrium.* Blood passes through an opening guarded by the *mitral valve* (or *bicuspid valve*) to the *left ventricle,* which is the part of the heart that pumps blood out through the aorta.

Blood flowing through vessels of the arterial system is under considerable pressure, and the vessels have resilience to expand and contract with the pulsating flow of blood. Vessels of the venous system lead from the capillaries back to the heart. Blood is under much less pressure here than in the arteries, because of the loss of energy from friction in the capillaries. Walls of veins are thinner and less muscular than arteries. Veins have one-way valves within themselves that aid in preventing a back flow of blood. The major veins of arms and legs are placed between large muscles of those appendages. When we use those muscles in normal activity they thicken at their middle portions and this squeezes nearby veins, thus helping push blood toward the heart. The overall efficiency of the human circulatory system as a transport device is demonstrated by the fact that a red blood cell can make an entire circuit around the body within about 30 seconds.

THE LYMPHATIC SYSTEM MAKES THE CIRCULATORY SYSTEM MORE EFFICIENT

It is estimated that about 1 percent of all blood plasma leaving the circulation at a capillary bed fails to return. This does not sound like much, but if it constantly occurs at all capillary beds, the loss from circulation seriously compromises total blood volume and flow rate. Vertebrate animals have a system of tubes that negate this loss: the *lymphatic system.* This system is in some ways parallel to the blood circulatory system, but it has major differences too.

At every capillary bed there is a set of *lymphatic capillaries.* Each of these has the thin, porous wall characteristic of capillaries. However, lymphatic capillaries are blind; that is, each is a cul-de-sac. Fluid (called *lymph*) entering a lymphatic capillary can move only away from the capillary bed. Lymphatic capillaries join together to form a *lymphatic vein* leaving a tissue. Eventually the lymphatic veins lead to a *thoracic duct* on the left side of the body and a *right lymphatic duct* on the other side. The thoracic duct is the principal one in that it collects lymph from most parts of the body. It empties lymph into the left subclavian vein near its junction with the right internal jugular vein, between the heart and the neck. The right lymphatic duct collects lymph from the right arm and the right side of the chest, neck, and head. Thus, all fluid that did not immediately return to the blood circulatory system at capillary beds is eventually returned to circulation and blood volume remains constant.

Lymph vessels have many one-way valves within their lumens. These valves keep lymph flowing toward the point of entry into the circulatory system. There is no heart in the lymphatic system to provide impetus for lymph flow. However, major lymphatic veins, like circulatory system veins, are located in spaces between large muscles (such as those of the arms and legs) and are squeezed whenever these muscles contract during normal movement of the body. Lymphatic veins of the chest region are similarly squeezed by muscles during breathing movements. The resulting flow of lymph is slow but steady if the body is normally active.

The lymphatic system provides a second useful activity, that of filtration. Interspersed along the major lymphatic veins are *nodes*. A node is an enlargement of the vein, within which there is a meshwork of connective tissue that can catch and hold objects such as bacteria or eukaryotic cells. Also within each node is a population of leukocytes (white blood cells) that can recognize and destroy certain cells that become trapped there. In humans, lymph nodes are particularly located in the groin, under the arms, and at the neck region. We are usually unaware of a lymph node until there is a bacterial infection upstream from it. In such a case, the node temporarily becomes distended and painful because of the accumulation of trapped bacteria and leukocytes within it.

BLOOD IS A COMPLEX COMBINATION OF CELLS AND FLUID

Let us now examine the material that circulates, the blood. If one were to centrifuge freshly drawn blood to separate its components by their sizes, one would find that slightly less than half of its volume, at the bottom of the centrifuge tube, is taken up by cells of several types. The remainder of the centrifuged material would be a straw-yellow liquid, the *plasma*.

Many Molecules Are Dissolved in Plasma

Over 90 percent of the volume of plasma is water. Dissolved in this water is a very complex mix of molecules. Most are proteins of several types; albumin, antibodies, and blood-clotting components are the major ones. Albumin serves to transport lipids and to help blood keep an appropriate osmotic value; the others will be discussed later in this chapter. Also dissolved in plasma are amino acids, sugars, hormones, nitrogenous wastes such as urea, carbon dioxide, and (to some extent) oxygen.

Three Structures Compose the Cellular Portion of Blood

When examined closely, the nonplasma portion of blood is found to be composed of three distinctly different types of cellular material: *red blood cells* (*erythrocytes*), *white blood cells* (*leukocytes*), and fragments of cells called *platelets.* All three types are manufactured in the marrow regions of large bones and released upon maturity.

The red blood cells comprise the vast majority of blood cells. They are highly specialized and carry the protein *hemoglobin*; about 25 percent of their cytoplasm volume is taken up by many copies of this oxygen-carrying molecule. Hemoglobin greatly enhances the ability of blood to carry oxygen into tissues needing it. It has been determined that the presence of hemoglobin allows a given volume of blood to carry 30 times more oxygen than it would have otherwise. Of course, it is necessary that hemoglobin hold oxygen *reversibly,* picking it up in an environment where it is plentiful (the lungs or gills) and dropping it off into the plasma when the red cell reaches a tissue with low oxygen concentration. Red blood cells are quickly worn out by their constant passage through the circulation. The typical cell has a lifetime of only four months, during which it probably makes about 350,000 circuits around the blood system. Aged and damaged red blood cells are recognized and destroyed in the *spleen,* an organ situated between the stomach and the diaphragm in mammals. The spleen also harbors many leukocytes that can attack foreign cells such as bacteria.

Not all animals use hemoglobin to carry oxygen. Blood of some Molluscs and Arthropods is blue due to a copper-containing analog of hemoglobin. Some Annelid worms use a greenish chlorine-containing molecule for the same function. The insect's blood has no pigment for absorbing and transporting oxygen. Actually, it needs none since it has branching tubes (*tracheae*) that circulate air within diffusible distance of all cells.

A second category of blood cells is the white cells, or leukocytes. These are much less numerous than red blood cells: one leukocyte per 700 red cells is the normal ratio. They originate from the same sort of bone marrow stem cells as the red cells, but differentiate into five slightly different varieties none of which carries hemoglobin. All leukocytes are active in recognizing and destroying foreign materials in the body. Some of them—the *neutrophils, monocytes, eosinophils,* and *basophils*—perform general infection-battling tasks. Additionally, eosinophils may be involved in allergic responses, and basophils are probably active in preventing inappropriate blood clotting. Cells of the fifth category, the *lymphocytes,* are much more specific in their targets and comprise the cells of the immune system. This is described in detail later in the chapter. Although the easiest way to find leukocytes for study is to examine a blood sample, most activity of these cells is outside blood vessels in the fluid surrounding cells.

The third category of nonplasma blood components is that of the *platelets.* These are fragments of cells that were produced in bone marrow. They carry vital materials used in forming a clot when a blood vessel has been damaged. The clotting mechanism is quite complex and cannot be described in all of its details here. Although clotting is essential when a breach occurs in a vessel, there must be some fail-safe aspect to prevent it from happening when it is not needed; otherwise, a normal vessel will become clogged. The initial trigger for clot formation is exposure of connective tissue normally located deep within the wall of a vessel. Exposure happens only if the wall is torn. Particularly, the protein *collagen*, if contacting blood, causes platelets to congregate around it. The first platelets to arrive become sticky and hold other platelets until they all form a temporary plug at the wound site. Then, a set of *clotting factors* is released from the platelets to act upon proteins in surrounding plasma. After a cascade of events involving at least 12 different clotting factors from both platelets and plasma, a final step occurs in which a plasma-carried protein called *fibrinogen* becomes converted into *fibrin*. This protein, as the name implies, forms a fibrous net-like structure capable of capturing red blood cells as they try to flow through the vessel wall gap. The combination of fibrin and red cells forms the final clot that prevents further loss of plasma and cells until the vessel wall can permanently heal. If any of the many clotting factors is missing or defective, the result is slowing of the clot-forming activity. Some clotting disorders are called *hemophilia*. The severity and type of hemophilia depends upon which factor is affected.

GAS EXCHANGE IS USUALLY ASSOCIATED WITH AN INTERNAL TRANSPORT SYSTEM

As was described in Chapters 5 and 6, most organisms must trade a pair of gases with the environment. Autotrophs such as plants take in CO_2 and release any excess O_2 that they produce in photosynthesis. Heterotrophs such as animals do the opposite as they perform aerobic respiration. Although not all organisms provide specialized surfaces for this exchange, most of those that do will also provide an intimate connection between those surfaces and an internal transport system. Indeed, internal transport is vital if an organism is devoting only a small portion of its body to gas exchange; otherwise, most of the body cannot be served by that specialized area. Thus, it is appropriate for us to sequentially examine internal transport and gas exchange.

Small or Thin Organisms Do Not Need Specialized Surfaces

Gas exchange is a phenomenon of diffusion across the plasma membrane. Both CO_2 and O_2 are capable of moving directly through such membranes if the surfaces are moist. For instance, after entry into a surface cell, O_2 can further diffuse into neighboring cells, although such movement is rather slow. Obviously, single-celled organisms have no difficulty performing direct exchange with their environment. Multicellular organisms whose bodies are thin enough can use cell-to-cell diffusion as their only means of distributing gases to and from their entire body surface. In general, any organism whose cells are all within about a millimeter of the surface can survive without providing either specialized gas exchange areas or an associated internal transport system.

Some organisms manage to obey this rule even though attaining large over-all size. For instance, the alga called kelp can grow to a length of 50 meters, yet all of its cells are very close to the surrounding water since it is extremely flat and thin. The Cnidarian animals commonly called jellyfish appear at first to be violating the one-millimeter rule since they can be several centimeters in thickness yet have no internal transport system. However, close examination reveals that most of the internal bulk of a jellyfish is noncellullar nonliving material that neither needs O_2 nor produces CO_2 as waste. All of its cells are a very short distance from the environment.

Leaves and Roots Are the Primary Gas Exchange Areas of Terrestrial Plants

A leaf, by its very shape, provides a large surface area for gas exchange. However, much of that surface is not available for this function, since it is covered with an impermeable waxy layer to prevent undue water loss. The points of entry and exit for gas exchange are the stomata scattered over the leaf surface (Fig. 8.12). Since there are thousands of stomata on the leaf surface, considerable opportunity for interaction is available. The interior of a leaf has much open space between cells, so movement of gases to and from stomata is available by simple diffusion.

Most cells of large stems are dead, so very little gas exchange is needed there. However, the bark of many stems has round, oval, or linear structures called *lenticels* (Fig. 8.16), where the cells are loose enough to permit the diffusion of gases. In cherry, plum, peach, and elderberry stems these are quite conspicuous. The constant expansion of older stems as they grow ruptures the bark and allows gases to diffuse through the cracks thus formed.

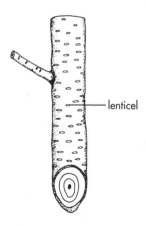

FIGURE 8.16.   *Stem with lenticels.*

Roots do not need gas exchange for photosynthesis, but their cells do carry out respiration to gain energy, so gas exchange is necessary at their surfaces. Since roots branch and rebranch to provide a large surface for absorption of water and minerals, they are also adapted to do efficient gas exchange with the soil. Because roots must do gas exchange, the quality of soil and its degree of aeration are important factors in the health of a terrestrial plant.

Deeply internal parts of a terrestrial plant are directly supplied with air via many connected open spaces. Oxygen entering at stomata can quickly move through these spaces to supply cells far from leaf surfaces.

Therefore, our conclusions for gas exchange in plants are two-fold. First, they require large surface areas to accomplish it efficiently. Second, they have many such surfaces over the body and therefore do not have to rely on their internal transport system (vascular tubes) to move gases to and from those surfaces.

A Typical Animal Has a Specialized Gas Exchange Organ

Most animals possess rather thick bodies with actively metabolizing cells far from their surfaces. Two major types of gas exchange organs necessary to keep all of the cells alive are the gills of many aquatic animals and the lungs of the non-Arthropod terrestrial animals. Both types have some features in common. First, they provide a very large membrane surface in a relatively small portion of the body. Second, they maintain moisture at the membrane surface to facilitate diffusion across it. Third, they include rich capillary beds just below the surface to connect with the rest of the body via the blood circulatory system. Fourth, they have mechanisms to protect the delicate surface membranes and capillaries from damage.

Gills Are Extensions into the Watery Environment for Aquatic Animals

Aquatic animals as diverse as worms, squid, fish, and newts use gills for gas exchange. Typically, a gill is a highly convoluted organ whose many branches and folds extend out from the core of the body into the watery environment (Fig. 8.17). Surrounding water not only provides a medium for gas exchange but also buoys up the gills to keep even the most delicate portions fully extended. Gills are vulnerable to damage in their extended position. Most animals protect their gills by providing some tough shield. The shells of a clam and the gill covers (called *opercula*) of a fish are good examples. This protection, however, restricts the flow of fresh water to a gill, so some means of *breathing* must be added. In the case of a fish, the animal repeatedly draws water into the mouth, then directs it through the left and right *gill slits* in the walls of the pharynx and toward the paired gills. Oxygen-poor water that has passed over the gill surface is expelled from the opercular opening to the outside. Such a one-way flow of water to and over the gill surfaces is very efficient.

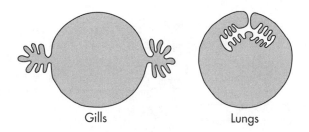

Gills   Lungs

FIGURE 8.17. *Generalized diagrams comparing gilled animals with lunged animals.*

Lungs Are Adaptations for Avoiding Water Loss from Terrestrial Animals

Vertebrate land-dwelling animals use internalized gas exchange structures, the *lungs*. The generalized diagram of such a body plan (Fig. 8.17) shows two features that make lungs better than gills for such an environment. One concerns the need for any terrestrial animal to conserve as much body water as possible. Gills would inevitably be exposed to dry air, causing rapid loss of water over their large surfaces. To cut water loss to an acceptable minimum, the internal lungs connect with the outside world by only a narrow air passage. Although the air deep in each lung is very humid, little of that evaporated water leaves during exhalation. Additionally, in many animals the nasal passage takes on a very convoluted shape to give airborne water molecules ample time to condense on its walls before leaving the body.

A second problem is that delicate gills would collapse under the force of gravity if removed from the buoyancy of surrounding water. This would lead to their being damaged as well as their effective surface area being severely restricted. Lungs are supported by their position inside the thorax.

The Internal Position of Lungs Necessitates Breathing Actions

Although it is highly beneficial to place lungs deep within the body, this creates the problem of how to provide fresh air and remove air laden with CO_2. The breathing mechanics of vertebrate animals provide this air motion. Breathing is accomplished in most cases by muscular action of the chest wall and *diaphragm* (Fig. 8.18). As the dome-shaped diaphragm contracts to become much flatter, the chest cavity above it undergoes an increase in total volume. This, in turn, causes air to rush into the lungs from the outside to equalize pressure within and without the lungs. The lungs are easily expandable under the force of this pressure. If more air is needed, rib muscles also participate in enlarging the chest volume, lifting the rib cage upward and outward. Exhalation occurs when the diaphragm goes back to

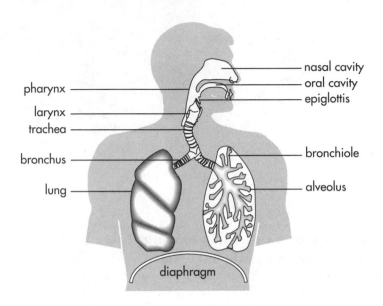

pharynx

larynx

trachea

bronchus

lung

nasal cavity

oral cavity

epiglottis

bronchiole

alveolus

diaphragm

FIGURE 8.18.   *The human air passages.*

its original position and elastic elements in the lung walls bring them back to their original size. The lungs themselves do not have muscles; all muscular work in breathing is accomplished away from the lungs. Air-breathing amphibians have no diaphragms, and their breathing is best explained as a swallowing action. Reptiles also lack a diaphragm and ventilate their lungs by various combinations of muscle contractions along the ribs and body flanks.

Lung Surface Area Can Be Huge

All lungs have considerable surface area to facilitate gas exchange. However, there is much variation among species in the internal shapes and in the degree of convolution within their lungs. Some amphibians have very simple lungs that are sac-like. Reptiles have more subdivisions and thus greater surface area. Birds have tiny *air capillaries* branching from larger tubes, to give a very large total surface. Mammals also have many branches into ever smaller tubes, but each of the smallest tubes expands into a cluster of spheres called *alveoli* (singular: *alveolus*). Each unit of several alveoli resembles a microscopically small cluster of grapes, except that the grapes are hollow air-filled spheres. The diagram of the human chest that is Figure 8.18 does not do justice to the expansion of surface afforded by alveoli, since each human lung actually contains about 150 million alveoli and the total internal surface area of an adult's lungs is about 80 square meters.

The Human Gas Exchange Apparatus Is Complex

The air inspired as a person breathes is moistened, cleaned, and moderated in temperature as it moves toward the lungs. During extremely cold or hot weather the temperature of inhaled air is brought closer to body temperature. Air moving through the passageway lined by moist epithelium picks up moisture and is thus more easily absorbed when it gets to the lungs. Dust and bacteria are removed from the air by trapping them in the mucus that is secreted by epithelial cells. A considerable amount is removed, as one can see, by blowing the nose or clearing the throat after working in a dusty place. In extreme conditions dust and bacteria are not completely removed. Certain industrial and mining operations are recognized as distinct occupational hazards in that large quantities of particulate matter like dyes and mineral dust get into the lungs and cause specific disorders. The lungs of people living in smoky cities may be blackened by the carbon of soot and smoke.

The passageway to the lungs of mammals (Fig. 8.18) consists of the *nasal cavity, pharynx, larynx, trachea,* paired *bronchi,* and *bronchioles.* Much of the nasal passage is lined by a ciliated epithelium that directs mucus toward the pharynx. The nasal cavity is separated from the oral cavity by the *palate* (roof of mouth) and enters the *nasopharynx,* which is the upper part of the pharynx, at the back of the nasal cavity. The nasopharynx is continuous with the *oral pharynx* at the back of the oral cavity. The oral pharynx also serves as a passageway for food to the esophagus. The pharynx leads into the larynx (voice box) through an opening called the *glottis.* Above it is a flap, the *epiglottis,* which by closing when one swallows prevents food from going down the air passage. It closes automatically, but choking occurs when it sometimes fails to work properly. The larynx is connected with the trachea, a large tube whose wall is reinforced by rings of cartilage that cause it to stand open. The trachea branches into two bronchi, which in turn branch into smaller bronchi. Like the trachea, their walls are reinforced with cartilage. Shortly after entering the lungs they branch into smaller and smaller tubes. The smaller ones are bronchioles. The trachea, bronchi, and to some extent the bronchioles have ciliated linings.

The absorbing parts of the lungs are the alveoli. They are lined by a single layer of epithelium (mostly flattened). Alveoli are closely packed, and adjacent ones are separated from each other by common walls consisting of their epithelial linings between which is a very thin layer of connective tissue richly supplied with blood vessels. Thin alveolar walls and a rich supply of capillaries enhance gas exchange.

Insects Use a Different Gas Exchange System

Not all terrestrial animals use lungs for breathing air. Insects have a system of many individual air-carrying tubes (*tracheae*) that branch into every part of the body (Fig. 8.19). Each trachea begins at an opening on the side of the body called a *spiracle.* The ends of the many-branched tracheae are extremely tiny and numerous tubes, the *tracheoles.* Each tracheole is fluid-filled to facil-

FIGURE 8.19.   *Tracheae of an insect. Walls are reinforced with rings of chitin, the same material used for the animal's exoskeleton.*

itate diffusion of gases into and out of nearby cells. Rhythmic muscle contractions of the body help move air in and out of the tracheal system. This direct delivery of air to all cells helps explain how insects can be very active even though they possess the relatively inefficient open style of blood circulation and lack oxygen-carrying pigments such as hemoglobin.

THE CIRCULATORY SYSTEM PROVIDES PATHWAYS FOR DEFENSIVE AGENTS

Animals (especially vertebrate animals) have multiple defense mechanisms to protect themselves from invasions by parasites and other potentially dangerous entities. Some of these are purely mechanical, such as the barrier provided by skin. Some are purely chemical, such as the presence of the bacteria-destroying enzyme lysozyme in tears, saliva, and other fluids. Some are activities that all body cells can initiate upon being invaded. For instance, any cell harboring viruses responds by producing the protein *interferon*, which weakens the infective ability of the viruses and prepares neighboring cells to fight off invasion. Additionally, cells damaged by *pathogens* (disease-causing organisms such as bacteria and viruses) can respond by releasing the chemical *histamine*. This causes increased blood flow into the area, which brings a large number of cells specialized for fighting the disease agent. It is this last activity that brings us to the topic of the remainder of this chapter. A complex and efficient system of disease-fighting cells resides in and moves through the blood circulation. These cells work individually and together to recognize foreign materials and to attack and destroy them.

Some Leukocytes Are Nonspecific Defensive Cells

Neutrophils and monocytes, two of the leukocyte types introduced earlier in this chapter, are able to ingest bacteria, foreign cells, and debris by the process of phagocytosis. They are nonspecific in this action, attacking any object they recognize as not being part of the body. Actually, monocytes are not able to do this, but when they mature they become very efficient nonspecific killers called *macrophages*.

Lymphocytes Direct Specific Immune Responses

Another class of leukocytes, the lymphocytes, provides most of the highly effective actions that are together called an *immune response*. This differs from the activities described above by its specificity. Each immune cell seeks out and helps destroy only a very particular foreign material. If a person were invaded by ten different pathogens, the immune system would mobilize at least as many different populations of lymphocytes, each of which would work on only one invading species. Each specifically recognized foreign material is called an *antigen*. Antigens are molecules or portions of molecules. They can exist alone or be surface components of invading cells or viruses. Although an immune response is somewhat slower than non-specific activities, it is more effective and long-lasting.

The Immune Response Is Actually Two Responses Acting in Concert

Lymphocytes differentiate into two distinct types, usually called T cells and B cells. When mature, both populations are capable of recognizing and acting upon antigens, but they do so in very different ways. Both populations become highly diversified in the kinds of antigens that their cells recognize and act upon. It has been estimated that a person's body has at least 100 million different kinds of immune-responsive T and B cells. B cells are most active in a response system called the *humoral immunity system* and T cells are most active in the *cellular immunity system*. However, as will be shown below, these systems sometimes work together to rid the body of a potentially dangerous antigen.

The Humoral System Uses Antibodies as Its Primary Weapon

B cells (also called *plasma cells* when mature) scatter into many places in the body after leaving the bone marrow. If molecules of a particular antigen appear in the body and spread to make contact with all of the B cells, they eventually touch the surface of one or more that are capable of manufacturing an *antibody* matching them in shape and capable of attaching to them.

Antibodies (also known as *immunoglobulins*) are proteins with the general shape shown in Figure 8.20. Each antibody molecule has a pair of identical tips that are shaped to match a particular antigen and hold onto it. Since there are millions of different B cell types, each capable of making an antibody with a slightly different shape at its tips, it is likely that any antigen entering the body will eventually come into contact with a B cell that has the correct antibody shape. When that happens, the B cell reproduces rapidly to make many identical plasma cells that together release into the bloodstream a large number of the matching antibody molecules.

The mere attachment of antibody to antigens is not sufficient to destroy the antigens, but this acts to mark them for future destruction or make them less likely to be damaging. If antibody coats the surface of a virus, it is unlikely that the virus will be able to bind to a body cell that it would otherwise be able to enter. The attachment of antibodies to the surface of a bacterium makes it an easier target for phagocytosis by nonspecific macrophages. The attachment of antibody to antigens on the plasma membrane of a foreign cell does not directly damage the cell, but it makes possible the attachment of a plasma protein called *complement*. When properly attached and activated, complement alters the cell membrane so that a hole opens. The result is loss of the cell's contents and death of the cell.

The Cellular System Responds to Antigens on the Body's Cells

Sometimes the surfaces of our own cells become altered, either by being infected by a foreign organism (virus, bacterium, parasite) or by becoming

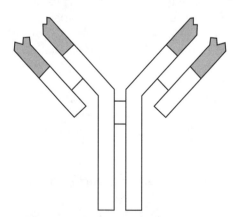

FIGURE 8.20.   *Diagrammatic view of an antibody molecule. The shaded portions represent highly variable regions that can be of proper shape to attach to specific antigens.*

cancerous. The T cells of the cellular immune system act to recognize and destroy such cells, thus protecting the rest of our cells from future damage. Three major types of T cell cooperate in this important activity. *Cytotoxic T cells* (also called *killer T cells*) are the variety that actually attack and kill such cells by phagocytosis. *Suppressor T cells* act at the end of an immune response to halt further activities (of both the T and B systems) after an infection has been successfully eliminated. Helper T cells get the immune response started. All T cells are made in bone marrow, then migrate to a gland in the upper thorax, the *thymus* (hence, *T* cell). In a way not yet fully understood, they mature within the thymus before leaving to patrol the body.

The sequence of events that involves the helper Ts is complex. It begins with a nonspecific macrophage ingesting some foreign or altered cell. As it does so, the macrophage decorates its own surface with antigen molecules that the ingested cell had on its surface. The macrophage can now be called an *antigen-presenting cell* (*APC*), because it then travels to helper T cells and presents its captured antigen to them. Just as there is a huge variety of B cells in the body, there are millions of different helper T cells, each with a different kind of receptor molecule on its membrane. Eventually, the APC finds a helper T cell whose membrane receptor shape matches the foreign antigen being presented. When such a match occurs, the chosen helper T cell rapidly reproduces to make many copies of itself. These newly produced helper Ts then search for the appropriate cytotoxic T cells that have surface properties enabling them to recognize and attack the foreign cells that the APC had ingested. When such a cytotoxic T cell is found, the helper T cell stimulates it to reproduce rapidly and search for the foreign cells.

Helper T cells also aid in activating the B cells of the humoral system, so they are vital in the response of both systems. Unfortunately, it is helper T cells that are the targets of the *Human Immunodeficiency Virus* (*HIV*) that is responsible for *Acquired Immune Deficiency Syndrome* (*AIDS*). If the virus destroys a large portion of the helper T population, the host person loses the ability to recognize and attack all other pathogens and is devastated by infections that normally are easily stopped. The victim is also prone to tolerating cancer cells that are normally recognized and destroyed by the cellular immune system.

The cellular immune system is capable of recognizing and attacking transplanted tissues, since they may be composed of cells with surfaces unlike one's own. If it is medically necessary to transplant a tissue or organ (or to transfuse blood cells), one of two precautions must be taken if the transplant is to be tolerated by the immune system. If a donor can be found whose cell surfaces are identical with those of the patient (as would be the case with identical twins), transplantation will elicit no immune response. Actually, an exact match may not be necessary. Certain cell-surface antigens (together called the *major histocompatiblilty complex,* or *MHC*) are much more likely to raise an immune response than others. If an organ donor can be found whose MHC antigens match those of the patient, this makes it likely that the transplant will be tolerated. The other technique used to help transplant success is to treat the recipient with drugs that suppress some or all of the immune system response. Of course, while this leads to tolerating the transplant, it also opens the patient to future infections.

Both Immune Systems Remember Antigens

When B cells and T cells multiply rapidly to make an initial response to an antigen's presence, some of those cells become *memory cells*. These remain after the antigen has been removed, although they are few in number and largely inactive. However, if this same antigen enters the body later (even many years later in some cases), these cells are available to get the immune response started much more quickly than in the first encounter. They give such a rapid and large response that the invading pathogen is unlikely to cause disease symptoms before being wiped out. A person who has previously had a *primary response* (the initial immune response) is said to be immune to the disease agent because of this protective *secondary response* being available.

We never seem to become immune to some disease agents such as the influenza virus. This can be explained by the ability of these organisms to mutate rapidly so that their surface antigens become different. If you fell victim to last year's influenza strain, you are immune to it but not to the new strain that has since developed by mutations.

Allergy Is Caused by an Unusual Antibody

Some people are genetically predisposed toward inappropriately responding to antigens by producing a set of symptoms called an allergic response. This is traced to their producing a form of antibody that not only attaches to antigens but also irritates one's own body cells. The attachment of this antibody to certain body cells causes them to respond by releasing a chemical called *histamine*. The action of histamine is to cause inflammation of the region—water release and blood vessel widening. It is good to bathe the area in an attempt to wash away the irritating substances, but doing so also causes the symptoms of allergy: swollen tissue, drippy nose or eyes, sneezing, and so on.

Some Diseases Are an Immune Attack upon One's Own Cells

It would be devastating if our immune system lost the ability to recognize the difference between our own cells and foreign cells and attacked both. This happens to a limited extent in some people, leading to damage of certain tissues. For instance, the degenerative disease rheumatoid arthritis is now known to be a destruction of connective tissue through inappropriate attack by the cellular immune system. This and a number of others are called *autoimmune diseases*. The list of known autoimmune diseases continues to grow as further investigation occurs. Other known or suspected autoimmune diseases and their targets are systemic lupus erythematosus (nucleic acids), insulin-dependent diabetes (certain cells of the pancreas), multiple sclerosis (insulating sheaths of neurons), rheumatic fever (heart valves), Graves disease (the thyroid), and myasthenia gravis (motor neurons).

STUDY QUESTIONS

1. Carefully distinguish between simple diffusion, facilitated diffusion, and active transport.

2. Why are biological membranes described as semipermeable?

3. If a cell is hypertonic to its aqueous environment, should it be expected to become more turgid? Explain.

4. What is receptor-mediated endocytosis?

5. Why are mosses limited in their size and their environment?

6. What happens at a root's cap? Its hairs? Its vascular cylinder?

7. Distinguish between the functions of xylem and phloem.

8. What is the role of the Casparian strip of a root's endodermis cells?

9. Explain how guard cells of a leaf's stomatal region can change shape. What are three ways this change is triggered?

10. Water movement against gravity in a tall tree is propelled by several forces acting together. Explain.

11. Why must active transport be used to drive sucrose into phloem of leaves?

12. Distinguish between open and closed circulatory systems. Name at least one animal that uses each.

13. Describe the important movements of blood plasma in capillary beds. Why are lymphatic capillaries necessary there?

14. How do some animals solve the problem of friction at capillary beds? Which vertebrate animals fail to solve this?

15. Trace the circulation of blood from the time it enters the right atrium until it returns to that same place. Distinguish between systemic and pulmonary circulation.

16. How are the lymphatic system and the immune system related?

17. What is the role of hemoglobin in internal transport?

18. What is hemophilia? How is it related to platelets?

19. At what places does a terrestrial plant accomplish gas exchange?

20. Compare and contrast lungs and gills. Why cannot a gilled animal survive on land?

21. What do the alveoli of lungs and the villi of the intestine have in common?

22. How do insects manage to perform efficient gas exchange without lungs, an efficient oxygen-carrying molecule, or a closed circulatory system?

23. Describe how the humoral and cellular immune systems cooperate to ward off a viral infection.

24. Distinguish between immune responses and the work of nonspecific macrophages.

25. Which immune system releases large quantities of antibody during an infection?

26. What is the medical importance of the major histocompatibility complex (MHC)?

27. Which immune cells are are primary targets of the human immunodeficiency virus (HIV)? Why is this particularly devastating?

28. Name three human diseases believed to be caused by autoimmunity.

9
REGULATION OF THE INTERNAL ENVIRONMENT

What do we know about:
—the advantages of having a steady internal temperature?
—why some plants thrive in extremely hot areas?
—why dogs pant?
—our ability to alter blood flow patterns for heating or cooling?
—whether dinosaurs were capable of regulating their temperature?
—how clams respond to change in the saltiness of water?
—how freshwater fish keep from losing all of their salts?
—how our kidneys respond to the need to save water?
—the brain's control of both internal temperature and salt content?
These and other intriguing topics will be pursued in this chapter.

HOMEOSTATIC MECHANISMS EXIST TO MAINTAIN OPTIMUM INTERNAL CONDITIONS

The ideal situation for any organism would be to have a constant internal environment just right to maintain health and perform tasks. Such an organism would have the optimum internal pH, ionic concentrations, and temperature, among many other qualities. Furthermore, that near-perfect organism would be able to change conditions to match changing needs and simultaneously maintain different conditions in different parts of the body. A term that was introduced earlier—*homeostasis*—describes many of the activities that such an organism would have to employ.

This chapter considers the fact that some organisms come much closer to the ideal than others in maintaining internal conditions. Two conditions in particular will be our focus: internal temperature and salt and water balance. Furthermore, since many animals carry on the removal of their nitrogenous wastes by the same organs used for salt and water balance, this will also be examined. It will be remembered that most of these toxic materials have their origins in the ammonia produced when amino acids are deaminated (Chapter 6).

MANY ORGANISMS ATTEMPT TO REGULATE THEIR INTERNAL TEMPERATURE

Temperature is a critical concern for living things. The rate of chemical reactions, including those inside a cell, is heavily dependent upon temperature. In general, the warmer the reactants the faster they interact. Organisms rely upon speedy chemical reactions to synthesize and tear down structures, move their parts, and so on. It would seem, then, that every organism should be trying to find the warmest possible environment. Indeed, we will discuss methods that organisms sometimes use to raise internal temperature.

However, two related facts must be considered in this context. First, an organism consists of molecules, many of them very large, complex, and precisely crafted. As temperature rises, these macromolecules eventually are at risk of being destroyed. Second, it must be remembered that one of the most important categories of large molecules is that of enzymes, which control most of the reactions within the body by acting as catalysts. If internal temperature gets too high, enzymes begin to break down (*denature*), and many critical reactions cannot then be regulated. Although there are some exceptions, most enzymes show at least some loss of catalytic ability at or above about 40°C (104°F). This temperature range is not often encountered in the external environment, but an active animal in a warm climate might easily generate enough internal heat to attain such a temperature in various organs.

In summary, organisms would benefit by having some mechanism for doing appropriate cooling and heating as their activities and the external environment dictate. As with any homeostatic device, a temperature regulating system needs some sensory device to monitor the internal condition, effectors to carry out appropriate corrections, and some method of communicating between these two. Let us now survey organisms for the presence or absence of such ability.

Plants Have Limited Temperature Regulating Ability

In the plant kingdom, the only homeostatic response available is that of *evaporative cooling*. As indicated in Chapter 3, this is the phenomenon of drawing off heat into the atmosphere as water evaporates over a surface. The large area provided by all of the leaves of a plant is ideal for taking advantage of such cooling during the heat of the day. Water is provided from the tracheal system of vascular plants and it evaporates from the stomata. As was mentioned in Chapter 8, this could be dangerous to the plant if it continued during drought times; a plant under the stress of low internal water closes its stomata even during daylight hours.

Some plant species have a permanent (rather than physiological) adaptation for living in very hot and dry environments. Known as C_4 plants, they use a slightly different set of biochemical reactions to make sugars during photosynthesis. The difference is that they can store carbon dioxide in the leaf (in the form of an organic molecule) and release it for carbon fixation after stomata are closed. When the ordinary plant (called a C_3 plant) is subjected to both heat and dryness during a day, it closes its stomata; consequently, it shuts off its supply line for carbon dioxide and stops carbon fixation. A C_4 plant also closes its stomata at such a time but can continue carbon fixation because it draws upon the carbon dioxide it previously stored within itself. Such plants (including corn, sorghum, crabgrass, and several thousand other species) flourish in tropical regions and in the hottest part of the summer in temperate regions.

The typical response of a plant to cold weather is to go into a state of torpor called *dormancy*. This is a condition in which the plant (or an embryonic plant, the seed) survives by dropping to a minimal activity level. Dormancy can be maintained by conditions other than cold; dry seeds remain dormant at any temperature. Breaking dormancy upon an increase of environmental temperature and/or humidity increase is necessarily a complex event. If temperature were the only factor, many plants would be "fooled" by an unusual warm period followed by return to winter conditions and would begin to produce flowers only to have them destroyed when winter conditions returned. To avoid this, many plants break dormancy only after a set of several conditions are met, including increase in the length of daylight hours.

Only Some Animals Can Regulate Their Temperature

In Kingdom Animalia, most species are classed as *ectothermic*, meaning that they passively take on the temperature of their surroundings. Such animals must avoid temperature extremes or suffer the consequences. Only the birds and mammals are truly *endothermic*, capable of maintaining a relatively constant body temperature independent of their environment's temperature. (However, some animals classed as ectothermic can perform some limited temperature control, as will be shown below.)

Two commonly used terms, cold-blooded and warm-blooded, are misleading and should be avoided. A "cold-blooded" ectotherm, if placed in a very warm environment, will have its blood (and all other tissues) matching that warmth. A "warm-blooded" endotherm will be able to maintain a cooler body temperature than its surroundings if they are at, say, 40°C.

Ectothermic Animals May Have Behavioral Tricks for Limited Regulation

Although an ectotherm cannot (by definition) thermoregulate within close tolerances by physiological means, it is not necessarily at the mercy of the environment. Many seek cool shelter during the hottest part of a day and move into the warmest places on very cool days. A rattlesnake sunning on a rocky shelf in early spring is placing itself in position to capture extra heat energy radiating from the rock. A lizard creeping into a desert cave during the middle of the day is doing the opposite. Some butterflies are seen on cool days sitting in sunny spots with wings spread at a particular angle that maximizes their ability to capture heat from the sun. Many such butterfly species have wing patterns with dark pigments on the upper basal portion of the wings, nearest the body. This and the basking behavior are believed to enhance the absorption of heat nearest the flight muscles of the body to warm them to a usable temperature. Honeybees keep the entire hive warmer than the environment in winter, by moving about in the hive to generate heat from muscular activity. They also bring water to the hive in hot weather, then circulate air over it by wing movements, thus providing a means of evaporative cooling.

Endothermic Animals Use Both Anatomy and Physiology

Mammals and birds, the true endotherms, have arsenals of physiological mechanisms to make themselves somewhat independent of the environmental temperature. Before describing some of these, one must point out two things. First, endotherms can regulate only over a certain range of environmental temperatures. Either extreme warmth or extreme cold will overwhelm even the best tricks and place an endothermic animal in jeopardy. Second, some of the mechanisms about to be described are activities that require the expenditure of energy. An animal must work to keep its temperature different from that of its surroundings. This is shown indirectly by the increased breathing rate of an animal placed under thermal stress (both heat and cold). No animal can be truly endothermic unless its body is well equipped to do gas exchange and to deliver oxygen and food molecules to the tissues involved in these energy-using actions. As mentioned in the previous chapter, it is not coincidental that endothermic animals are those with efficient four-chambered hearts and the best lungs.

Evaporative Cooling and Thermal Radiation Are Used to Lose Heat

Just as vascular plants perform evaporative cooling on their leaf surfaces, endothermic animals under heat stress place water on their body surfaces for the same function. This can be in the form of sweating, panting, or simply licking the body surface. Panting animals (birds and some mammals) have a long tongue that is well provided with vessels to bring hot blood from the body core to the moist surface. They add the behavioral aspect of rapidly moving the tongue and breathing over its surface to enhance evaporation of salivary water.

Some heat loss can also occur by direct radiation from the body surface. An animal needing to cool off changes its pattern of blood flow to send more into vessels near the skin surface. This is accomplished by ceasing nerve messages to muscles in vessel walls just under the skin surface, thus allowing them to relax and increase vessel diameter. Called *vasodilation*, this change leads to an increase in blood flow from the body core to regions where heat transfer can occur. Some animal species genetically adapted to living in very hot climates have specific portions of their bodies enlarged to provide more surface for radiative cooling. For instance, warm-climate rabbit species have much longer ears than related species living in arctic regions.

Anatomical, Physiological, and Biochemical Mechanisms Increase Body Temperature

Endotherms use a number of features to raise their temperature when necessary. Certainly one strategy is to avoid loss of whatever heat the body has generated. An animal can reverse the vasodilation described above, performing *vasoconstriction* of vessels near skin surfaces. It can be genetically equipped to have a rounded body shape with few protruding parts that increase surface area. Some mammals (such as polar bears and whales) living in extreme cold seasonally store fat under the skin as insulation. Both bird feathers and mammalian hair shafts can be raised by muscular action to increase the layer of dead air between skin and the environment, acting to decrease heat loss.

Two activities are employed by endotherms to increase internally generated heat rather than conserve already produced heat. One is to increase muscular activity. This can be in the form of overall movements, such as running, or very local movements known as shivering. The latter involves rhythmic contraction of muscles in a way that does not lead to overall movement of body parts. Since any muscle contraction event involves release of some heat as ATP is broken down, the active animal benefits in cold weather. The other major way to increase heat production is called *nonshivering thermogenesis*, which acts at the biochemical level within the entire body. Under hormonal stimulation (mainly thyroxin and epinephrine) there is a general increase of all metabolic reactions, resulting in as much as double the basal rate of heat production within cells. Nonshivering thermogenesis, unlike the other mechanisms described here, is a slow response to continued cold, requiring days of continued exposure to low temperature.

Endotherms Must Integrate Their Responses to Changing Temperature

It can be seen that many mechanisms are available to carry out temperature change in an endotherm. It becomes obvious that there must be two additional portions of the system if the appropriate actions are to be used at any moment. First, there must be some way to measure internal temperature. In mammals, temperature-sensitive neurons both near the skin surface and deep within the body send messages to a region of the brain, the *hypothalamus*. Some of these neurons switch on only when low temperature is encountered, others when high temperature occurs near them. The hypothalamus itself also contains temperature-sensing cells, which can override the others. This is not surprising, since brain cells are more likely than others to encounter difficulty in their performance if their temperature changes significantly.

The second necessary portion of this (or any homeostatic) system is analysis and signaling to get the correct response started. The hypothalamus integrates the messages from all regions of the body and then sends appropriate nerve signals to peripheral tissues and organs, causing combinations of the actions described in preceding paragraphs.

Were Dinosaurs Endothermic?

Modern reptiles (lizards, snakes, etc.) have only limited capacity to regulate their internal temperature, mostly through behavioral actions such as moving to warmer or cooler places within an environment. However, there is some indirect evidence that their ancient relatives, the dinosaurs, may have been much closer to being truly endothermic. In 1994, a report was published showing that the ratio of certain isotopes of oxygen in well preserved bones of the dinosaur *Tyrannosaurus rex* is in a pattern similar to that found only in the bones of modern endotherms. While the isotope ratio in an ectotherm's skeleton varies according to whether the bone is near the core of the body or near the skin, the ratio is constant in all bones of an endotherm, showing that the entire body is maintaining nearly the same temperature. Throughout the *T. rex* skeleton, according to the report, the isotope ratio is constant. However, it should be pointed out that at this writing there are many experts on both sides of this issue and much more evidence will be needed before everyone is convinced one way or the other.

ORGANISMS MUST KEEP THE PROPER INTERNAL SALT AND WATER CONCENTRATIONS

Any cell must have an appropriate concentration of materials dissolved in the water of its cytoplasm if it is to avoid serious problems. Foremost among these problems is the gain or loss of too much water by the process of osmosis. Cells gaining much water risk bursting their plasma membrane. This is not likely in plants, since each cell can expand no more than the rigid cell wall around it, but it is a major hazard for organisms of Kingdom Animalia. We will examine animals as they cope with this and other problems of salt and water balance. It will be shown that animals generally fall into one of two categories: *osmoconformers* that passively match whatever environment they are in and *osmoregulators* that do something to maintain the ideal internal balance independently of their environment.

Some important descriptive words introduced in Chapter 8 should be brought back to our attention now. This set describes relative concentrations of dissolved material, comparing one place with another. A *hypertonic* area is one with more solutes (and therefore less solvent, such as water) than the area being compared. *Hypotonic* is used for the reverse, and *isotonic* indicates no difference in concentration between the compared areas.

Many Marine Invertebrate Animals Are Osmoconformers

Saltwater of the oceans is isotonic to the cytoplasm of most invertebrate animals living in it. These animals, such as sponges, clams, marine worms, and lobsters are osmoconformers, but the sea is usually very constant in its tonicity and conforming to it is not dangerous. However, if such an animal wanders into an estuary (an area where freshwater is mixing with saltwater, as at the mouth of a river), it can find itself in trouble, losing too many salt ions to the hypotonic area and gaining water. Some marine animals living in zones with changing salt content close their bodies to water contact when encountering less salty water. For instance, a clam in such a situation might close its shell halves until it senses that its surroundings have become sufficiently salty again.

Marine Fishes Perform Osmoregulation to Remain Hypotonic

The bony fish species (class Osteichthyes) have tissues with only about a third the salt concentration of the surrounding sea. To maintain their hypotonicity, they must either close their bodies to any salt and water movement or constantly export salts while importing water. They cannot do the former; their gills continually present a large surface area of membranes across which diffusion can occur. Therefore, they drink much seawater to regain the water being lost at the gills, then move the concurrently gained salt ions to the gills where active transport mechanisms pump the salts out to the environment.

The cartilaginous fishes (sharks and their relatives, of class Chondrichthyes) also have a lower internal salt concentration, but they maintain this in a quite different fashion. They are able to retain more of their nitrogenous wastes (primarily urea) than other animals can tolerate. Since urea acts osmotically like salt, the blood and tissue situation for a shark is one of slight hypertonicity to the sea. The excess water gained by being hypertonic is efficiently removed in urine made at the kidneys.

Freshwater Animals Osmoregulate to Remain Hypertonic

Any animal living in a stream, pond, or lake faces the problem of being surrounded by a very hypotonic environment, a situation tending to cause much salt loss and water gain. All freshwater animals must be osmoregulators. A strategy that many such animals use is to make most of the body surface impermeable. For instance, crayfish and other freshwater Arthropods have a thick exoskeleton over much of the body, and expose only the gills as sites of water and salt interchange. Then, there must be active (energy-using) methods for removing water and bringing in salts. Gills of freshwater animals are equipped to perform active inward transport of ions. Excess water is usually removed in a very dilute urine made in large volume.

Terrestrial Animals Must Osmoregulate to Retain Both Water and Salts

Dry land (and more importantly the air above it) is perhaps the most challenging environment osmotically, since the difference between cells and the air is so extreme. Consequently, terrestrial animals employ a number of devices to make as much as possible of the body impermeable to water loss. For instance, Arthropods such as spiders and insects coat their surfaces with a thin layer of wax, which repels the movement of water. As described in Chapter 8, terrestrial animals inevitably lose some water over gas exchange surfaces, but keep this to a minimum by placing these areas deep within the body with only small exterior openings. Many animals perform efficient reabsorption of water from fecal material before it is excreted from the intestinal tract (review: Chapter 7, functions of the colon).

A major avenue for water loss from terrestrial vertebrate animals begins at the kidneys. Here, urine must be continuously manufactured to be a watery vehicle for carrying nitrogenous wastes from the body. As described in detail below, the vertebrate kidney is well equipped to respond to the need to retain water when necessary, by manufacturing a more concentrated urine. The kidneys are also sites of salt retention.

Terrestrial Arthropods also must remove nitrogenous wastes but are even more efficient at doing this with minimal water loss. Insects have a set of two devices that results in the release of an essentially dry waste material. First, they convert their wastes to *uric acid*, a material that can be brought to dryness and excreted in crystalline form. Second, they collect the uric acid from the body fluids into a set of many tiny dead-end tubes called *Malphghian tubules*. Each of dozens of such tubules empties its collected uric acid and water into the intestinal tract, where an efficient reabsorption of water can be accomplished before ejection of the wastes.

Of course, an important aspect of water balance is the behavioral response to becoming too dry: finding water to drink. Only the most efficient water-retaining Arthropods (such as those inhabiting furniture and book bindings) can live without drinking and/or capturing the water within food. Those few not needing to include bulk water in their diets obtain enough as metabolic water. This is water made during the chemical reactions of oxidative phosphorylation.

The Mammalian Kidney Acts in Both Osmoregulation and Waste Removal

For all vertebrate animals, the kidneys are the most active sites of both osmoregulation and the processing of nitrogenous wastes. Although the specific osmoregulatory needs of a particular animal will dictate the exact kidney anatomy and activity, a good representative is that of the human.

As shown in Figure 9.1, the paired kidneys are connected to other units of the complete excretory system. In overview, each kidney produces the waste-containing fluid, *urine*, and releases it into a tube, the *ureter*. The ureters dump their contents into a temporary storage bag, the *urinary bladder*. Then, whenever the bladder is full, it can be emptied to the outside via a single tube, the *urethra*. Once urine has been released from the kidneys, its composition cannot be altered, so it is to the kidneys that we must look for regulation of the amount and quality of urine.

The functional units within a kidney are the *nephrons* and their associated blood vessels and connecting tubes. There are about 1 million nephrons in each human kidney, tightly packed together. Each of them operates in approximately the same fashion to produce urine, so it is useful to examine just one nephron in detail. Figure 9.2 indicates the anatomy of a nephron and nearby structures. The nephron and collecting duct are tubes, each of whose walls are made of a single layer of living epithelial cells. This must be kept in mind as we follow the conversion of blood plasma into urine.

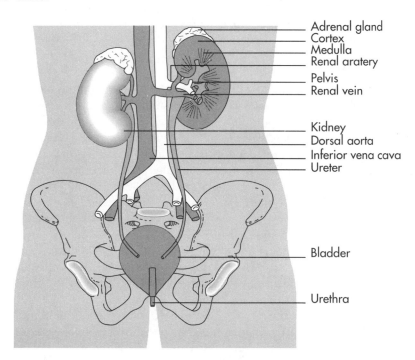

FIGURE 9.1. *The human excretory system for production and removal of urine. (From John W. Kimball,* Biology, *6th ed. Copyright © 1994 Wm. C. Brown Communications, Inc., Dubuque, Iowa. All Rights Reserved. Reprinted by permission.)*

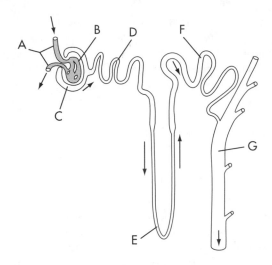

FIGURE 9.2. *The nephron and associated structures. A: an arteriole. B: the glomerulus. C: Bowman's capsule. D: the proximal convoluted tubule. E: the loop of Henle. F: the distal convoluted tubule. G: a collecting duct. Portions B-F constitute the nephron. Arrows indicate direction of flow. Many capillary vessels lie over the nephron and collecting duct; these are not shown here.*

As the renal artery enters the kidney, carrying blood rich in the nitrogenous waste material urea, it branches and rebranches until each nephron is supplied by one arteriole. That arteriole splits into a very small and compact bundle of capillary vessels, called a *glomerulus*. As at any capillary bed, some plasma leaks out. Red blood cells traveling in the plasma cannot leave the glomerulus vessels. Thus, even at this first stage an alteration of blood is occurring. This is called *filtration*. Much of the modified plasma immediately passes through the membranes of the nephron at the region called *Bowman's capsule*. Note how closely this area of the nephron surrounds the area where plasma is available. As plasma crosses cell membranes to enter the nephron, most of its dissolved proteins are filtered out and reabsorbed back into the blood vessels of the glomerulus. Of course, this is a valuable activity, since material entering the nephron may eventually exit the body with urine. Proteins should not be given the opportunity to be lost.

The second portion of the nephron is a twisted tubular area called the *proximal convoluted tubule*. This is an area very busy in removing several materials from the nephron to avoid their loss. This is called *reabsorption*. Glucose, amino acids, and vitamins are recognized and removed by active transport. If these materials are in normal amounts in blood, the proximal convoluted tubules will be successful in reabsorbing virtually all of them. However, if they are in significantly greater than average concentration, the reabsorption process will be overwhelmed and some will be allowed to leave the body in urine. This is what happens in uncontrolled diabetes, where blood glucose is in excessively high concentration because it cannot enter body cells.

Also at the proximal convoluted tubules there is active transport of some (but not all) of the Na^+ and Cl^- ions present in the fluid. It makes the area surrounding this portion of the nephron hypertonic to the interior of the tubule. Therefore, much water follows passively, by osmosis. Water reabsorption is very important, since a huge amount starts the trip through the nephron. It is estimated that about 200 liters of water enter the nephrons of a person each day. Since the total blood volume in the body is much less than that, it is obvious that much of the water is recaptured before it can leave the kidneys. Approximately 80 percent of water reabsorption occurs at the proximal convoluted tubule.

The liquid, now quite different from plasma but not yet urine, passes next into a long straight portion of the nephron, the loop of Henle. It descends, then returns in a parallel portion of the loop. During movement through the loop, some more modification occurs. Sodium chloride is actively transported out, creating a very salty neighborhood. That, in turn, removes more water by osmosis.

The last portion of the nephron that the fluid traverses is the *distal convoluted tubule*. Here, some more active transport of salt occurs, followed by passive osmosis of water. A new activity also occurs: *secretion*. This is an addition into the fluid in the nephron, rather than a removal. Potassium, hydrogen, and ammonium ions are actively transported into the fluid if they are in higher than optimal concentration in blood. Hydrogen ion secretion is a major homeostatic event, since it regulates the pH of blood. Once H^+ ions are pumped into the fluid, they are trapped there by attachment to other ions that cannot leave.

The fluid entering the collecting duct can truly be labeled urine, but one further act of processing occurs in the duct. Under certain circumstances, the cells of the collecting duct become very permeable to water movement. Since the collecting duct lies parallel to the loop of Henle, and since the area around the loop is very salty at all times, water molecules are prone to leave the urine and reabsorb back into the tissue between the loop and the duct. The loop transports NaCl in such a pattern that the area around it is saltiest near the bottom, at the hairpin turn. Thus, as urine flows down the collecting duct, it moves into ever saltier conditions and is likely to lose water all along this parallel pathway. The longer the loop of Henle, the more opportunity there is for water loss from the urine of the nearby collecting duct. This is why the kidneys of desert mammals have extraordinarily long nephrons, providing the opportunity to save a maximum amount of water.

The cells of the collecting duct are not always equally permeable to osmotic movement of water. They become most permeable under the influence of a hormone, *antidiuretic hormone (ADH)*, released from the pituitary at the base of the brain. When certain cells of the hypothalamus sense that blood salt concentration is high (meaning perhaps that total blood volume is becoming low), they send a nerve message to the nearby pituitary. The response of the pituitary is to release ADH, which travels via blood to the cells of the collecting duct. They, in turn, respond by become much more permeable to water, thus saving more water from being lost in urine and placing it back into blood circulation. This is a very sensitive homeostatic device for keeping blood volume at the optimal value. Although most of the water that entered the nephron at Bowman's capsule has been removed before the collecting duct is reached, this last bit of water removal is a matter of fine tuning that can be very important to the welfare of the body.

The degree of reabsorption of Na⁺, the most abundant salt ion, is also under hormonal control. The cortex region of the adrenal glands (Fig. 9.1) releases the hormone *aldosterone* when it senses that the blood Na⁺ level is dropping below the optimum value. The target cells of this hormone are the distal convoluted tubule and the collecting duct; they respond to its presence by increasing the reabsorption of the ion from urine.

Thus, it can be seen that each of the 2 million nephrons and their associated collecting ducts are major homeostatic devices, retaining most water and salts but also capable of changing this retention when body conditions dictate. Figure 9.3 summarizes the selective removal of materials at various portions of the apparatus. The urine that leaves the kidneys is water with much dissolved urea (the toxic nitrogenous waste that must be excreted) and low concentrations of several ions. A physician examining urine can find evidence

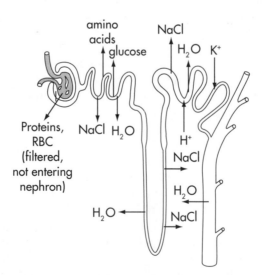

FIGURE 9.3.   *Summarizing diagram of the modifications occurring as fluid moves from blood and through the nephron and collecting duct.*

of several systemic diseases or of kidney damage if it is found to contain significantly high concentrations of any other materials.

STUDY QUESTIONS

1. What is the biochemical hazard of becoming too warm?
2. How do vascular plants keep their leaves cool in the heat of the day?
3. How does a C_4 plant such as crabgrass grow better than a C_3 plant such as fescue grass during the heat of summer?
4. Define the term *ectothermic*, and indicate why it is a better term than *cold-blooded*.
5. What is an example of an ectothermic animal? Of an endothermic animal?
6. Describe behavioral actions that some ectotherms use to exercise partial temperature control.
7. Under what circumstances would a person be likely to vasodilate the vessels just below the skin surface?
8. How does the hypothalamus control temperature in your body? What role does it play in controlling kidney function?
9. What is the risk of being an osmoconformer? Name a few such animals.
10. How can sharks be nearly isotonic to their surroundings, despite having lower salt concentrations in their blood?
11. Describe the role of a freshwater fish's gills in osmoregulation.
12. Would a freshwater fish be expected to have more concentrated urine than a typical terrestrial animal, or less concentrated? Explain.
13. How are insects well adapted for living in dry environments?
14. Distinguish among these kidney activities: secretion, filtration, and reabsorption. Where does each occur?
15. Why is it necessary to pull most of the water out of a human's nephrons?
16. How is the nephron active in maintaining proper blood pH?

10
COORDINATION
AND MOVEMENT

What do we know about:
—*the structure and function of neurons?*
—*how some neurons are specially equipped for speedier message transmission?*
—*methods used by neurons to communicate with each other?*
—*how instinctive behavior differs from learning?*
—*the functions of various brain regions?*
—*the molecular basis of muscle contractions?*
—*how hormones make their presence known at target cells?*
—*the control of endocrine glands by the brain?*
—*interactions among the coordinating chemicals of plants?*
These and other intriguing topics will be pursued in this chapter.

ORGANISMS MUST SENSE THEIR ENVIRONMENT AND RESPOND TO IT

One of the characteristics of all living things is the ability to sense the environment (both external and internal) and then to make appropriate and timely responses to whatever has been analyzed. This was seen in Chapter 9 and in many portions of previous chapters. It is time to examine in detail the methods by which the various portions of an organism gather information and communicate with each other during analysis and response. It will be seen that plants communicate only by a rather slow dispersal of chemicals, but that animals have two systems—a slow chemical one and a much quicker system based on nerves. We will also examine how organisms move their parts, since this is a common response to either chemical or nerve stimulation and is part of the story of coordination.

COORDINATION BEGINS WITH A STIMULUS

When changes in the environment can register strongly enough and last long enough to produce a response, they are called *stimuli*. Some stimuli of the environment are light, sound, temperature, gravity, contact, chemicals, and electricity.

Receptors are portions of a body (or a cell) specialized to receive specific kinds of stimulus. Special sense organs are very different from each other in their sensitivity to environmental factors. What acts as a stimulus to one receptor or organism may not be a stimulus to another. However, some receptors respond strongly to the stimulus for which they are specialized but also weakly to other stimuli. For instance, if the retina of a person's eye is hit by physical pressure (such as a blow), it responds to that stimulus by sending nerve messages to the brain by the same routes as when it is hit by light. It requires much more physical stimulation than light stimulation, but it does respond. Since the same neurons are being stimulated, the brain interprets the message of a blow as light: the person experiences flashes of light.

THE RESPONSE TO A STIMULUS IS ACCOMPLISHED BY SPECIALIZED REGIONS

The kinds of response that organisms can make are related to their own body equipment. The responding part is the *effector*. Most animal effectors are either *muscles* or *glands*. In considering responses one ordinarily thinks about those that are rapid and visible muscular actions. Glandular responses are much more subtle. Endocrine glands (to be discussed later in this chapter) secrete chemicals called *hormones* in response to a variety of stimuli; for example, fright produces a reaction of the adrenal glands and visual, contact, or mental stimuli can produce sex-related hormonal secretions.

To a smaller degree there are *chemical, electrical, photic,* and *chromatic* responses by animals. The production or discharge of irritating secretions that serve defensive purposes is sometimes a response to stimuli. The bee and other stinging insects discharge stinging chemicals in response to stimuli. Upon proper stimulation the electric eel is able to discharge a considerable amount of electricity. The photic response, a discharge of light, is known in insects, Molluscs, fish, and others. It is also termed *bioluminescence*. Chromatic responses, which are color changes, are brought about by changes in surface cells known as *chromatophores*. Among the many animals that make such responses are the chameleon and the frog.

Some responses, being internal, are not so readily identifiable. A case in point is permeability changes in the plasma membranes of cells, such as was described in Chapter 9 in the response of kidney cells to antidiuretic hormone.

Thus far, reference has been made to responses that are *activated*, but potential responses may also be *inhibited*. As will be discussed below, some neurons release chemicals that inhibit the ability of effectors to respond to activating messages.

Sudden responses in plants are uncommon. Like animals they have responses of motion, secretion, permeability changes, and bioluminescence. Motion, of course, is not directed locomotion of the whole organism from place to place. Secretions are numerous, some of them having economic uses. The carnivorous plants even secrete enzymes when they are stimulated by the capture of animals suitable as food.

Many organisms have directional responses. If the organism is more or less permanently attached to a substrate (e.g., a plant rooted into soil or a coral animal attached to the colony's skeleton), the response is properly called a *tropism*. This takes the form of orientation of the body toward some stimulus, such as growth of a plant stem toward light. If a motile animal moves its entire body, the response is a *taxis* (plural: *taxes*). Fish often respond by heading into water currents, and moths often fly toward a light. Descriptive prefixes may be added to the word taxis to show the kind of stimulus involved; for instance, *phototaxis*, *geotaxis*, *thermotaxis*, and so on. Taxes are responses not only of animals but also of unicellular motile organisms.

NEARLY ALL ANIMALS COORDINATE SENSING AND RESPONDING BY USE OF A NERVOUS SYSTEM

The sponges (phylum Porifera) are the only major animals having neither nerve cells for perception and transmission nor muscles for reaction. The most primitive development of specialized nerve cells is in phylum Cnidaria. The nerve cells of these animals (Fig. 10.1) are connected in a *nerve net*. A stimulus received by a cell is transmitted in all directions from it and is thus diffuse. As one might expect, the response is also diffuse. A strong enough stimulus applied at one point may make the animal contract into a tight ball, as if it were being stimulated from all directions. Note the lack of any central processing area, such as a brain.

Nearly all other animals have a more centralized nervous system, in which many neurons connect in such a way as to send sensory information to a *central nervous system (CNS)* where the integration of information can occur and appropriate messages can be sent out to effectors. Usually this CNS has its most controlling elements at the anterior of the body, close to the many sensory devices also located there to analyze new territory as the animal explores.

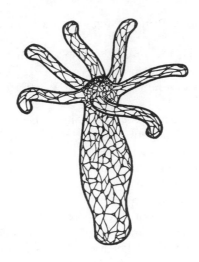

FIGURE 10.1. *Nerve net of hydra, a typical Cnidarian animal. (From* General Biology *by Stauffer, D. Van Nostrand Company, Inc.)*

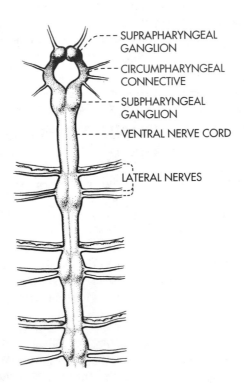

FIGURE 10.2. *Nervous system of* Lumbricus terrestris *(earthworm). (From* General Biology *by Taylor and Weber, D. Van Nostrand Company, Inc.)*

This *cephalization* is most pronounced in the vertebrate animals with a distinct head including a *brain*.

In simpler animals like earthworms (Fig. 10.2) and insects, clumps of nerve cells called *ganglia* (singular, *ganglion*) occur at the anterior end of the CNS. It is arguable whether this congregation is sufficiently large and controlling to be called a brain. That term is more unambiguous when referring to the portion of the CNS that has distinct control over all other nerve function and that cannot be damaged without placing the entire organism in great peril. Ganglia may occur in other parts of the same body or in higher animals as well.

In animals of Phylum Chordata, the central nervous system consists of a brain and *spinal cord* (really a tube), both of which are dorsal. All nervous tissue outside the CNS is considered the *peripheral nervous system*. It must be remembered, however, that the nervous system is a highly integrated association of cells and that this division into central and peripheral portions is for convenience of discussion only.

NEURONS ARE THE BUILDING BLOCKS OF A NERVOUS SYSTEM

The functional unit of any nervous system is the nerve cell or *neuron* (Fig. 10.3). It consists of a *cell body* that contains the nucleus and many other organelles commonly found in cells and from which extends one or more processes known as *nerve fibers*. The fibers are of varying lengths but may be a meter or more long in large mammals. Nerve fibers normally transmitting messages toward the cell body are *dendrites*, and those transmitting away from it are *axons*. Most neurons have several rather short dendrites and one long axon. Impulses travel from neuron to neuron in a chainlike fashion.

Neurons Vary in Shapes and Functions

Neurons may be classified according to the number of fibers they have. There is a *unipolar* type whose cell bodies are concentrated in the dorsal root of the spinal nerves. Actually, the single fiber attached to its cell body is divided a short distance from it, with one part extending to the peripheral region and another to the central nervous system. Another type is *bipolar*, as in the retina of the eye. Bipolar cells have one dendrite and one axon. Most book illustrations show the *multipolar* type, where there are several dendrites and one axon.

Another way to classify neurons is according to where they send their messages and how they are connected. Those that receive a stimulus and send their message toward the central nervous system are *sensory neurons*. Those that send a message from the central nervous system to an effector (such as a muscle fiber) are *motor neurons*. Usually, sensory and motor neurons are connected via one or more neurons entirely within the central nervous system, known either as *association neurons* or *interneurons*.

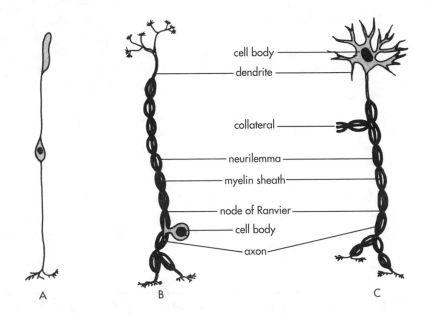

FIGURE 10.3. *Three types of neurons: A: bipolar neuron of retina. B: unipolar sensory neuron. C: multipolar motor neuron. Drawings are not in correct proportions.*

A third way to distinguish neurons is by the effect they have upon the cells to which they connect. An *excitatory* neuron causes some positive response, while an *inhibitory* neuron inhibits some response or causes a cessation of activity.

Axons may or may not be surrounded by insulating sheaths in their portions outside the central nervous system. As usually prepared and observed under the light microscope, the sheaths look two-layered. The outer layer is called the *neurilemma* and the inner one the *myelin sheath*. Actually, they are one and the same. By using the electron microscope it is possible to show that the myelin sheath is a tightly wrapped portion of a *Schwann cell* closely adhering to the axon. The infolded part of this cell is wrapped around the axon like a roll of paper. The portion of the Schwann cell occupying the outermost layer (and thus keeping some of its original structure, including a nucleus) is the neurilemma. The sheath has gaps along its length known as *nodes of Ranvier*. These gaps indicate the boundaries of adjacent Schwann cells. The presence of the myelin sheath and its nodes insulates axons that may lie very close together and also speeds the nerve message's speed of movement. More will be said about this below.

The Nerve Message Is Both Chemical and Electrical

The nature of a nerve impulse is not simple. It has some characteristics that are chemical and others that are electrical. The description that follows is necessarily simplified and the reader is referred to specialized texts in neurophysiology.

In essence, the message moving along a neuron is a change in the relative electrical charge on its surface. An explanation of this must consist of four parts: (1) what the charge is at an unstimulated portion of a neuron and what creates this charge, (2) what it is that changes when that portion is stimulated, (3) how that change can be moved from one end of a neuron to the other end, and (4) how two neurons (or a neuron and an effector) can communicate this message.

The Resting Neuron Maintains a Positive Charge on the Outside of Its Membrane

The plasma membrane of a neuron (and, indeed, of any animal cell) is equipped to maintain certain inequalities in the concentrations of ions. The concentration of sodium ions (Na⁺) is higher on the outside surface than in the cytoplasm just beneath the membrane (Fig. 10.4). Potassium ions (K⁺) show the opposite distribution. Dissolved proteins with negative charges, and other negatively charged large molecules, are richly represented in the cytoplasm. There exist channels through which K⁺ can passively diffuse toward the outside, but there are also active transport potassium pumps that bring much of it back into the cell. Actually, these pumps do double duty, ejecting Na⁺ as they bring in K⁺. This *sodium-potassium pump*, as it is called, is somewhat more efficient in working with Na⁺ than with K⁺. There are also pro-

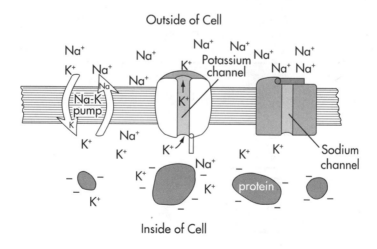

FIGURE 10.4. *Ionic conditions leading to the establishment of a resting potential across the neuron membrane. Although both potassium and sodium channels exist, only the potassium channel is open, allowing passive loss of that ion to the outside.*

teins built into the membrane that could act as channels for the passive diffusion of Na⁺, if they were open. However, as the neuron rests (unstimulated), these channels are not functioning.

The result of all of these conditions is that there is more of the combination of Na⁺ and K⁺ on the outside of the membrane than there is of these two ions on the inside. Now, speaking electrically, if there is more positive charge in one place than in another, the less positive area can be said to be negative in comparison to the more positively charged area. To enhance this difference there are the negatively charged proteins always held inside the cell. Therefore, the net result of active transport plus the absence or presence of functioning channels for passive diffusion is that the outside surface is slightly positive compared to the cytoplasm at the inside surface of the membrane. Figure 10.5 A represents this condition at one point along a neuron.

Voltage is the measure of such a charge difference, and an alternative way of indicating voltage is to use the term *potential difference*, or just *potential*. The voltage across the membrane is measured at approximately 70 millivolts (70/1000 of a volt); this is sometimes called the *resting potential* of a neuron.

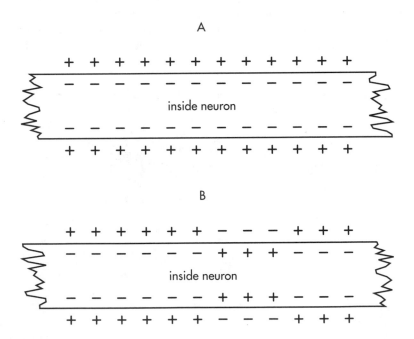

FIGURE 10.5. *Net charge distribution at a point along a neuron. A: resting (unstimulated) condition. B: just after stimulation has occurred at the point marked by the arrow.*

Stimulation Leads to Reversal of Charge at the Membrane

If a portion of a neuron's membrane is impacted with a sufficient stimulus (electricity, chemicals, pressure, and the like), a series of changes suddenly occurs. The first change is that previously closed Na^+ channels begin to open, allowing this ion to begin flowing inward by passive diffusion. Each entering positive charge further stimulates the gates to open, and soon a number of Na^+ ions can flow into the cytoplasm. Of course, this changes the relative charges inside and outside, making the outside much less positive than it had been. In response, the previously open K^+ channels close, trapping this ion inside the cell. The effect of the inward flow of positive charges can be graphed, as shown in Figure 10.6. So much movement happens that the relative charge condition actually reverses from the prestimulus situation, with the outside becoming negative to the inside (Fig. 10.5 B). The inside goes from -70 millivolts to $+35$ millivolts, all in the short span of as little as 1/1000 of a second. This change is sometimes called the beginning of an *action potential*, since it is a change in potential difference (voltage) signaling an action of stimulation. It can be accurately stated that the action potential is the message of the neuron.

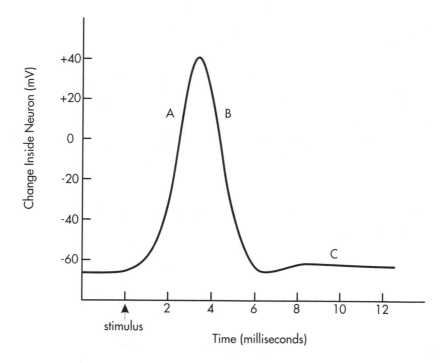

FIGURE 10.6. *A typical action potential. A: change resulting from inward flow of Na^+. B: change resulting from outward flow of K^+. C: period when Na^+ and K^+ are being transported back to their original positions. Actual voltage change and time span varies from neuron to neuron.*

This new condition does not last long. The change in charge itself triggers the reclosing of Na$^+$ channels and the reopening of K$^+$ channels. Potassium rushes out, leading to a reversal of the charge again. The second portion of the action potential, as shown on the graph, is a drop back to nearly the same degree of negative charge as there was before stimulation.

However, the neuron is not ionically the same as it was before being stimulated. There is much K$^+$ outside and Na$^+$ inside, just the reverse of the condition of the unstimulated neuron. At this point, the Na-K active transport pump acts to bring the stimulated region back to resting condition. The pump had operated continuously during the action potential, but had been overwhelmed by the diffusion of the two ions. Now, with the Na$^+$ channels again closed, there is effective pumping to get both ions back to their original concentrations. This recovery period is slower than the action potential, requiring about 5/1000 to 10/1000 of a second. During much of that time any restimulation attempt will fail to elicit a second action potential.

From this description, it can be seen that the action potential is an automatic series of events if a large enough stimulus is applied to get them started. It follows, then, that an action potential at a particular neuron will always be the same size (same number of millivolts of change), no matter how much more powerful the stimulus might be beyond the threshold level. How, then, does the body distinguish varying degrees of stimulation? How do we know that a finger has been pinched just a little on one occasion and very hard on another, if both cause each stimulated neuron to send the same sized action potential toward the brain? Two methods are used by organisms to distinguish sizes of stimuli. First, each neuron can vary the rate of sending its messages (action potentials). The brain can interpret a higher frequency as meaning a larger stimulus is being applied. Second, a larger stimulus probably activates more individual neurons in the area being affected. The brain counts the number of neurons simultaneously sending messages.

A Neuron Can Send a Wave of Action Potentials Along Its Length

All of the above describes a very localized response to a stimulus. Since the function of neurons is to send messages from place to place, it is necessary to move the message of the action potential along the length of a neuron. Although a single action potential has a very short lifetime (Fig. 10.6), it stimulates the adjacent membrane region before it disappears. That is, the response to an externally applied stimulus can itself become a stimulus for a new action potential further along the neuron. The new action potential can do the same thing to the next adjacent membrane section, and so on for the length of the membrane. The overall effect is as if the original action potential flowed continuously along the membrane. If the origin of stimulation is in the middle of a fiber (not a normal event), two waves will be generated, moving in opposite directions to the two ends of the neuron. The more likely thing to happen with a neuron in its natural setting is that a stimulus is applied at one tip of the neuron and the flow of action potentials is from there to the cell body to the other extreme end of the neuron.

The speed of transmission, although not nearly as high as that of electrons along a copper wire, is still impressive. Depending on the animal and the neuron type, transmission rate can be from a few centimeters per second to about 100 meters per second. Two factors influence transmission speed. One is that a larger fiber diameter increases speed. This is the only device available for invertebrate animals, where some species develop *giant axons* to carry messages quickly over long distances. The other device concerns the myelin sheath described earlier (Fig. 10.3). This tight wrapping of Schwann cell membranes around an axon provides an alternative route for transmission of a stimulus by a method called *saltatory* (jumping) *conduction*. Here, action potentials occur only at the tiny gaps between insulation units, the *nodes of Ranvier*. When an action potential forms at one node, its stimulating effect jumps over the myelin sheath (through the liquid on its surface) to the next node where it initiates a new action potential (Fig. 10.7). The neuron membrane between two nodes is not stimulated. Compared to the many regenerations of action potentials needed for continuous conduction along the entire neuron surface when no sheathing is present, this method is much faster. Only vertebrate animals have myelinated axons.

Neurons Are Connected to Each Other and to Effectors at Synapses

A nervous system requires communication between neurons. Adjacent neurons (end to end) do not usually quite touch each other, although their membranes are very close together. The region that includes the ends of two neurons plus the tiny gap between them is called a *synapse*. Since the two membranes are not in contact, it is impossible for an action potential of one

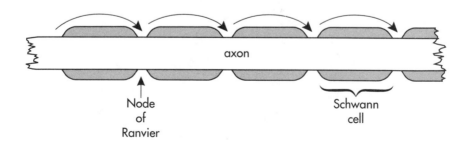

FIGURE 10.7. *Saltatory conduction of an impulse along a myelinated axon. Arrows indicate the route taken between nodes. Action potentials occur only at exposed regions (at the nodes).*

to stimulate directly the making of an action potential at the other. Rather, a purely chemical transmission occurs at the synapse.

At a synapse both neurons are quite specialized, one to send a chemical message and the other to respond (Fig. 10.8).

The synaptic end of a sending neuron is an area where a number of vesicles are stored, each containing molecules of the neurotransmitter molecule. The arrival of action potentials propagated along the fiber stimulates a migration of *synaptic vesicles* toward the *synaptic cleft* (the fluid-filled gap between neurons) and exocytosis of their contents into the cleft. Diffusion occurs throughout the narrow cleft, and neurotransmitter molecules soon make contact with the membrane of the second neuron. Here are located receptors attached to closed ion channels. When the proper stimulus (the attachment of a neurotransmitter molecule) is received, each receptor responds by causing its ion channel to open. The resulting flow of ions into the neuron triggers the

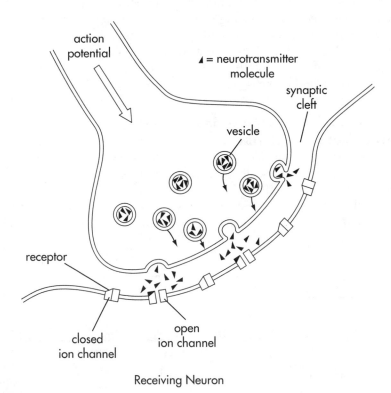

Receiving Neuron

FIGURE 10.8. *A typical synapse. The neuron portion at the top is equipped to make and release neurotransmtter; the receiving neuron's membrane has receptors to match the shape of arriving molecules of the neurotransmitter at ion channels.*

production of a new action potential, which can then travel over the surface of this neuron. Since this all hinges upon diffusion through an intercellular space (the cleft) and chemical recognitions, there is a significant time period between the extinguishing of one action potential and the beginning of a new one at the other neuron. If a pathway of connected neurons between two parts of the body has many synapses to be traversed, this significantly increases the total time needed to get the message sent.

A necessary feature of any synapse is some way to clear out already released neurotransmitter molecules, so that the stimulation will not continue after it should have ended. In some cases (such as when acetylcholine is the neurotransmitter) there is an enzyme in the synaptic cleft and at the receiving neuron's surface that specifically destroys the neurotransmitter of the synapse very soon after it is released. Some other synapses perform the same function of removal by having a mechanism for absorbing the neurotransmitter back into the cell that released it.

A feature of the synapses described above is that they transmit in only one direction. Only the sending end can make and release a neurotransmitter; only a receiving end is equipped with receptors and ion channels to respond appropriately. This one-way feature helps in the integration of the nervous system.

Well over 60 different neurotransmitters are known to exist, each made and released by a specific type of neuron. The first to be discovered, *acetylcholine*, is released at the specialized synapse between a motor neuron and an effector such as a muscle fiber. The muscle fiber's plasma membrane responds just as a neuron's would, generating an action potential that is propagated along the fiber surface. Acetylcholine is also used between neurons in the brain and elsewhere. It is an example of an *excitatory neurotransmitter*, since its action is to initiate an action potential. Another excitatory neurotransmitter is *norepinephrine*, released at synapses of the autonomic nervous system. An example of an *inhibitory neurotransmitter* is *gamma-amino butyric acid (GABA)*, released at certain synapses within the central nervous system. The action of this and other inhibitors is to increase the resting potential of the receiving neuron's membrane, making it more difficult to start an action potential even if a stimulus comes to it.

NERVES ARE BUNDLES OF PARALLEL NEURON FIBERS

Communication between the central nervous system and peripheral regions is in pathways called *nerves*. These are bundles of parallel fibers, either dendrites or axons, enclosed by connective tissues and serviced by the circulatory system. The cell bodies of these fibers are located in ganglia or the central nervous system.

Nerves may be classified into several categories. On the basis of function, they are *sensory, motor,* or *mixed.* Depending on where they are connected to the central nervous system, they may be *spinal* or *cranial* (the former entering the CNS at the spinal cord, the latter at the brain). There are 31 pairs of spinal nerves in humans, all of which are of the mixed type. There are 12 pairs of cranial nerves, which are totally sensory, totally motor, or mixed. For example, the first, the olfactory, is entirely sensory; the third, the oculomotor, is entirely motor; and the seventh, the facial, is mixed.

THE REFLEX ARC IS THE SIMPLEST USEFUL NERVOUS SYSTEM PATHWAY

The simplest kind of response of the nervous system is an unconscious one known as a *reflex.* It follows a pathway of message transmission, known as the *reflex arc,* from the place of stimulus reception to the place of effector reaction. The simplest reflex pathway is composed of a sensory neuron and a motor neuron, but most reflex arcs also include an association neuron connecting the other two within the CNS (Fig. 10.9).

Reflex arcs may have attached pathways that link with areas of conscious perception in the brain. With such connections, other responses may accompany (follow) the reflex. However, reflex arcs by themselves produce involuntary reactions to stimuli, such as the lifting of the arm when one's finger has encountered a painful stimulus. Of course, such a response is appropriate, even though it has not been preceded by conscious thought and decision-

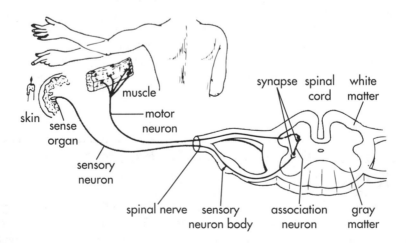

FIGURE 10.9. *The reflex arc.*

making. Reflexes regulate many vital activities such as heartbeat and secretion of enzymes. Many, like the one regulating heartbeat, have to be involuntary, as reflexes are, so that the effector can operate during sleep. Reflexes may be classified as *somatic* (their effectors are skeletal muscles) or *visceral* (having effectors that are the smooth muscles of the intestine and other areas).

THE SPINAL CORD IS A CENTER OF INVOLUNTARY COORDINATION

In a cross section of the spinal cord (Fig. 10. 9) one can see a butterfly-shaped area of *gray matter* surrounded by *white matter.* The gray matter contains a concentration of cell bodies, whereas the white matter consists of ensheathed axons and dendrites. Paired spinal nerves connect the spinal cord with the peripheral region.

Near the spine, spinal nerves have one trunk that contains both sensory and motor fibers. Toward the peripheral region they have branching connections with the tissues. Just before joining the spinal cord, the trunk of each nerve branches into two segments known as the *dorsal* and *ventral roots.* The dorsal root, the pathway of sensory neurons in mammals, has an enlargement, the *spinal ganglion*, which is a concentration of sensory cell bodies. These sensory cells are unipolar (Fig.10.3 B). The single extension from the cell body divides into two fibers, one a dendrite connected to the peripheral region, and the other an axon extending into the gray matter of the spinal cord. The ventral root is the pathway of motor neurons. It does not have a spinal ganglion because both its cell bodies and dendrites are in the gray matter of the spinal cord; therefore, the ventral root consists of axons of motor neurons. Unlike the sensory neurons, motor neurons are multipolar (Fig. 10.3 C). Association neurons connect the sensory and motor neurons in the gray matter. The reflex arc just described is sufficient for the reflex, but other association neurons have connections in the brain that relate the reflex to conscious events.

THE AUTONOMIC NERVOUS SYSTEM IS NOT USUALLY UNDER CONSCIOUS CONTROL

Visceral reflexes have pathways that are in what is described as the *autonomic nervous system*. It is in no sense an independent system but is a functional unit integrated with the central nervous system. Its effectors are in the eyes (controlling iris contraction), several internal organs, and blood vessels (causing vasoconstriction). Responses are both muscular and glandular. Sensory neurons of visceral reflexes carry their impulses into the brain or spinal cord. These do not usually have associations that lead to conscious activity. As in somatic reflexes, they are linked by association neurons to motor neurons.

Chains of no fewer than two autonomic motor neurons lead to the effectors. The body of the first neuron is in the brain or spinal cord; the second is in a ganglion outside the spine. Ganglia of the autonomic system occur parallel to the spine and are connected to the spinal nerves as well as to one another. Some of the ganglia (thoracolumbar) are close to the spinal cord and others (craniosacral) are near or within the organs they innervate.

Each effector organ served by the autonomic system (with the exception of sweat glands) is innervated by two antagonistic sets of motor neurons. One set of neurons is known as the *sympathetic system* and the other as the *parasympathetic system* (Fig. 10.10). The sympathetic system is centered in the thoracolumbar region and the parasympathetic in the craniosacral regions. One set may be an inhibitor and the other a stimulator; or one may be a dilator

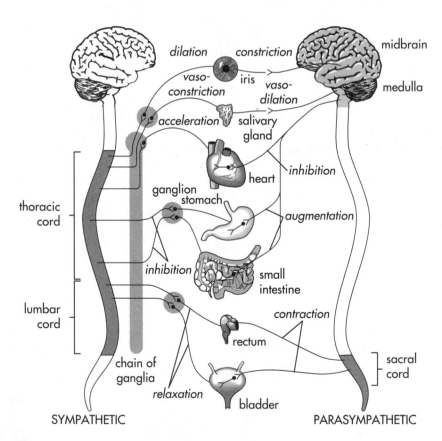

FIGURE 10.10. *The sympathetic and parasympathetic systems. (Adapted from* Biology, Its Principles and Implications *by Harden. Copyright © 1961 by Harden. Used with permission of W. H. Freeman and Company.)*

and the other a constrictor. While the two systems are antagonistic in any specific organ, they do not always have the same influence on all organs they innervate. For example, the sympathetic system excites the heart and relaxes the stomach.

INSTINCTIVE BEHAVIOR CAN BE BOTH INVARIANT AND COMPLEX

The term *instinct* is used to denote genetically determined (rather than learned) behavior that is more complex than any one of the simple reflexes described above. Indeed, some instinctive behavior is very complex, such as the building of a hive by honeybees or maternal behavior by a bird. Lower animals have numerous fascinating examples of unlearned responses, but even the most highly developed animals may rely on some instinctive activities.

Although the hallmark of instinct is the ability of the animal to perform it without being taught, this does not mean that instinct cannot be modified by learning. For instance, although nest building is instinctive to a bird, the quality of building may improve as a bird repeats its task the second and third times. Instinct and learning must often combine to get a particular job done. A solitary wasp has an instinctive ability to dig a burrow, into which she lays her egg. However, she must learn the geographical placement of this burrow, for her next task is to go off at a distance, capture and paralyze an insect, find her way back to the burrow and place the prey's paralyzed body in the burrow as food for the larva when it hatches.

The generally invariant quality of instinctive behavior suffices for an animal in nearly all situations. However, under unusual circumstances, instinct can lead an animal into trouble. For instance, there is an adaptive advantage in the instinctive behavior directing caterpillars to crawl upwards (an example of negative geotaxis). In their normal setting, caterpillars benefit by this behavior, always moving upward toward tender leaves at the tip of branches. However, if they are put in a vertical tube they will crawl to the top as directed by instinct and starve rather than go down to the bottom where food has been placed.

LEARNING IS THE PREDOMINANT BEHAVIOR OF MAMMALS

Although even the simplest animals (and even some Protozoa) can demonstrate some learning, the use of this is most developed in the vertebrate animals. *Learning* is formally defined as a relatively long-lasting change in behavior resulting from previous experience. Significant learning depends upon the ability to store memory in the nervous system, a function associated with certain regions of the brain. To demonstrate learning, an animal must be able to compare the present situation to the memories of similar past situations, choose the closest-matching memory, and then perform the appropriate response based upon that memory.

One kind of learning is *classical conditioning*. It is a learning experience involving the association of simultaneous events. One of the events is the stimulus of a natural reflex, the other an event repeated with it. The classic experiment that Pavlov performed with a dog is a good example. The dog, by nature, salivated (a simple reflex) when given food. The experimenter repeatedly rang a bell when food was given the dog. Finally, he could omit the food but make the dog salivate at the sound of a bell.

A form of learning that may be confined to the mammals is that which is called either *reasoning* or *insight*. In this variety, the animal can (at least sometimes) perform the correct response to a situation without relying on previous experience. That is, an animal might develop an innovative solution to a problem that it had never faced before. Another way to look at this is to say that the animal mentally considers many possible solutions to a problem and then chooses the one that seems to be most likely to work before physically attempting it.

Not even all mammals can use reasoning to solve certain problems. For instance, a dog tethered to a tree is often seen straining toward an object such as food that is just out of reach because the tether is wrapped around another tree and cannot extend to its full length. A person or chimpanzee in the same circumstance would quickly analyze the situation and walk away from the food, around the restraining tree, then successfully back to the food. A task for behavioral biologists is determining how many other mammalian species are truly capable of reasoning at even this simple level.

THE MAMMALIAN BRAIN CONTROLS MOST BODY FUNCTIONS

As one examines many animals, classifying them from simplest to most complex, one observes that animals with larger brains show a greater degree of brain control over the rest of the nervous system. This culminates in the mammals, in which even a small amount of damage to the brain is likely to be life threatening.

The largest part of the human brain is the *cerebrum* (Fig. 10.11), which overcaps the rest of the brain. It is much convoluted and divided into right and left hemispheres. Gray matter is on the outside and covers a zone of white matter beneath. It functions in such activity as learning, memory, consciousness, and language. The cerebrum also functions in controlling the voluntary motor functions, such as willed body movement.

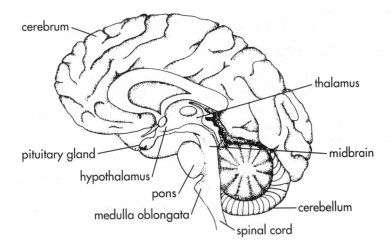

FIGURE 10.11. *The human brain.*

The next largest part of the brain is the *cerebellum*, which is underneath the cerebrum, near the spinal cord. It consists of two hemispheres and is very much wrinkled. It coordinates senses related to locomotion. Damage to this area leads to loss of fine movement control although gross, jerky movement is still possible. Ventral to the cerebellum is the *pons* (bridge), which has connections with the two hemispheres of the cerebellum.

The *midbrain* region is a center for certain vision and hearing functions. For instance, blinking, pupil response to light conditions, and adjustment of the ear to sound volume are controlled here.

The *thalamus*, located just above the midbrain, connects most sensory input neurons to the cerebrum. As it provides this relay some interpretation occurs so that the various signals can be sent to proper portions of the cerebrum for further analysis.

The *hypothalamus*, quite close to the thalamus, is a major center for homeostatic mechanisms. Located here are sensory cells for temperature, blood pressure, blood glucose level, blood salt concentration, and a number of other conditions. It is active in the experiencing of some strong emotions such as rage and pleasure. It is the target for some addictive drugs, such as cocaine. It releases hormones that affect the pituitary (see below), thus indirectly exerting control over a number of peripheral endocrine glands.

The *medulla* (or *medulla oblongata*) is the part of the brain merging into the spinal cord. Transmission pathways to and from the higher centers pass through it, and it is the center of autonomic reflex control. Among the reflexes centered here are regulation of breathing, heartbeat, blood pressure, temperature, visceral glands, and muscular action of much of the digestive tract including swallowing, coughing, and vomiting.

MUSCLES ARE THE PRIMARY EFFECTORS OF AN ANIMAL'S BODY

Having examined the nature of nerve messages in animal bodies, we now turn to how effectors respond to those messages. Although various glands respond to nerves, the most usual effectors are muscles of various types. Thus, it behooves us to examine the structure and responsive movements of muscles in some detail.

The characteristics of three types of muscles found in higher animals were given in Chapter 4 (see Fig. 4.15). The variety that is under voluntary control and is used to move the body by pulling on bones is the one called skeletal muscle. As we examine how muscles contract, this type will be our primary object of focus.

A Muscle Is Composed of Smaller Units Called Fibers

The size of a muscle is determined by the number of its functional units, *muscle fibers*. Each fiber is an elongated structure (Fig. 10.12) that was formed embryologically by the joining together of a number of cells. Each cell lost its plasma membrane boundary with its neighbors but retained its nucleus. The general name for such a multicellular functional unit is *syncytium*; its unified anatomy helps it to react in a single, coordinated action. The plasma membrane surrounding each muscle fiber is called the *sarcolemma*; it is this membrane that receives a neuron's message by being part of a synapse called the *neuromuscular junction*. The response of a muscle fiber to a neuron's stimulation is to shorten its length; i.e., to contract. The number of fibers contracting in a muscle determines the degree of contracting that the entire muscle undergoes.

Close examination of a single fiber reveals much internal detail. The fiber is a bundle of long units called *myofibrils* (Fig. 10.12). Skeletal muscle is also called striated muscle because each myofibril has repeating dark and light striations (bands). Dark regions appear that way because they are areas of more densely packed materials. These materials are *myofilaments* composed of proteins of two sorts. Thin myofilaments are composed mostly of actin; the thicker ones are made of *myosin* (Fig. 10.13). In a relaxed myofibril (not yet stimulated to contract), there is partial overlap of myosin and actin filaments. Actin filaments are attached to cross-structures called *Z lines*. The pattern repeats along the length of a myofibril, with Z lines acting as boundaries of repeating units called *sarcomeres*.

A Sarcomere Contracts by Sliding Actin and Myosin Filaments Over Each Other

A contraction event is, in essence, a sliding movement of thin filaments across thick filaments which, in turn, causes Z lines at the edges of each sar-

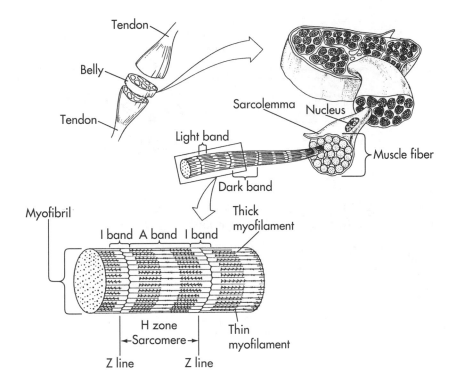

FIGURE 10.12. *A muscle is composed of a number of muscle fibers, each of which is composed of many myofibrils. The pattern of overlap between thick and thin myofilaments within each myofibril produces a repeating pattern of dark and light bands. (Adapted from Mitchell et al.,* Zoology, *Benjamin/Cummings, 1988.)*

comere to be pulled closer together (Fig. 10.14). This is the result of a sequence of events at the heads of myosin molecules. Using energy derived from ATP, each myosin head attaches to a specific part of the nearby thin filament, pulls it, and then releases to repeat the process a little farther along the filament. Many myosin heads exist on each myosin filament, so the progress gained by each head is not lost by back-movement of the thin filament as each head releases and reattaches for another pull.

Contraction Is Controlled by the Presence or Absence of Calcium

All of the structures described above are constantly present in a muscle fiber, as is ATP. What, therefore, triggers contraction? What signals the end of a contraction event? The key component initiating the sliding of the thin fila-

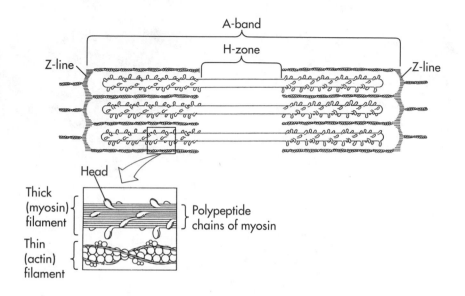

A-band

H-zone

Z-line

Z-line

Head

Thick (myosin) filament

Polypeptide chains of myosin

Thin (actin) filament

FIGURE 10.13. *Close views of the myofilaments within a sarcomere. The protruding heads of individual myosin molecules can interact with actin molecules of the thin filaments. (Adapted from Mitchell et al.,* Zoology, *Benjamin/Cummings, 1988.)*

ment across the thicker myosin filament is calcium ions (Ca^{++}). If they flood into the area of a sarcomere, they interact with the thin filament, changing it so that it can form connections with projecting myosin heads. This, in turn, makes it possible for nearby ATP to be broken down and release energy for movement. Conversely, removal of calcium causes loss of this bridge and cessation of contraction.

The control of calcium presence occurs at a complicated network of membranous bags that closely overlie the myofibrils (Fig. 10.15). Called the *sarcoplasmic reticulum*, it can retain calcium ions within itself by active transport, thus keeping contractions from happening. However, if stimulated, the sarcoplasmic reticulum becomes very permeable to calcium because specific membrane channels open. Calcium pours out by diffusion and bathes the sarcomeres.

A Motor Neuron Initiates Contraction

We come now to how the contraction events within a muscle fiber can be controlled to allow body coordination. As described above, contraction occurs only when calcium channels open at the sarcoplasmic reticulum deep within

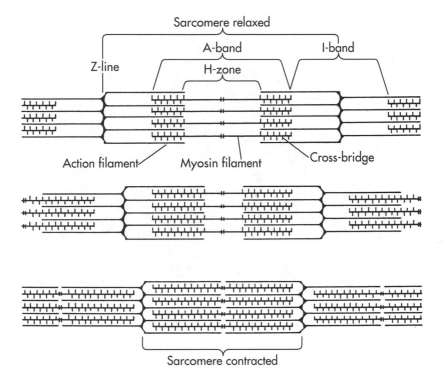

FIGURE 10.14. *The relation of myosin and actin filaments in a sarcomere that is relaxed, then contracted. As actin filaments are pulled across myosin filaments, attached Z lines are pulled closer together. (From Mitchell et al., Zoology, Benjamin/Cummings, 1988.)*

a muscle fiber. It happens in response to the arrival of action potentials at that place. Where do these action potentials come from? To understand this, one must begin by examining the connection between the tip of a motor neuron and the surface of the muscle fiber. At the synapse formed there, a neurotransmitter is released to stimulate the sarcolemma. A sarcolemma can respond to arrival of neurotransmitter molecules just as if it were the membrane of a neuron: it begins and propagates its own action potential. Scattered over the sarcolemma are many inward extensions called *T (transverse) tubules* (Fig. 10.15). These are continuous with the sarcolemma, so each action potential dives down the T tubules into the interior of the fiber. There, since T tubules lie directly beside portions of the sarcoplasmic reticulum, action potentials can influence its calcium channels and stimulate the release of calcium to bathe the sarcomeres.

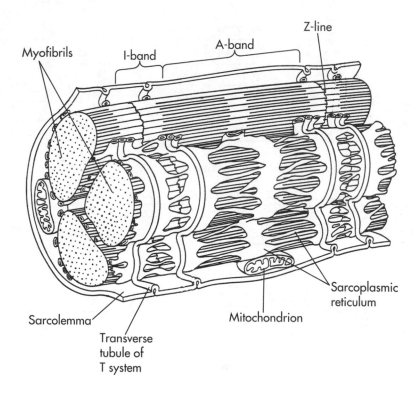

FIGURE 10.15. *The sarcoplasmic reticulum overlies myofibrils, controlling the presence of Ca⁺⁺ at the sarcomeres. (Adapted from Mitchell et al.,* Zoology, *Benjamin/Cummings, 1988.)*

HORMONES ALSO ACT TO COORDINATE BODY ACTIVITIES

The nervous system is not alone in coordinating and influencing activities of the animal body; it has an ally in a variety of potent chemical secretions known as *hormones* that produce striking effects. A hormone is a chemical that is secreted from one cell type, travels to one or more other cell types via body fluids, and has specific effect(s) upon the target cells for the function of coordinating body activities.

Since a hormone's movement to target cells depends upon both diffusion and the bulk flow of carrying fluids, this system operates much more slowly than the nervous system. In most animals, the medium of movement is circulating blood. Very little hormone is needed for some target cells to begin responding, so an effect might begin within a few seconds of initial release of certain hormones. However, a neuron sending a message from one end of a large mammal to another would be on the order of several hundred times faster.

Actually, a neurotransmitter released at a synapse within a nervous system is very like a hormone: made in one cell, diffusing through fluid to another cell, and causing a specific effect upon reaching the target. Two differences exist: neurotransmitters have to move only over extremely short distances (entirely within a synaptic cleft) and they do not circulate through the entire body since they do not enter the bloodstream.

Many, but not all, animal hormones are made in and secreted from epithelial tissues called *endocrine glands*. These tissues release their products into blood vessels. (Another variety, the *exocrine glands*, release nonhormone materials only to specific areas via ducts. For instance, the salivary glands release saliva into the mouth by way of ducts.) Since endocrine glands are intimately involved in many hormonal actions, the field of their study is called *endocrinology*.

Two Types of Animal Hormones Work Differently at Their Target Cells

Hormones must be placed into two structural categories when the discussion centers around how they have their effect in target cells. The hormones that are steroids (such as testosterone, progesterone, and estrogen) can move easily right through the plasma membrane since they are lipids. After entering a cell, a steroid hormone moves to the nucleus, where there can be a receptor protein that matches it and binds to it. The hormone-receptor combination has the proper shape to attach to a particular portion of the nucleus's DNA. The attachment is to an area of DNA that acts as a switch for activating one or more genes, causing them to release their messages and make the cell's response.

The other hormones are, of course, nonsteroids. All that have been discovered are either single amino acids (sometimes modified to the form called amines) or polypeptides of several to thousands of amino acids. They cannot enter a target cell, being blocked by the plasma membrane. Nevertheless, they have effects within the cell, so there must be some way that they exert their influence without entering. This is accomplished by use of a *second messenger* that relays the message of the hormone. The first step is attachment of the hormone to a specific receptor protein that is embedded in the plasma membrane's outer surface. The act of attachment triggers a series of events at the membrane that culminates in activation of an enzyme. The enzyme catalyzes the synthesis of a second messenger within the cell's cytoplasm. The second messenger is then responsible for one or more changes within the cell, either at the gene level (to make new gene products) or upon gene products already present. Several second messengers have been found in various target cells, but the most understood one is *cyclic AMP*, a variant of the nucleotide adenosine monophosphate.

With both varieties of hormone, it is necessary to explain how some body cells are responsive as target cells for a particular hormone and others are not, since a hormone moves throughout the blood circulatory system and therefore contacts virtually all cells. The key is the presence or absence of receptor molecules (within the nucleus for steroid hormones and on the cell surface for nonsteroid hormones). As each cell differentiates during development, part of its specialization can be the placement of such receptor proteins. A single cell can have several different receptor types, each capable of responding to a particular hormone.

Hormones Control a Wide Variety of Animal Activities

Nearly all animals rely upon hormonal coordination for the sorts of changes that do not have to be immediate. Here is a small sampler to show the variety of hormonally controlled events in nonhuman animals.

—Metamorphosis in insects and other Arthropods, such as change from caterpillar to butterfly.
—Metamorphosis in amphibians, such as transition from tadpole to adult frog.
—Color change in insects and other Arthropods, matching the environment's color.
—Shell growth in Molluscs.
—Sexual maturation and reproductive timing in many animals.

Most Human Endocrine Glands Are Controlled by the Hypothalamus and Pituitary

Humans (and all mammals) rely upon a wide variety of hormones for some vital coordinating functions. The following paragraphs introduce some of the more important systems, but are not intended to be comprehensive.

Arguably the most important gland of the body is the *pituitary*, located at the base of the brain just below (and physically connected to) the hypothalamus (Figs. 10.11 and 10.16). It is a master gland in that it controls the hormonal output of other glands. However, the pituitary itself is under the control of the nearby hypothalamus. The hypothalamus secretes chemicals generally called *releasing hormones*, which travel through tiny blood vessels to the pituitary and stimulate it. The hypothalamus can also exert a negative influence on the pituitary via *release-inhibiting hormones*.

A compound gland originating in the embryo from both a part of the brain and a part of the mouth cavity, the pituitary secretes at least nine hormones. It consists of three anatomical regions: anterior, intermediate, and posterior. (The intermediate part is a thin zone of cells whose functions are not well understood). The *anterior lobe*, originating from the roof of the embryonic mouth, produces hormones that activate the thyroid, adrenals, gonads, and mammary glands (breasts). They are known as *thyroid-stimulating hormone, adrenocorticotropic, gonadotropic hormone,* and *prolactin*, respectively. It also produces a hormone with effects at many places in the body, *growth hormone.* Hypersecretion of this hormone results in gigantism if it occurs during the normal growth period, or acromegaly (disproportionate growth of only

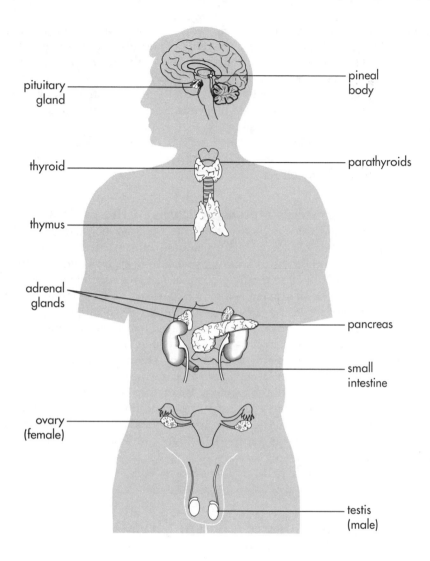

FIGURE 10.16. *Locations of endocrine glands in humans.*

some body parts) if it occurs in adolescence or adulthood. Hormone release from the anterior lobe of the pituitary is under control of the hypothalamus.

The posterior lobe, of neural origin, produces *oxytocin* and *antidiuretic hormone* (also called vasopressin). The former causes contraction of smooth muscles, especially the uterus, and is important in childbirth. It is sometimes administered to induce or speed up labor. It also is the trigger for breast milk *release* during infant suckling. (The anterior pituitary hormone prolactin stimulates the *production* of milk.) The other posterior lobe hormone, antidiuretic hormone, raises blood volume by increasing reabsorption of water in the region of kidney nephrons (Chapter 9). Both of these hormones are actually manufactured by cells of the hypothalamus. They travel to the posterior pituitary in vesicles, moving through neurons that connect the hypothalamus to the lobe. When these neurons are stimulated from the hypothalamus, they release the contents of the vesicles stored at their tips. Such hormones are sometimes called *neurohormones*.

Pituitary hormone release is under the tight control of feedback systems, with constant monitoring of their levels (or the levels of the hormones released by peripheral endocrine glands).

The Thyroid Controls Metabolic Rates of the Body's Cells

The *thyroid* gland is located close to the trachea just below the larynx. It consists of two lobes connected by an isthmus, which makes it roughly H-shaped. The gland originates as an outgrowth of the digestive tract but loses all connections with it before the time of birth. It secretes two hormones, *thyroxine* and *calcitonin*. Thyroxine is a modified amino acid containing iodine. It functions to increase the speed of metabolism in nearly all body cells. Hypothyroidism decreases metabolism to the extent that it causes physical and mental sluggishness, a tendency toward obesity, and an alteration of skin texture. When thyroxine deficiency occurs in early childhood, it results in mental retardation, small size, and inability to become sexually mature. Since iodine is a part of the thyroxine molecule, it cannot be produced in sufficient quantities when the element is deficient. The gland enlarges as if to compensate and becomes a goiter. Hyperthyroidism increases metabolism, thereby causing greater consumption of food, energetic muscular activity, loss of weight, insomnia, and bulging eyes.

Calcitonin, the other thyroid secretion, functions in relation to the parathyroid glands, discussed next.

The Parathyroids Control Calcium Levels

There are usually four small *parathyroids* located on the surface or imbedded in the tissue of the thyroid. This intimate association sometimes complicates surgery when the thyroid has to be removed. The glands secrete *parathyroid hormone*, which causes release of calcium from bones and increases its retention at the kidneys. A deficiency of this hormone results in excess deposition of calcium in the bones and removal of calcium from the blood and tissues. This, in turn, increases the sensitivity of muscles so that they overrespond in cramps, twitching, or convulsive activity. Unless parathyroid hormone is supplied, death quickly follows. Oversecretion produces the opposite effect of taking calcium from the bone and increasing it in blood and tissues. When this happens, bones soften and muscles become much less responsive. The thyroid hormone calcitonin acts antagonistically to parathyroid hormone, inhibiting removal of calcium from bones if too much is sensed in blood.

The Adrenals Respond to Stress

The *adrenal glands* are compound glands located anterior to the kidneys. Each has an outer zone known as the *adrenal cortex* and an inner one known as the *adrenal medulla*. The cortex secretes several hormones that are steroids manufactured from cholesterol. Their combined action controls sodium and potassium levels in blood by promoting their reabsorption at nephrons of the kidneys. The control of these ions, in turn, helps determine blood volume (via osmotic balance) and blood pressure. Deficiency of cortical hormones results in a syndrome known as Addison's disease. The patient suffers a general decline in muscular strength, a lowering of blood pressure, disturbance of digestion, and a bronzing of the skin.

The adrenal medulla secretes two related hormones called *epinephrine* (also known as adrenaline) and *norepinephrine* (noradrenaline), which are modified from amino acids. Norepinephrine, it will be remembered, is also a neurotransmitter at synapses of the autonimic nervous system. They are released in accelerated quantities when one is stimulated by anger or fear. Their influence of epinephrine on the conversion of glycogen to glucose, increase of metabolic rate, increase of heart rate, enrichment of blood supply to muscles, and several related changes is useful in preparing the body to cope with stress or injury. Either of these hormones can be artificially administered in certain medical situations; their stimulating action makes them useful to restart the heart after it stops or to increase blood pressure. Their relaxing action on smooth muscles makes them useful in the treatment of asthma.

A Portion of the Pancreas Controls Carbohydrate Metabolism

In addition to being a digestive gland (an exocrine function), the pancreas produces two hormones (an endocrine function). Both hormones are secreted by patches of cells known as *islets of Langerhans*. In each islet there are some cells called *alpha cells* and others called *beta cells*. Alpha cells produce the hormone *glucagon*, which influences the change of glycogen in the liver to free glucose molecules for distribution throughout the body. It also acts in the liver to produce glucose from other stored materials including amino acids and fatty acids. Thus, its overall effect is to increase blood glucose.

The second pancreas hormone, made by beta cells, is *insulin*. This acts to remove glucose from blood by changing the permeability of all body cells and allowing glucose to enter them for metabolic use. Insulin also inhibits the conversion of amino acids and fatty acids into glucose. Thus, glucagon and insulin act as mutual antagonists with reference to the concentration of glucose in blood. Neither alpha nor beta cells are controlled directly by the hypothalamus-pituitary axis; rather, they independently monitor blood glucose level and respond accordingly.

When there is an insulin deficiency, glucose increases in the bloodstream. This condition, with associated symptoms, is the disease *diabetes mellitus*. In extreme cases of deficiency, a coma results. The opposite extreme also produces a coma known as insulin shock. This state may accidentally occur in those who take insulin injections for diabetes. Type I diabetes, also called insulin-dependent diabetes, results from a decrease of beta cells in the pancreas, probably caused by an autoimmune attack upon them. It is treated by artificially supplying insulin via injections. Type II diabetes, also called noninsulin-dependent diabetes, is the much more common form of the disease, accounting for about 90 percent of all cases. Type II is caused by loss of insulin receptors on target cell surfaces, not by loss of beta cells or lack of ability to produce insulin. Consequently, insulin injection will not help this variety. Type II diabetes usually occurs later in life than Type I, and is much more likely to happen if one is obese. It can usually be controlled by changing the diet and losing weight.

Reproductive Glands Make a Variety of Hormones

Both *ovaries* and *testes* produce a complex of hormones to sustain sexual processes vital to reproduction and secondary characteristics that play supporting roles. Let us examine the condition in females first. The ovaries are covered with an epithelium that generates *follicles* within the tissue beneath. A follicle is a combination of a developing egg cell and a cluster of supporting and nourishing cells around it. The follicle cells around each egg are responsible for producing and releasing two types of steroid hormones known as *estrogen* and *progesterone*.

The anterior pituitary controls the production of these hormones and release of eggs from the ovary. It does so by releasing *follicle stimulating hormone (FSH)*, which stimulates the maturing of a follicle. The follicle cells, in turn, gradually increase their production of *estradiol*, which is the primary estrogen. Its general presence is responsible for the onset of sexual changes at puberty and its absence for the regressive changes at menopause. It therefore maintains feminine characteristics. It also acts on the uterus wall on a repetitive monthly basis, causing it to thicken with a rich supply of blood vessels in preparation for supporting an embryo. During the first half of each menstrual cycle, estradiol is produced in increasing amounts and reaches peak quantities just before the time of ovulation. This large output of estradiol causes the pituitary to release a spike of a second hormone, *luteinizing hormone (LH)*. The message of that hormone is to release a mature egg from the follicle, a process called *ovulation*. Thus, the combined effects of FSH, estradiol, and LH in the first half of the monthly reproductive cycle are to cause maturation and release of an egg plus the preparation of the uterus to nurture an embryo that will be produced if that egg is fertilized.

Cells at the follicle change after the egg is released from it. They fill the follicular space by rapid proliferation and are collectively known as the *corpus luteum*. The corpus luteum secretes high levels of both progesterone and estradiol. Together, these inhibit the further release of FSH and LH, thus temporarily holding back the development of any other follicles. Progesterone and estradiol also continue the preparation of the uterine wall for supporting a pregnancy.

If a pregnancy occurs, some cells of the embryo itself begin to send signals back to the ovaries. The signal is in the form of yet another hormone, *human chorionic gonadotropin (HCG)*, whose effect is to maintain the presence of a corpus luteum. This, in turn, continues to support the uterine wall, keeping it from breaking down as a menstrual flow of blood and tissue from the body. If no pregnancy occurs, the degeneration of the corpus luteum causes a drop in progesterone level, followed by menstruation and a repeat of the entire cycle.

In the male the primary sex-determining and controlling hormone is *testosterone*, made in the paired testes. This hormone is produced by *interstitial cells* located among the seminiferous tubules whose walls produce sperm cells. Testosterone has multiple roles. Before birth, its presence directs the development of testes and supporting tubes. If it is absent then, these same tissues develop into ovaries and their accessory structures. At puberty, testosterone is responsible for the final development of male reproductive structures and of secondary sexual characteristics such as overall body growth, prominent facial hair, and deepening of the voice. From then on, testosterone's role is maintenance of these characteristics and stimulation of the production of sperm cells.

Other Hormones Exist in the Human Body

Although it could be argued that the hormones described above have the most profound activities within the body, there are quite a few others with somewhat more circumscribed functions. One is *thymosin*, produced by the *thymus gland* located just below the thyroid in the lower neck and upper thorax. It will be remembered that the thymus is essential in the maturation of the T cells of the cellular immune system (Chapter 8). The specific role of the several polypeptides that are together called thymosin is still obscure.

Several hormones help coordinate the activities of the stomach and small intestine. *Gastrin* is produced by stomach cells when stimulated by the entry of food, and its action is to induce the secretion of hydrochloric acid into the lumen of the stomach. As partially digested food reaches the small intestine, its acidity triggers the release of the hormone *secretin*, which acts on the pancreas to cause a release of alkaline (basic) liquids into the small intestine. This, of course, brings the food back to nearly neutral pH. Simultaneously, secretin increases the rate of bile production in the liver. Fatty acids in the food at the small intestine induce the release of another hormone, *cholecystokinin*, from intestinal cells. This acts on the gallbladder, causing it to contract and squirt stored bile into the intestine. It also stimulates the pancreas to release its several digestive enzymes. All of these hormones act to ensure that the proper conditions and digestive aids will be placed into the intestinal tract only when there is food present.

A rather enigmatic gland is the *pineal*, which is located in the brain, posterior to the thalamus. On a daily cycle, it secretes the hormone *melatonin*.

Nearby brain regions receive information about light conditions around the body and seem to communicate this information to the pineal body to regulate the release of melatonin. The hormone may be involved in coordinating the body functions that depend on the 24-hour-day cycle. Melatonin may also be involved in triggering the onset of puberty.

PHEROMONES ARE COORDINATING CHEMICALS SENT BETWEEN ANIMALS

There are known to a number of animal secretions that strongly influence other members of the same species in such ways as attracting mates, marking territories and routes, alerting with alarm signals, limiting populations by inhibiting reproduction, and regulating social systems. Known as *pheromones*, these substances are mentioned with hormones only because they are secretions that influence other members of a species as hormones influence hormone-sensitive cells within individuals.

HORMONE-LIKE CHEMICALS PROVIDE COMMUNICATION AND COORDINATION IN PLANTS

Plants have no nervous systems for receiving and coordinating responses to changes in the environment. Nevertheless, they have their own way of receiving stimuli and of reacting to them. Since plants have no muscles and their cells are mostly bounded by walls, there can be no sudden contraction of parts to bring about movement. Plant responses, then, are confined to much slower movements produced by growth or by variations in water content of cells. Since plants do not grow throughout their entire bodies as do animals, hormone effects are typically observable only in the growing parts.

Some plant physiologists object to using the word hormone in conjunction with the various chemicals to be discussed below. This is based on the fact that they do not seem to act upon target cells by altering gene settings or changing biochemical reactions, as animal hormones do. Since the plant analogs have most of their effect by influencing the rate of cell division, the direction of cell division (or direction of cell elongation after division), or by causing the aging and death of cells, the preferred term for them might be plant growth regulators. Whatever one calls them, there seem to be five major types of these chemicals. It will be seen that several of these interact to produce effects, and that the plant seems to measure their relative concentrations as a cue for responding.

Plants React to the Environment with Specific Growth Patterns

Growth responses may be toward or away from the direction of the environmental stimulus. As introduced early in this chapter, directional responses of this type are called *tropisms*. Prefixes to the word tropism indicate the nature of the stimulus. Examples are geotropism (gravity), phototropism (light), thigmotropism (touch), and hydrotropism (water).

Roots grow down in positive response to gravity while stems grow up in negative response to gravity. Seeds may be sprouted in any position, but roots and stems orient themselves very quickly.

Many plants conspicuously turn their leaves or flowers in definite positions with respect to the sun. The sunflower is so named because the head faces the sun. Equally sensitive, if not more so, is the pennywort (genus *Hydrocotyle*), a small herb frequently found in large, almost pure stands in some eastern coastal areas of the United States. The degree of its response is seldom equaled by any other plant, for every leaf blade in entire stands of the plant follows the sun.

Contact responses are characteristic of climbing plants. The growth of climbing roots, in English ivy for instance, is probably a contact response. Many plants have tendrils that help support them. They are always thigmotropic. Well-known examples of tendril-bearers are peas, vetch, and grapes.

Directional growth responses of roots to water are known to exist but are not as common as might be expected. True, roots may be more abundant in soil well supplied with water when compared to dry soil, but the case is one of providing a more favorable environment for growth. To be tropistic, roots must actually bend in the direction where more water is available.

Auxins Induce Cell Elongation and Other Responses

Phototropism, as described above, is caused by unequal cell elongation on two sides of a stem. Specifically, if the cells on the side away from light elongate more than those nearest the light, the effect is a bending of the stem toward the light (Fig. 10.17). A class of organic chemicals called *auxins* is responsible for greater elongation on one side. The most active auxin is called *indoleacetic acid*, or *IAA*. It is made in shoot tips, young leaves, and seeds. IAA quickly migrates from cells being hit by light toward the darker area of a stem. This migration is not through the vascular system, but cell-to-cell by an active transport mechanism. Upon arriving at darker cells, the effect of IAA is to make their cell walls more flexible so that they can stretch in one dimension.

Auxins have other effects too. Artificially adding IAA to a root leads to significant increase in its elongation rate. But this is a concentration-dependent activity: very high concentration of IAA at a root actually inhibits its elongation. By contrast, similarly high concentrations applied to stems lead to their elongation. In some plants, the auxins that form in seeds help stimulate the development of the fruit surrounding the seeds. Some plants, such as toma-

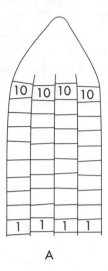
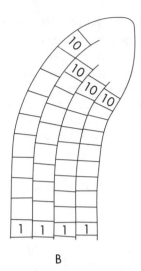

A   B

FIGURE 10.17. *Schematic drawing of the stem-bending effect of differential cell elongation. A: a stem in which all cells are of approximately the same vertical
length. B. the same stem after elongation of the cells in the rows on the left
of the stem. The resultant bending to the right is equivalent to the effect
caused when auxin migrates to the left side of a stem under influence of
light hitting the right side.*

toes and cucumbers, can actually be induced to produce fruits without first
being fertilized, just by spraying with auxin.

Cytokinins Induce Cell Divisions

A second category of plant growth regulators is that of the *cytokinins*,
which act to stimulate the production of new cells. They are manufactured in
roots, embryos, and fruits. It is believed that an interaction between cytokinins and auxins helps determine the pattern of a plant's shape: whether it
grows straight up or branches into a rounded shape. The general rule seems
to be that cytokinins moving from the roots of a plant induce much growth of
side stems, but only if the concentration of cytokinin is relatively higher than
that of auxin emanating from the main shoot tip. As a plant grows larger, with
proportionally more cytokinin-producing roots, it begins to branch out rather
than growing straight up.

Gibberellins Have Multiple Effects

Gibberellins comprise a class of over 70 compounds, all of which stimulate
stem elongation. They interact with auxins in subtle ways. They also enhance
the size of fruits and help seeds break dormancy and begin to develop into

adult plants. In the latter capacity, they probably act as the link between specific environmental cues (such as warmth, light, humidity) and the initiation of new biochemical activities needed to start seedling growth. This is shown by the ability of artificially applied gibberellins to start embryonic development even if those cues are not received.

Abscisic Acid Sends Negative (Inhibitory) Signals

Abscisic acid's effects all seem to be inhibitory rather than stimulatory. It is necessary to have such a material in a plant, because there are times and situations in which it is in the plant's interest to go into dormancy. Abscisic acid is responsible for stopping the growth of buds as winter approaches. It also influences the development of a protective covering over the buds at this time. Abscisic acid in seeds helps keep them dormant until arrival of favorable conditions for seedling development. In this context, abscisic acid and cytokinin are antagonists. That is, only a high concentration of the latter will overcome the inhibition of the former. The same sort of balance occurs at buds, where the positive effect of a large amount of gibberellin negates the inhibitory effect of abscisic acid when it is time to break dormancy and resume growth. Abscisic acid also acts to close the stomata of leaves whenever the plant is under the stress of too much water loss. The hormone was given its name in the mistaken belief that it is responsible for the loss of leaves (abscission) as cold weather approaches. That function is now known to be performed by the next hormone to be described, ethylene.

Ethylene Acts to Promote Necessary Aging Changes

Some portions of a plant must undergo degenerative changes in the normal course of the organism's life. Such changes include the ripening of fruit and the periodic loss of leaves. The hormone triggering some of these changes is *ethylene*, the only one that is a gas. Ethylene is made in fruit and diffuses throughout the interior by moving through air-filled spaces between cells. Its effect in fruit is to cause it to soften and (sometimes) become sweeter and thus more attractive for animals to eat. (Of course, this benefits the plant, since animals carry off the fruit and thus disperse the seeds buried within it.) Ethylene, being gaseous, also diffuses out of each fruit. Grocers wanting to avoid early ripening of their produce can periodically flush ethylene from stored fruit by adding CO_2 to the storage bin. Conversely, if it becomes desirable to hasten ripening, one can place several fruits in a tightly sealed bag to retain the emitted ethylene in high concentration.

Ethylene is also responsible for leaf drop from deciduous trees as cold weather appears in fall. This is caused by *abscission,* a weakening of the stalk at the junction between leaf and stem. Microscopic examination of this region shows that the specific effect of ethylene is to thin and weaken the walls of parenchyma cells in a narrow strip across the stalk. When wind blows, the pressure of movement causes the stalk to break at that point. As days shorten and environmental temperature drops, an antagonistic interaction between ethylene and auxin shifts in favor of the ethylene and the abscission changes occur. Leaf loss is an adaptive strategy for these plants. During winter they lose the ability to gain water from frozen soil and thus benefit by also losing the regions where transpirational water loss occurs—the stomata of leaves.

Plant Reproduction (Flowering) Is Linked to an Internal Clock

A critical change in a plant's life is the series of complex developmental activities that culminate in the building of a flower and its reproductive tissues. In many plants, this occurs at a particular season of the year, determined largely by the ratio of darkness to light during a 24-hour period. Use of this environmental cue by a plant requires that it have some method of counting; that is, it needs an internal clock.

Certain plants, such as violets that bloom in the early spring or asters that bloom in the late fall when the days are short, are called *short-day* (or *long-night*) plants; others, like the morning glories that bloom only in the summer when the days are long, are called *long-day* (or *short-night*) plants. Still other plants, like the tomato, are indifferent to the length of day and bloom continuously through the growing season.

The relation between the blooming of plants and length of exposure to light is *photoperiodism*. When the phenomenon was first discovered in the 1920s, it was believed that such plants counted daylight hours. More recent studies indicate that the supposed relationship to light was a case of getting the cart before the horse, that it is really the length of uninterrupted darkness that is important rather than the length of light. This can be shown by interrupting periods of darkness with quick bursts of light striking a plant, or vice versa. If it is a short-day plant, the interruption of its long night retards the initiation of flower development just as if it had experienced a short night and long day. In contrast, short periods of darkness interjected in the middle of a long light exposure do not retard flower development in long-day plants. Thus, both types of plant seem to be counting periods of uninterrupted darkness, rather than of light.

Although much work has been done to determine how a photoperiodic plant gets the job done, much of the clock mechanism is still a mystery. A protein called *phytochrome* has been much examined in this context, since it is capable of changing its form as it goes from darkness to light. However, it is now believed that phytochrome is involved in setting this clock each day, but is not the primary counting molecule. Total understanding of the clock mechanism in such plants would be of much interest beyond the realm of botany, since organisms of all kingdoms show the ability to change their activities based on counting the passage of time. For instance, humans have many biochemical and physiological actions recurring on 24-hour, monthly, and other cycles. Our understanding of sleep disorders, jet lag, and menstrual cycles would be enhanced if we understood how our bodies count time within cells and tissues.

STUDY QUESTIONS

1. What are the two most common effector types in animals?

2. What is a ganglion? How is it similar to a brain? How is it different?

3. Distinguish between axons and dendrites. Which type stimulates muscle fibers?

4. What are the roles of Na^+ and K^+ channels in establishing a resting potential? In initiating an action potential?

5. What kind of nervous system is found in the Cnidarian animals? How do animals with this kind of system respond to stimuli?

6. Describe the anatomy of a typical neuron.

7. How can the brain receive varying information of stimulus strength, if an action potential is always the same size?

8. What are the dual roles of the myelin sheath? How is this sheath related to Schwann cells?

9. How is it ensured that a synapse sends messages in only one direction across itself?

10. What are neurohormones? Where are they produced? What function do they perform?

11. How are nerves different from neurons?

12. "Sympathetic and parasympathetic nerve systems are antagonistic to each other." What does this mean?

13. Trace an action potential from the central nervous system to the place where it stimulates contraction of a sarcomere, describing what happens at each place.

14. How does an instinct differ from a simple reflex? Are instincts always invariant?

15. Which animals can perform reasoning?

16. What are conditioned responses?

17. Which part of the brain is the center of autonomic reflex control?

18. What is the role of calcium in controlling muscle contraction?

19. Distinguish between endocrine and exocrine glands.

20. Why do some hormones need the help of second messengers? What are their roles?

21. What organs of the human serve dual purposes, one of which is endocrinal?

22. How are the hypothalamus and the anterior pituitary functionally related? Describe the same for the hypothalamus and the posterior pituitary.

23. Why is the pituitary known as the master gland? What is its embryological origin? What are its target glands?

24. What are symptoms of hypothyroidism? Of hyperthyroidism?

25. What hormone regulates calcium metabolism?

26. Briefly describe the locations and functions of these glands:
    —thymus
    —adrenal cortex
    —parathryoids

27. Distinguish between Type I and Type II diabetes. Which is treatable by insulin supplement?

28. What hormones are produced by the islets of Langerhans? How do they interact to control one set of activities?

29. Cite an example of how two plant growth regulators interact to get a job done.

30. What is the action of indoleacetic acid in causing a stem to bend toward a light source?

31. Describe the role of ethylene in fruit ripening.

32. What plant growth regulator is most active in keeping a seed dormant?

33. How do we know that long-day plants are actually counting night-time hours?

34. What is the value obtained by trees that lose all of their leaves in winter?

11
HEREDITY

What do we know about:
—*the interest in heredity demonstrated by our ancestors?*
—*why Mendel succeeded?*
—*how genes are arranged on chromosomes?*
—*why sex is so common among organisms?*
—*predicting what your offspring might look like?*
—*why so many human blood types occur?*
—*how an organism's sex is genetically determined?*
—*the benefits and perils of inbreeding?*
These and other intriguing topics will be pursued in this chapter.

WE HAVE ALWAYS BEEN INTERESTED IN GENETICS

For generations people have observed that children may closely resemble one or both of their parents in physical and behavioral characteristics. Considering the fact that these are such common observations, not only in humans but in all living organisms, it is surprising that a breakthrough in our understanding of the mechanics of the process was so slow in coming. A very old misconception that held back the development of this field was that the determinant of heritable physical characteristics has the character of a liquid, and that sexual reproduction results in an irretrievable flowing together of this liquid in a manner not unlike the mixing of two colors of paint. This idea remains with us in the expressions bloodline, blood kin, or blue blood. However, as will be demonstrated in this chapter, genetics is now one of the fastest growing areas of science, and one that is rich with both theoretical and practical implications.

Heredity is the transmission of characteristics from parents to offspring. The beginning of any real understanding of how inheritance works dates from Abbot Gregor Mendel's papers published in 1865 and 1866 in the *Proceedings of Natural History Society of Brunn*. To be more exact, one may say that it dates from the independent rediscovery of hereditary laws first found by Men-

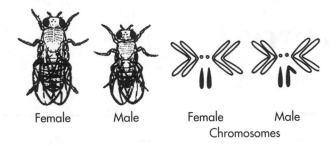

Female   Male   Female   Male
                          Chromosomes

FIGURE 11.1.   *Drosophila melanogaster and its chromosomes.*

del and the recognition of the true value of his work in 1900 by Correns (Danish), De Vries (Dutch), and Tschermack (Austrian). Since then developments have been rapid and widespread, first in the area of predicting the results of matings, then in understanding the molecular basis for these predictions.

Fortunately, heredity works much the same way in many living organisms, thus making possible the application of knowledge gained from one plant or animal to similar problems in another. Such transfer of knowledge has been particularly useful in studying humans. To find answers to human hereditary problems is difficult. In the first place, a generation is too long for geneticists to gather data as fast as they would like. They could hardly observe more than three generations during a lifetime. How much easier it is to experiment with the fruit fly *Drosophila* (Fig. 11.1), in which a generation is as little as ten days long, or on a crop plant like corn, in which a generation can be produced outdoors in warm months and another in the greenhouse in winter. Another obstacle is the impossibility of mating humans experimentally in all sorts of combinations just to learn about the resultant offspring. The usual approach to the problem is to experiment where possible and apply the information to observable traits in human families as they occur naturally in our society. Regardless of the difficulty of this procedure, an understanding of human heredity is emerging.

GREGOR MENDEL BEGAN THE MODERN UNDERSTANDING OF HEREDITY

Mendel's approach to the problem accounts for the great success of his undertaking. Whereas others concerned themselves with the general resemblance of an offspring to its parents, Mendel dealt with only one characteristic at a time—for example, height. This greatly enhanced his ability to recognize the inheritance mode. Furthermore, Mendel was careful to calculate the proportions of offspring in each generation that showed one characteristic or the other, a method that was very important in guiding him to important conclusions. This will be illustrated below.

Mendel was able to establish that—

1. Information for determining characteristics is transmitted as units.
2. This information maintains its identity from generation to generation (segregation).
3. The information for some characteristics is dominant over that of others.
4. Some characteristics are inherited independently of others (independent assortment).

BOTH HEREDITY AND ENVIRONMENT CAN INFLUENCE AN ORGANISM'S APPEARANCE

Not all of the observable characteristics of an organism are attributable to inheritance. The expression of some genetic information can be strongly influenced by the environment around the organism, or even the internal environment of its cells. For instance, some characteristics of the fruit fly appear only when the animal develops in a warm climate. In other cases, there is a subtle interplay between genetic instructions and environmental influence, in which the *degree* of genetic action is moderated by the organism's environment. A human example of this is how adult height is determined generally by parental genetic contributions (a range is set) but specifically by the kind and amount of food available in childhood.

GENETICISTS PRECISELY DEFINE THE TERMS *TRUE-BREEDING* AND *HYBRID*

The term *true-breeding* is used in describing a group of individuals without hereditary variations. Proof of this situation is obtained if two very similar organisms are mated and the resulting offspring are very similar to each other and to the parents. Actually there may not be any sexually produced individuals exactly like one another. The pure condition is closely approached in domesticated animals and cultivated plants in the so-called breeds or varieties. They are produced by selecting certain desired characteristics, getting them all together in an individual, and keeping them together by close breeding. In another sense, true-breeding may be used to describe a pair or a group of individuals alike in one or a few characteristics, like true-breeding for flower color or eye structure.

A *hybrid* is an offspring resulting from mating two genetically different individuals. The degree of difference can be great or small. Hybrids have been produced by crossing individuals of different species or even genera. Crosses of this type are rare, and the hybrids resulting from the crosses will have one or more significant defects. Of this type is the mule, resulting from the mating of a male donkey with a female horse. The mule is not biologically perfect in that it is very rarely fertile. Infertility of the offspring is sufficient proof that the two parents were of different species. The term *hybrid* is also used to describe individuals resulting from crosses between parents differing only in one or a few pairs of contrasting characteristics. An example of a hybrid of this type results from a cross between a tall and a short parent.

GENES ARE THE MOST IMPORTANT UNITS OF HEREDITY

Because of the influence of genes, all individuals develop according to a predetermined pattern. *Genes* are the fundamental units of heredity, which store and release information on how to build and control the cell. In all organisms studied thus far, genes are composed of nucleic acids (usually DNA). Each gene occupies a specific position along one of the linear structures called chromosomes. The methods by which a gene stores and then releases information are topics to be pursued in the next chapter.

CHROMOSOMES ARE THE PHYSICAL SITES OF GENES

Strings of genes, plus supporting material, are *chromosomes*. As described in Chapter 4, chromosomes are usually located in the nucleus of a cell. An exact number is characteristic of a species: horse, 60; earthworm, 32; fruit fly, 8; and corn, 20. The actual number of chromosomes is of limited significance since out of the enormous number of species many very different species have the same number. For example, most humans, a particular tropical fish species, and the privet hedge all have 46 in each diploid cell. The types of genes on the chromosomes make the difference. Sometimes the number can be helpful in determining relationships among species. Chromosomes are not easily visible as individual units in the nuclei of all cells at all times. They can clearly be seen with the aid of a light microscope only when cells are dividing. At other times, one can use an electron microscope to see them, but they tend to be tangled together and are thus not easily counted and characterized.

EACH PARENT CONTRIBUTES A COPY OF EACH GENE

Body cells usually have two sets of chromosomes and therefore two sets of genes, one contributed by each parent. When the two sets are brought together at fertilization, the resultant offspring will be endowed with an equal complement of genes from each parent. The interaction of genes from the two sets determines the heritable characteristics of the individual. Recent work indicates that it sometimes (but rarely) matters which parent contributed each version of a gene. That is, in a few known animal cases the same gene will have effect if contributed by one parental sex, but be inhibited in its expression if it was inherited from the other parent. This phenomenon is called *parental imprinting*. All of these interactions mean that, rather than being an exact copy of just one parent, the new individual has a mixture of observable characteristics similar to those of one parent, or the other parent, or intermediate between the parents. It is this mixing to produce new combinations in each generation that fuels the process of evolutionary change in populations, a topic pursued in Chapter 15.

GAMETES ARE HAPLOID CELLS SPECIALIZED FOR SEXUAL REPRODUCTION

To understand fully some phases of inheritance, it is necessary to consider how eggs and sperm, each with a single set of chromosomes, originate from cells with two sets of chromosomes. In animals, the *primordial germ cells* from which these gametes originate are distinguishable from other body cells early in the development of the embryo and migrate to become incorporated into the testes and ovaries. Within those organs, they are known as *spermatogonia* and *oogonia*. They develop into sperm or eggs when the animals reach a certain stage of maturity. Mitotic cell division constantly replaces the precursors of human sperm cells, but the number of cells available to become eggs in women remains constant from early in embryonic life.

The manner in which chromosome reduction occurs is best understood by contrasting mitosis with meiosis (see Chapter 4). The real difference becomes obvious in prophase. In mitosis the chromosomes, each composed of two chromatids, behave independently, with homologous chromosomes—similar ones from each parent—having no spatial relationship with each other. Then, at the beginning of anaphase, the two chromatids of each chromosome separate and move toward opposite ends of the cell. Each of the resultant two cells (after division by cytokinesis) has the same number of chromosomes as the original cell had. By contrast, in the first portion of meiosis the chromosomes that are homologous align themselves side by side in a pairing process called *synapsis*.

As the first meiotic division proceeds, paired chromosomes separate and move toward the poles. Each chromosome is still composed of two chromatids. Mitosis and meiosis should be further compared at this point, for in the former the individual chromosomes consist of single units, whereas in the latter each of the chromosomes consists of two chromatids (Fig. 4.11). But since the two chromatids of this pair are identical in the genetic information that they carry, and since there is one copy of each gene per chromatid, the number of copies of any particular gene in a cell after the first portion of meiosis is still two, the same number as any nondividing body cell has. Thus, the first portion of meiosis has been reductional in the sense that there is now only one copy of each chromosome (the cell is haploid), but the crucial genetic condition remains unchanged—there are still two copies of each piece of genetic information.

It is therefore necessary to have a second splitting of cells in meiosis (refer again to Figure 4.11). In this division, each chromosome splits, with its two chromatids moving in opposite directions. After cytokinesis, each of the resulting cells has one chromatid of each type (now called a chromosome) and thus, has only one copy of each gene. Such a haploid cell is genetically ready to function as a gamete: when it unites with another gamete in sexual reproduction, the resulting diploid cell will have two of each gene, the normal situation. In animals meiosis produces cells that then mature into sperm or eggs; in seed plants, and in many other plants as well, meiosis results in spores (review life cycles of plants).

A MONOHYBRID CROSS SHOWS THE CONCEPT OF DOMINANT AND RECESSIVE ALLELES

One of the experimental crosses reported by Mendel involved flower color in the pea plant. In this organism, color is determined by a single gene, which can be found in two forms. Another way of saying this is that the gene exists as two *alleles*. The capital letter R may be used to represent the allele for red flower color and the lower-case letter r for white flower color. Since Mendel followed the inheritance of a single gene at a time, the controlled mating described here is often called a *monohybrid cross*. A true-breeding red flowered pea plant (RR) crossed with another true-breeding red one produces only red offspring, and a true-breeding white flowered pea (rr) crossed with another true-breeding white one produces only white offspring. (Both of these true-breeding organisms can be called *homozygous*, a term referring to possession of only one of the alleles of a gene.) Furthermore, the crossing of a pure red with a true-breeding white produces all reds (Fig. 11.2).

Obviously the allele for red expresses a red flower color; alleles that act in this way are called *dominant*. On the other hand, the allele for white flower color is not expressed; such alleles are called *recessive*. The hybrid red plant (Rr), also called a *heterozygote*, is just as red as the pure red (RR); however, r maintains its individuality and could later express itself in combination with another r. The individuality and manner of expression of recessive genes ex-

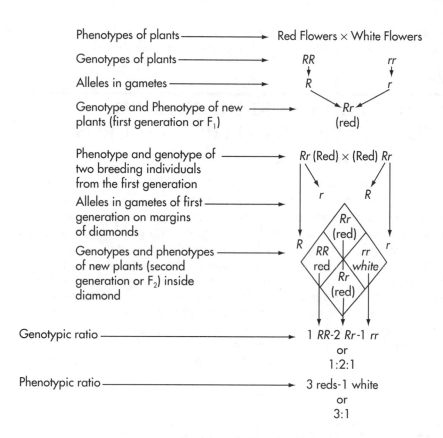

FIGURE 11.2. *Simple cross of pea plants involving dominance and recessiveness.*

plains how a characteristic may appear unexpectedly without having expressed itself in recent generations.

Figure 11.2 also shows a second mating that Mendel arranged. When he crossed two heterozygous (Rr) plants, the next generation included both red and white flowering plants. This indicated to him that the r allele had not been lost or changed while residing in a heterozygous plant, even though it had been dominated over. Indeed, the r allele could be separated from the R allele during meiosis in a heterozygous plant, with each allele coming to reside in a separate gamete. Because of this, Mendel's monohybrid cross is said to show that alleles segregate from each other in meiosis, and the phenomenon is known as Mendel's *Law of Segregation.*

Remember that one virtue of Mendel's work was that he quantified his results. The generation in which white flowers reappeared showed a ratio of red to white very close to 3:1. The box in Figure 11.2 shows why this ratio would be expected if gametes of two heterozygous plants came together in a random fashion. This ratio is a good indicator of any cross involving two alleles of a gene, in which one is completely dominant over the other.

SOME TERMS MUST BE DEFINED: PHENOTYPE, GENOTYPE, TESTCROSS, AND BACKCROSS

Note that the above-mentioned 3:1 ratio is of flower appearance, not of the various combinations of alleles present in the plants. A general term for the observable appearance of an organism is *phenotype*; the term for the genetic basis of this appearance is *genotype*. Thus, red is a phenotype, and Rr is one of two genotypes (along with RR) that would cause the red phenotype to occur. In this case one cannot distinguish between RR and Rr organisms by looking at them. They can, however, be determined by crossing with a recessive rr (white). If RR is crossed with rr, all offspring will be Rr (red); if Rr is crossed with rr, half will be Rr (red) and half will be rr (white). The crossing of an individual having the dominant trait with a recessive to find out the genotype of the dominant one is known as a *testcross*. Since in this case the testcross involves going back to one of the parents (rr) in the original cross, it can also be called a *backcross*.

SOME ALLELES ARE NOT SIMPLY DOMINANT: THE CONCEPT OF INCOMPLETE OR PARTIAL DOMINANCE

Not all alleles of genes interact in the completely dominant and recessive fashion that Mendel saw with pea flower color. For instance, when a true-breeding red four-o'clock plant is crossed with a true-breeding white one, the resulting generation has flowers that are all intermediate between the two parents, pink (Fig. 11.3). That is, using the R and r symbols for the red and white alleles, the phenotype of the Rr (heterozygote) offspring is a blend of the phenotypes of the RR and rr parents. Such a situation, where neither allele is dominant, is called either *incomplete dominance* or *partial dominance*. One is tempted to suppose that this result contradicts our assumption that genes are inherited as discrete units that do not irretrievably intermingle. However, when two pink four o'clocks are crossed (Fig. 11.4), the reappearance of white flowers and the 1:2:1 ratio of red, pink, and white show that the same segregation of alleles has occurred here as in Mendel's peas. The only difference between the two cases is in how alternate alleles interact to produce the phenotype when thrown together in one organism.

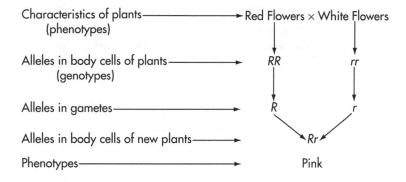

Characteristics of plants —————————→ Red Flowers × White Flowers
(phenotypes)

Alleles in body cells of plants ———————→ RR   rr
(genotypes)

Alleles in gametes —————————————→ R   r

Alleles in body cells of new plants ——→ Rr

Phenotypes ————————————————→ Pink

FIGURE 11.3. *A case of incomplete dominance: a cross between a true-breeding red flowered four o'clock and a true-breeding white flowered four o'clock.*

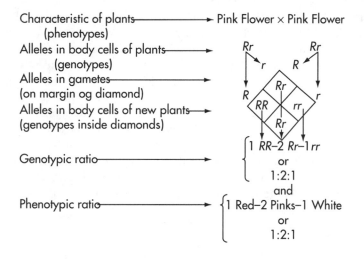

Characteristic of plants ——————→ Pink Flower × Pink Flower
(phenotypes)
Alleles in body cells of plants ———→ Rr   Rr
(genotypes)
Alleles in gametes ———————————
(on margin og diamond)   R   r
Alleles in body cells of new plants —→   RR   rr
(genotypes inside diamonds)   Rr

Genotypic ratio ————————————→ 1 RR–2 Rr–1 rr
or
1:2:1
and
Phenotypic ratio ———————————→ 1 Red–2 Pinks–1 White
or
1:2:1

FIGURE 11.4. *Cross between two pink-flowered four o'clocks.*

SOME CHARACTERISTICS EXIST AS MORE THAN TWO ALTERNATIVE FORMS: MULTIPLE ALLELES

In examples shown so far, there were only two alleles for a single characteristic, such as color. Some characteristics can be determined with three or more alternative alleles. Of course, only a maximum of two of them would be

TABLE 11.1. GENE COMBINATIONS AND BLOOD GROUPS IN HUMANS.

GENE COMBINATIONS (Genotypes)	BLOOD GROUPS (Phenotypes)
$I^A I^A$	A
$I^A I^B$	AB
$I^A i$	A
$I^B I^B$	B
$I^B i$	B
ii	O

present in any particular individual, on its two homologous chromosomes. Human ABO blood groups are determined by a combination of any two of the three known alleles, located at the same positions on the chromosomes. These three are often designated I^A, I^B, and i. There are six possible combinations of the three alleles (Table 11.1).

From the table, it is evident that both I^A and I^B are dominant over the expression of allele *i*. However, when I^A and I^B are both present in a person's cells, neither one is dominant over the other, and both are completely expressed (as the AB phenotype). This is yet another variation on the theme of dominance and recessiveness. When both alleles are able to make recognizable products, with neither influencing the other, this is a case of *codominance*.

This knowledge is cited as evidence in paternity cases taken to court. To prove that a man is the father of a child is impossible because many men have the same combination of genes determining a blood type. It is sometimes possible, however, to prove conclusively that an accused man cannot be the father. A case in point is a mother with type O blood who has a child with type A blood. Being of the genotype ii, she can produce only gametes with *i*, and

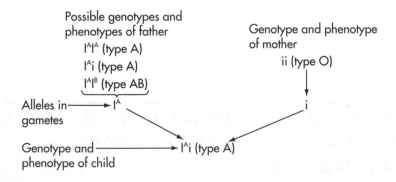

FIGURE 11.5. *Blood type inheritance when mother has type O and child has type A blood. The allele I^A from the father may come from one of three genotypes.*

the child must have either $I^A I^A$ or I^Ai (Table 11.1). The first combination is impossible since the mother has no I^A allele. The child's genotype is therefore IAi, the mother contributing i and the father I^A. To supply I^A the father could be either $I^A I^A$, $I^A i$, or $I^A I^B$. Genotypes $I^A I^A$ and $I^A i$ are type A; genotype $I^A I^B$ is type AB, meaning that any man with type O or type B blood could not possibly be the father (Fig. 11.5).

SOME ALLELES ARE LETHAL FOR THEIR ORGANISM

Sometimes combinations of alleles result in death of the individual in the embryo stage or shortly thereafter. Most of these alleles appear to be recessive and can be detected when there is considerable variation from expected offspring phenotype ratios. It would be expected in a simple cross between two individuals who are heterozygous, as Aa X Aa, that 25 percent of their offspring would be of the recessive phenotype (aa genotype). If the number of offspring from such a mating is continually reduced to approximately three fourths as large as the average and if all the offspring have the phenotype characteristic of either the AA or the Aa genotype, the results indicate that the 25 percent recessives are being killed early in life, perhaps in the embryo stage. A good example of such lethality is that of albinism in plants, where the inability to produce chlorophyll is lethal. Such plants cannot survive longer than the time needed to exhaust the food stored in the seeds.

SEX DETERMINATION CAN OCCUR IN A NUMBER OF WAYS

In most animals, sex is determined by a specific pair of chromosomes. In humans, diploid cells of females carry 22 pairs of chromosomes (*autosomes*) that are not sex determining and one pair of chromosomes (*sex chromosomes*) carrying genes to determine the female phenotype. These latter are designated the X chromosomes, so a female can be said to be an XX individual. Males have only one X chromosome; the other member of the sex chromosome pair differs in appearance and in the genes that it carries. This chromosome is designated Y, and a human male can be said to be an XY individual. A rather small portion of the Y chromosome has been identified as containing one or more genes that are responsible for controlling embryonic development of male characteristics. If this chromosome region is present in an individual's cells, male characteristics will develop, regardless of the presence or absence of the rest of the Y chromosome or the

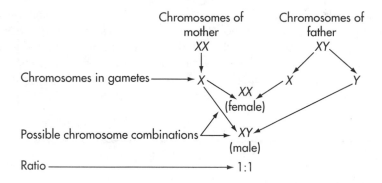

FIGURE 11.6.   *Sex determination in humans.*

number of X chromosomes. In the fruit fly *Drosophila melanogaster*, the same XX and XY conditions are normally found in females and males respectively. However, in this species it has been shown that an entirely different role is played by these two chromosomes in sex determination. A few groups of animals, including birds and some insects and fish, show the reverse condition, with the cells of normal males containing two copies of the same chromosome (designated ZZ) and normal females containing two non-identical sex chromosomes (designated ZW). Some animal groups have cells in which no visibly distinguishable chromosome differences exist between the sexes.

SEX LINKAGE REFERS TO GENES ON A PARTICULAR CHROMOSOME

The X chromosome carries thousands of genes that have nothing to do with sex determination. Since only females (in species described above) carry two copies of this chromosome, and since the Y chromosome does not carry the same genes as are on the X, any *sex-linked* genes (those on the X) will have an inheritance pattern quite different from genes found on autosomes. For instance, consider a sex-linked gene that helps human eyes perceive colors. A recessive allele of this gene can cause the condition called red-green color blindness. In the following examples (Fig. 11.7) one may observe how sons get the defect from their mothers and daughters get it from their fathers. The letters C (normal) and c (color-blind) stand for the alleles of this gene, while Y stands for the sex chromosome without either allele of this gene.

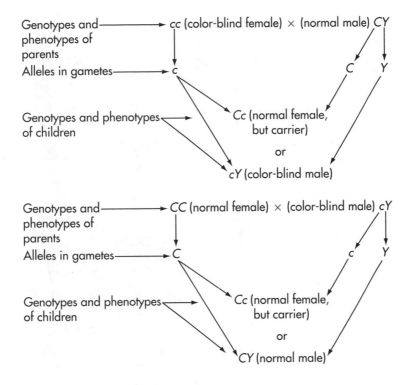

FIGURE 11.7. *Sex linkage. Crosses between color-blind female and normal male, and between normal female and color-blind male.*

TWO GENES CAN BE STUDIED SIMULTANEOUSLY: THE DIHYBRID CROSS

Whenever genes for different characteristics are located on different chromosomes, they occur in random combinations in the gametes. This is due to the random assortment of independent chromosomes occurring during meiosis. When the gametes are combined by chance in the fertilization process, definite and predictable proportions result. This can be illustrated by crossing a true-breeding black, short-haired rodent of some hypothetical species with a true-breeding brown, long-haired mate (Fig. 11.8). If black and short hair are both dominant, and the two genes are not linked together on the same chromosome, all offspring of this cross should be heterozygous (BbSs) and have black short hair. When two such animals are then mated, many genotype combinations are possible. Figure 11.8 shows all of these and the phenotypes, in the diamond diagram (also called a Punnett square). Any single

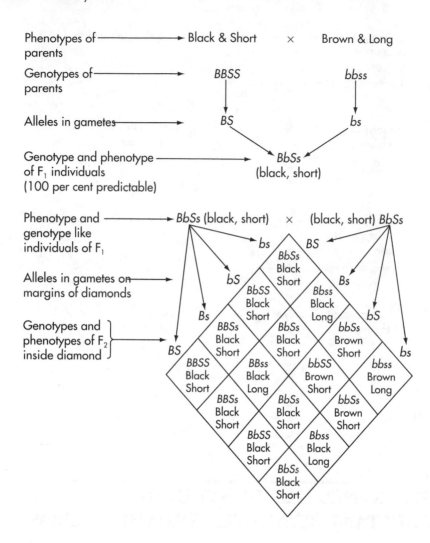

Phenotypes of parents ——→ Black & Short   ×   Brown & Long

Genotypes of parents ——→ BBSS   bbss

Alleles in gametes ——→ BS   bs

Genotype and phenotype of F₁ individuals (100 per cent predictable) ——→ BbSs (black, short)

Phenotype and genotype like individuals of F₁ ——→ BbSs (black, short)   ×   (black, short) BbSs

Alleles in gametes on margins of diamonds ——→

Genotypes and phenotypes of F₂ inside diamond ——→

bs   BS

BbSs Black Short

bS   Bs

BbSS Black Short   Bbss Black Long

Bs   bS

BBSs Black Short   BbSs Black Short   bbSs Brown Short

BS   bs

BBSS Black Short   BBss Black Long   bbSS Brown Short   bbss Brown Long

BBSs Black Short   BbSs Black Short   bbSs Brown Short

BbSS Black Short   Bbss Black Long

BbSs Black Short

FIGURE 11.8. *Dihybrid cross between a hypothetical black, short-haired rodent and a brown, long-haired one. Black and short are dominant.*

individual is unpredictable, but the percentage possibility of each combination is predictable.

The following phenotypic ratio is taken from the diamond:

9 black, short
3 black, long
3 brown, short
1 brown, long

Such a 9:3:3:1 ratio in this generation is a good clue that the two genes under examination are (a) each acting in a simple dominant-recessive manner, and (b) located on different chromosomes. Gregor Mendel was the first to study pairs of genes and to ascribe their inheritance behavior to their being part of independently acting units (now called chromosomes). This phenomenon is often termed *independent assortment.*

CROSSING-OVER IS A TOOL TO STUDY LINKAGE AND MAKE A CHROMOSOME MAP

Some rodent species show hair inheritance patterns exactly as described above. However, if one examined a group of actual crosses involving guinea pigs, where the matings were as described in Figure 11.8, it might be found that the phenotypes from the second cross were 749 with black, short hair; 2 with black, long hair; 3 with brown, short hair; and 240 with brown, long hair. Obviously the ratio is not 9:3:3:1. For the moment, let us ignore the two observed phenotypes that occur in very small numbers, and focus only on those with black, short hair and brown, long hair. They are in the approximate ratio of 3:1. The explanation of ratios like these is that the two genes are linked to the same chromosome like beads on a string (Fig. 11.9, example A). Therefore, they cannot independently assort during meiosis, and can produce only two types of gametes (BS and bs) in equal numbers. In this case black and short alleles are linked together on one copy of the chromosome whereas brown and long alleles are linked together on the other copy of the same chromosome in each diploid cell of a BbSs animal. Figure 11.10 illustrates such a cross.

The explanation just given still does not account for the two offspring with black, long hair nor the three with brown, short hair. In some way, during the production of sperm and eggs, the two genes became scrambled in relation to each other; otherwise all gametes would have carried BS and bs. The rearrangement happened in prophase of the first portion of meiosis, while chromosome pairs were very close together in synapsis (Fig. 4.11 B). At this

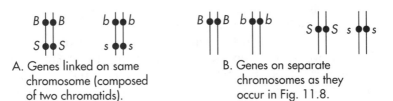

A. Genes linked on same chromosome (composed of two chromatids).

B. Genes on separate chromosomes as they occur in Fig. 11.8.

FIGURE 11.9. *Location of two genes in a doubly heterozygous organism: A. if linked; B. if on separate chromosomes.*

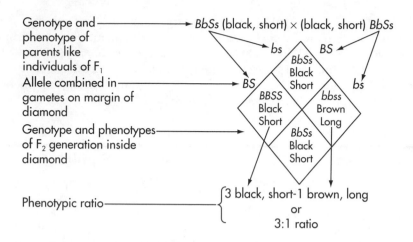

FIGURE 11.10. *A dihybrid cross involving linkage. Second generation phenotypic ratio is 3:1 rather than 9:3:3:1.*

time, some portions of two chromatids exchanged places with corresponding segments of others and produced gametes with Bs and bS. This phenomenon, known as *crossing-over*, could happen as illustrated in Fig. 11.11. If a gamete with the *Bs* or *bS* combination should unite with a gamete with bs, individuals with black, long hair and brown, short hair would result. Crossing-over is a naturally occurring activity, which can help a species by rearranging genes and thus leading to novel combinations of phenotypes in a population. This, in turn, makes a population more diverse and thus more likely to be able to cope with changes in climate, food availability, and so on.

Positions of many genes are now known exactly in a number of organisms. One of the first discovered ways to construct chromosome maps was by studying linkage and crossing-over. Should the sequence of hypothetical genes A, B, and C be desired, the percentages of crossing-over between A and B, B and C, and A and C are tested. For example, let us say that after many crosses 6 percent of offspring are found to show the result of crossing-over between A and B, 4 percent by crossing-over between A and C, and 2 percent by crossing-over between B and C. Crossing-over frequency indicates the distance between genes: the closer together two linked genes are, the smaller the chance that random crossing-over will occur somewhere between them (Fig. 11.12). The linear arrangement of these three genes would have to be A-C-B, since no other arrangement would make logical sense. If one adds the percentages given above, 2% + 4% = 6%, but no other pair of numbers adds up to be the third number. (To prove this, try reconciling these percentages with any other sequence of these three genes.)

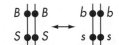

A. Two chromosomes pairing (synapsis). Each chromosome composed of two chromatids

B. Normal separation of chromosomes

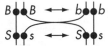 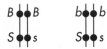

C. Chromosome mates separating but two chromatids tangled

D. Chromosomes separated after crossing-over

E. New chromosomes (composed of single chromatids) separated after aligning themselves at equator in division following synapsis. One goes to each of the four gametes

FIGURE 11.11. *Behavior of chromosomes during and after synapsis. The middle chromatids in sketch C are stretched out of proportion to better illustrate what is happening.*

A. Easier to cross between genes here

B. Harder to cross between genes here

FIGURE 11.12. *Spacing of genes on chromatids as related to opportunities for crossing-over.*

SOME CHARACTERISTICS ARE DETERMINED BY TWO OR MORE PAIRS OF GENES

The inheritance of some characteristics is by the additive interaction of two or more genes (each with its alleles). When two or more genes add their influence in a somewhat equal manner to give a single phenotype, this type of inheritance is called *quantitative inheritance*. This is likely to be happening if one observes a large number of phenotypes for a single characteristic, ranging from one extreme to the other and with little difference between any two when all of the phenotypes are arranged in order. For instance, Herman Nilsson-Ehle found a range of wheat grain colors, from white to deep red. Between these extremes are three intermediate shades of red. The existence of a total of five varieties of a phenotype can be explained by postulating that two genes, each with two alleles, are interacting to produce wheat color. The genotype AABB causes production of one of the extremes, deepest red, while aabb leads to white grain. A double heterozygous individual, AaBb, is exactly intermediate between the extremes in both genotype and phenotype. The remaining observed phenotypes are produced by the other genotypes possible when two genes are considered (AaBB, aaBb, etc.). In some other gene systems more than the above-cited five phenotypes are counted in a population; this usually indicates that more than two genes are adding their effects. A number of human characteristics, including skin color, adult height, and fingerprint patterns, are believed to fall into this category of inheritance. In some cases of quantitative inheritance, environmental factors can be as influential as any single allele in determining final phenotype. For instance, human adult height is determined both by several genes inherited from one's parents and by the amount and kind of food available during childhood and adolescence.

INBREEDING HAS BOTH VALUES AND HAZARDS

Inbreeding, the practice of mating among closely related organisms, is common; in fact, it is the method used in obtaining and maintaining true-breeding groups with desirable traits. This is the most used method to obtain such highly prized organisms as fast race horses, fancy show dogs, and high-yield grains. As for inbreeding in humans, the results are the same as in any other living organism. If characteristics that are determined by recessive alleles are known to be in a family, close marriage results in a greater likelihood of their being expressed in the phenotype of the resulting children. (Remember that both parents must possess at least one copy of the recessive allele if the child is to be homozygous recessive and thus express the recessive phenotype.) If the characteristics are good, the children will receive the benefits; if they are bad, they will suffer the consequences. Characteristics rarely found in the general population sometimes occur with much greater frequency in small societal groups that have chosen, for religious or other reasons, to marry only among themselves.

THERE ARE MANY OTHER MODES OF INHERITANCE

The phenomena discussed thus far occur with many genes and many organisms. However, a large number of variations on these themes are known.

For instance, some genes enhance or inhibit the action of others. In some cases, a single gene produces several quite different effects in various parts of the body. Certain genes are known to move to new positions on their chromosome, or even to jump from one chromosome to another. Any good text specializing in genetics will describe these and many other variant modes of inheritance.

STUDY QUESTIONS

1. As a subject for genetic studies, why is *Drosophila* sometimes preferred over the human?

2. Why did Mendel make substantial progress in understanding hereditary laws whereas others had failed?

3. What was Mendel able to prove as a result of his work with the garden pea?

4. What is the haploid chromosome number of humans?

5. How does synapsis of chromosomes during meiosis allow rearrangement of genes?

6. What is the difference between a monohybrid and a dihybrid cross?

7. What are contrasting versions of a gene called?

8. Define the following: genotype, phenotype, dominance, and recessiveness.

9. What are the genotypic and phenotypic ratios in the generation produced by mating between two heterozygotes, if one is observing a monohybrid cross where incomplete dominance is occurring? Would one expect a different phenotypic ratio if the gene shows complete dominance of one allele over the other allele? What would that phenotypic ratio be?

10. In a case where T is dominant over t, how is it possible to determine whether the genotype is TT or Tt?

11. How many alleles determine human blood groups A, B, AB, and O? Is it possible to prove by blood types that a particular man is the father of a child? Explain.

12. Explain how sex is determined in humans.

13. If a color-blind mother and a normal father have a boy and a girl, what will be their phenotypes?

14. What important information is obtained from the study of crossing-over?

15. What are the pros and cons of inbreeding?

16. Work to the second generation a cross between a true-breeding tall, hairy individual and a true-breeding short, hairless individual. Assume that tall

and hairless are dominant and located on separate chromosomes. What are the appearances and ratios of organisms in the second generation?

Answers to problems:

9. If incomplete dominance: 1/4 are "dominant" phenotype, 1/4 are "recessive" phenotype, and 1/2 are an intermediate phenotype.
   If complete dominance: 3/4 are dominant phenotype, 1/4 are recessive phenotype.

13. All boys are color-blind, all girls have normal vision (but are heterozygotes)

16. First generation: all are tall and hairless (heterozygotic for both genes). Second generation: 9/16 are tall and hairless, 3/16 are short and hairless, 3/16 are tall and hairy, 1/16 are short and hairy.

12
THE MOLECULAR BASIS FOR HEREDITY

What do we know about:
—*how DNA holds all of the information needed to build a person?*
—*how DNA can direct its own reproduction?*
—*the relationship between DNA and proteins?*
—*the unity of all organisms on the gene level?*
—*how a cell controls which genes will release their information?*
—*cancer-causing genes?*
—*the revolutionary development of genetic engineering techniques?*
—*curing genetic diseases by gene therapy?*
—*creating new organisms by gene manipulation?*
These and other intriguing topics will be pursued in this chapter.

MOLECULAR GENETICS IS A FAST-GROWING AREA OF BIOLOGY

No field of biology, or indeed of science, is growing faster or showing more potential for impact upon society than that which deals with the chemical workings of genes. It may be surprising to those born since the last quarter of the twentieth century that the underpinnings of this remarkable field were laid much more recently than those of most areas of biology. The chemical complexity of the chromosome, along with a lack of necessary biochemical techniques, retarded fruitful research until the 1940s. With ever increasing speed, a series of discoveries through the 1960s provided many of the facts that are the basis for this chapter. Then, just when many thought that the field was thoroughly understood, another set of experiments led to a resurgence: genetic engineering was born in the mid-1970s. This set of techniques provides both new insights into basic knowledge of the gene's workings and very practical applications for medicine, agriculture, and many other endeavors. With a set of tools invented in the 1980s and 1990s, biologists are now attacking such difficult projects as mapping the entire set of human genes, treating inherited diseases by replacing defective genes, and understanding evolutionary relationships by comparing genes of both living and extinct organisms.

NUCLEIC ACIDS ARE IDEAL FOR HOLDING INFORMATION

Chapter 3 provided a starting point for discussion of the chemical nature of the gene, and the section entitled Nucleic Acids should be reviewed. Remember that any nucleic acid is a polymer of linked units called nucleotides. When it was discovered that the nucleic acid of the chromosome, DNA, contains four distinctly different nucleotide types (differing by their nitrogenous base: adenine, cytosine, guanine, or thymine), it became obvious that the linear sequence of the nucleotides along one strand of the double-stranded DNA can hold a great deal of information. This is analogous to the linear sequence of letters in a written sentence, with the limitation that only four letters could be used throughout. For instance, consider the two strings of nucleotides shown below. Let each nucleotide be signified by a single letter, corresponding to the initial letter of the nitrogenous base that it includes (e.g., C represents a nucleotide carrying the base cytosine).

—C—T—T—**G**—C—A—T—A—A—C—G—G—
—C—T—T—**A**—C—A—T—A—A—C—G—G—

These two sequences differ by only one base (the fourth from the left), but this would be sufficient to make a difference in their messages. Think of the huge number of unique messages that could be manufactured by substituting at multiple base positions within a string of thousands of nucleotides; even an alphabet limited to four letters could signify a great deal of information. It is estimated that the set of chromosomal DNA within a single human cell contains at least six billion pairs of nucleotides! Two things are obvious: every person holds a great deal of information in the DNA of the nucleus, and the chance that two randomly selected persons would have exactly the same information is infinitesimal.

DNA REPLICATION IS ESSENTIAL FOR TRANSMITTING INFORMATION FROM GENERATION TO GENERATION

When James Watson and Francis Crick provided the first accurate model of DNA (in 1953), they ended their report by pointing out that the double stranded feature of the molecule provides a mechanism for a critical function that any information-storing molecule would have to be able to do: replicating itself. It had been known for many years that the first action of a cell preparing to reproduce is to double each chromosome, in order to provide a complete set of genetic instructions for each of the two cells that will be made. If DNA is the actual molecule holding these instructions, it must be able to double, and to do so very precisely.

The basis for exact DNA replication lies in the way the pairs of nucleotides on parallel strands interact to hold the strands together (Fig. 12.1). *Hydrogen bonding*, a stickiness between a hydrogen atom and another atom caused by the attraction of a slight positive charge on the hydrogen to a slight negative charge on the other atom, is responsible for holding together the pairs of nitrogenous bases that reach toward each other from their respective strands. Although any single hydrogen bond can be easily broken and reformed, the aggregate of thousands of such bonds along the DNA is sufficiently strong to keep the molecule's two strands from being easily separated.

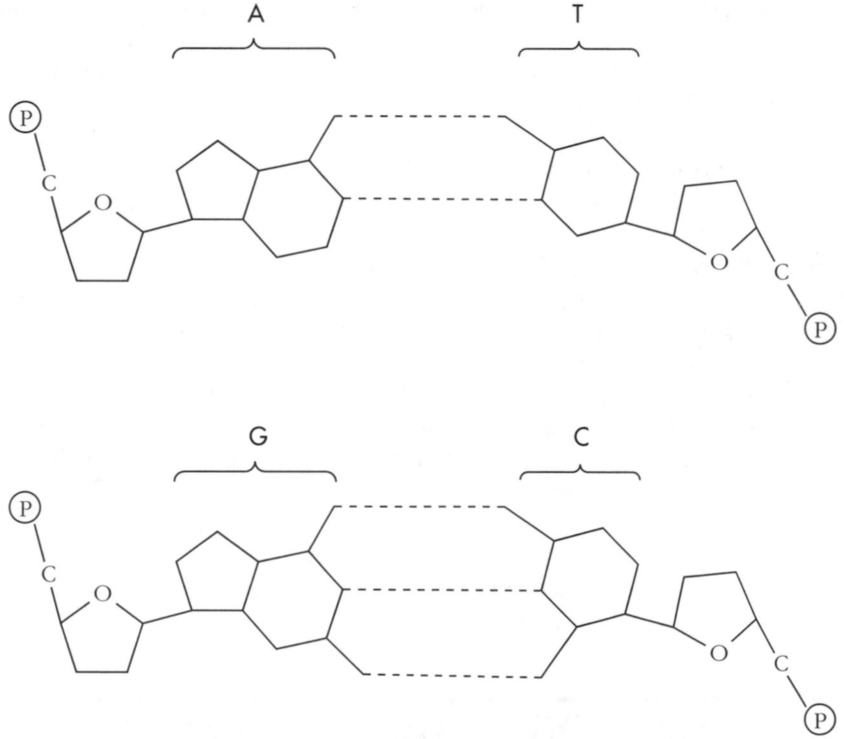

FIGURE 12.1.   *Specific base pairing to link parallel strands in DNA. Dotted lines represent hydrogen bonds. Both types of base pairs (A-T and C-G) are illustrated.*

In order for hydrogen bonding to occur, there must be an exact matching of the bases that are participating. They must be shaped so that they have pairs of oppositely charged atoms in the proper positions. Of the four bases that are in a DNA molecule, there are only two combinations that have the proper size and composition to accomplish this bonding. As seen in Figure 12.1, these combinations are adenine with thymine (A-T) and cytosine with guanine (C-G). This phenomenon is called *specific base pairing*, and it makes possible DNA replication and several other critical activities discussed elsewhere in this chapter.

As a cell prepares to reproduce, each of its chromosomes expands from its usual compact conformation, to make the DNA more accessible. Then, simultaneously at many sites (if it is a eukaryotic cell), the DNA uncoils from its helical shape, becoming more like a ladder. With the aid of an enzyme, hydrogen bonds are broken and the two strands pull away from each other (Fig. 12.2). This allows the string of bases along each strand to act as a model for building a new strand. With the cell providing free (unbonded) nucleotides as building materials, each base of the original DNA directs the placement of a nucleotide across from it, according to the base pairing rules (A-T, G-C). Then the enzyme *DNA polymerase* catalyzes the linking together of newly placed nucleotides to form a continuous strand. This proceeds along both of the original DNA strands (in opposite directions), until two complete double strands are formed. Finally, each of the double-stranded molecules that has been made coils into DNA's characteristic helical shape.

Note that each of these molecules has one of its strands intact from the original molecule and one strand that has just been synthesized by linking together free nucleotides. Therefore, this method of DNA replication is called *semiconservative replication*, meaning that half of each DNA molecule was conserved from a previously existing DNA molecule. Also note that each of the two molecules thus formed is identical to the other one (in the sequence of the bases along its length) and also identical to the DNA molecule that was its predecessor. This was made possible through the precise modeling done by each of the original strands, using the specific base pairing phenomenon. Of course, this is the most critical feature of the entire process, since each of the two cells that is soon to form must have the entire set of genetic instructions that the parent cell had. After DNA replication is accomplished, the cell goes into the steps of mitosis and cytokinesis described in Chapter 4.

THE DEFINITION OF GENE HAS EVOLVED

The idea of the gene has undergone considerable evolution. Before much was known about the nature of DNA, a gene was defined as the fundamental unit of heredity, located on a chromosome, and detectable by the presence of allelic forms that showed up in the phenotype. We can now refine this def-

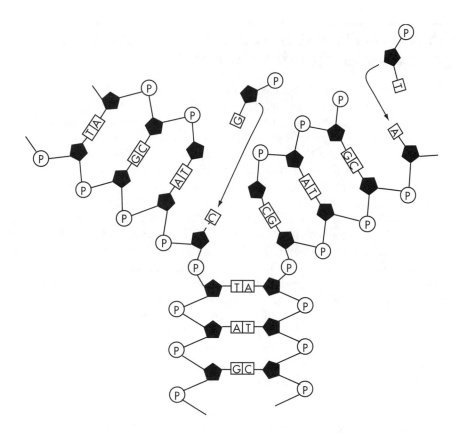

FIGURE 12.2. *Replication of DNA. Original double-stranded formation is at bottom. Free nucleotides are arriving to continue the elongation of two new strands. Note specific base pairing of A with T and C with G, as the new strands are formed.*

inition by adding that a gene is composed of DNA (except in certain viruses, where it is RNA) and that it carries instructions for influencing the phenotype of a cell in a very specific way: it tells the cell how to build a particular polypeptide. As described in Chapter 3, a single polypeptide (a polymer of amino acids) is sometimes sufficient to act as a fully functional protein, either enzymatically or structurally. However, many proteins are formed by linking together two or more polypeptides. If such a protein requires at least two different polypeptides, then its instructions are held by more than one gene. For example, human hemoglobin, the protein that carries oxygen in red blood cells, is composed of four polypeptides. They are of two types. Thus, there are two different hemoglobin genes, working together to direct the building of this one protein.

DNA'S NITROGENOUS BASES CARRY THE MESSAGE OF THE GENE

What is the nature of the message built into DNA that directs the building of a polypeptide? Remember two facts described above. First, DNA is a long string of nucleotides. Second, a polypeptide is a long string of amino acids. Through a series of ingenious experiments performed in the 1950s and 1960s, it became apparent that these two strings are colinear. That is, a sequence of small groups of nitrogenous bases arranged along one strand of DNA somehow directs the placement of a corresponding sequence of amino acids in a polypeptide (Fig. 12.3).

Furthermore, there is enough information in a group of only three bases to code for the placement of an amino acid. Each of these three-base segments of DNA is termed a *codon*. Eventually, the meaning of each of the 64 possible arrangements of bases was determined. Not only is each of the 20 naturally used amino acids signified by a codon, but many of these amino acids are found to have synonyms. For instance, the codons CCT, CCA, CCC, and CCG all direct the placement of the amino acid glycine. Additionally, 4 of the 64 represent punctuation, directing the polypeptide-building machinery to either begin or end its activity much as uppercase letters and periods function in the reading of sentences. The specific action by which the codons' messages are used to make a polypeptide product will be a topic a little later in this chapter.

A flow chart showing how a cell's function is controlled begins with the inheritance of the genes from a preceding cell. Then, at the appropriate time, a protein is manufactured under the direction of one or more genes. If this protein is purely structural in function, the effect of the gene(s) is immediate. If the protein is an enzyme, the effect is secondary. That is, the enzyme catalyzes a particular reaction, using chemicals provided by the cell, and it is the end products of the reaction that produce the final effect to alter the cell. To appreciate this chain of events more fully, one must examine two processes—the reading of a gene's instructions to build a polypeptide, and the control of this reading so that it occurs only when the cell needs the polypeptide.

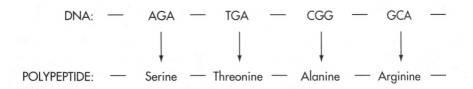

FIGURE 12.3. *The colinearity of DNA bases with the polypeptide amino acid arrangements that they signify. Note that each amino acid's placement is directed by a set of three bases.*

A SERIES OF STEPS CARRIES THE MESSAGE FROM GENE TO POLYPEPTIDE

It is best to consider this as a sequence of two distinct processes. First, a gene is stimulated to produce a version of its message, in the form of ribonucleic acid (RNA). This version, in turn, directs the building of a polypeptide.

The First Steps Are in the Process Called Transcription

Obviously, a gene is a very valuable entity, which should not be altered or lost during the life of a cell. In eukaryotic cells, it is almost always sequestered within the nucleus. Polypeptide synthesis occurs at cytoplasmic organelles, outside the nucleus, so a mechanism must be provided to get the message of a gene out to the cytoplasm. Furthermore, if the gene is to have impact upon the cell's phenotype, copies of the polypeptide for which it codes should be made in large quantity over a short period of time. Both of these functions are served by producing many copies of the gene's message in the form of RNA. The conversion of the message from DNA to RNA is called *transcription*.

As described in Chapter 3, RNA is a single strand of nucleic acid, not unlike one of the two strands of a DNA molecule. One should be reminded of an important difference between DNA and RNA—the latter never contains the nitrogenous base thymine. Instead, RNA uses a slightly different material, uracil. As will be shown later, RNA sometimes does hydrogen bonding, following the base pairing rules. However, because it uses uracil rather than thymine, there is a modification of the rules, as follows: uracil pairs only with adenine (U-A).

Figure 12.4 shows transcription to produce a molecule of RNA. The portion of a DNA molecule that constitutes one gene can be uncoiled and opened up, while neighboring areas of the DNA remain in their double-stranded helical form. Since the two halves of this DNA portion have temporarily ceased to be bonded with each other, each is theoretically free to direct the building of a

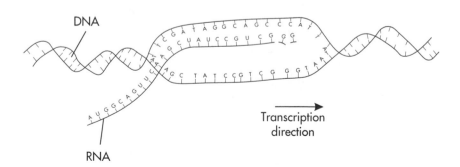

FIGURE 12.4. *Transcription. One strand of DNA is directing the placement of nucleotides to build a molecule of RNA. RNA polymerase is not shown here.*

new strand of nucleic acid, just as occurs during DNA replication. However, the enzyme directing transcription, *RNA polymerase*, selects only one of the DNA strands to act as a model; the other strand does not participate. It does this by recognizing and temporarily binding to a small portion of one strand, called the *promoter.* This region is not part of the gene to be transcribed, but is quite near one end of that gene. Using free nucleotides of the variety that can compose RNA, and obeying base pairing rules, RNA polymerase then catalyzes the building of a molecule of RNA, beginning at the end of the gene nearest the promoter. As soon as one copy of RNA has been made, it is released to be used. The same opened gene area can then repeat its modeling, sequentially producing many identical copies of the RNA.

The variety of RNA that carries the entire sequence necessary to produce a polypeptide is called *messenger RNA* (abbreviated mRNA). However, other types of RNA must also be provided to the cell as it builds polypeptides by linking together amino acids. These RNAs are also produced by transcription, under the direction of other portions of DNA. One of the helping RNA's is called *ribosomal RNA* (rRNA). Along with a number of proteins, rRNA forms the physical structures called ribosomes (Fig. 12.5), upon whose surfaces polypeptides can be manufactured. Ribosomes are considered by many cell biologists to be the most complex organelles, despite their diminutive size (a single cell might contain thousands to millions of ribosomes). A complete ribosome is formed by the union of two subunits, each composed of specific strands of rRNA and proteins.

Another RNA variety that plays a part in the synthesis of polypeptides is called *transfer RNA* (tRNA). The smallest of the RNAs (only 75 to 90 nucleotides long), its function is to carry single amino acids to a ribosome, to be attached to a growing polypeptide chain. There are enough tRNA varieties in a typical cell, differing slightly in their shapes, so that each of the 20 different amino acids can be carried by its own specific tRNA.

Each type of tRNA differs from all of the others in two critical ways. First, its shape ensures that only one of the 20 amino acids can be attached to it.

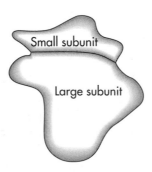

FIGURE 12.5.   *A somewhat stylized view of a ribosome formed by the joining of two subunits. Each subunit is composed of both rRNA and proteins.*

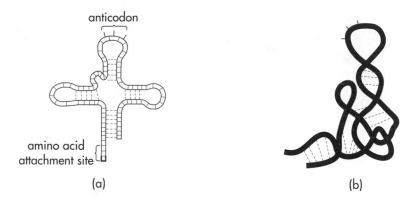

FIGURE 12.6. *A tRNA molecule. (a) A diagrammatic view. (b) A more accurate view.*

Second, it has a unique sequence of three bases, at the region farthest from the point at which the amino acid becomes attached (Fig. 12.6). This group of bases is called an *anticodon*, a term whose meaning will become clear as the process of building polypeptides is explained below.

The Second Steps Are in the Process Called Translation

Follow a sequence of activities that together will result in the production of many copies of a gene's final product, a particular polypeptide. The entire multistep process is called *translation*. A messenger RNA molecule moves through the nuclear membrane (if the cell is eukaryotic) into the cytoplasmic portion of the cell. There, either floating in the cytosol or attached to endoplasmic reticulum, is found the smaller of the two ribosome subunits; the mRNA strand attaches to its surface. (The two subunits are not attached to each other at this time.) This first activity of translation is sometimes labeled *initiation*. It is important for subsequent events that the mRNA becomes attached at its correct end, not at the other end or somewhere between the ends. What helps to select this end? A cytoplasmic enzyme comes to the attachment site and recognizes only a particular short sequence of mRNA's nitrogenous bases which effectively identifies that end of the strand. Figure 12.7a shows the configuration that results from the initiation steps. Note that the size of the small ribosomal subunit determines that only two of the mRNA codons can lie over the ribosome surface simultaneously, and that the remaining codons must wait their turn to come onto the ribosome. The ribosome surface regions that lie under the two codons are sometimes labeled A and P, as the figures show.

Once the two codons have been placed on the ribosome, the first selection of an amino acid can be accomplished. In random fashion, tRNA's bounce into the region of the mRNA codon at the P site. Remember that each tRNA carries its particular amino acid. If any arriving tRNA has an anticodon whose bases cannot form hydrogen bonds with the bases of the mRNA codon, that one does

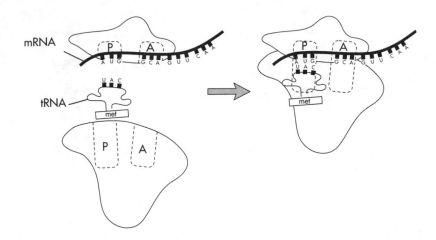

FIGURE 12.7a. *Initiation of translation. A tRNA carrying the amino acid methionine arrives and attaches at the P site, then the large and small ribosomal subunits unite.*

not remain. Eventually, a tRNA with the correct anticodon arrives, in the proper orientation to allow base pairing, and it becomes part of the configuration shown in Figure 12.7a. It is this matching of codon with anticodon that has selected a particular tRNA, and therefore a particular amino acid, to remain at the P site of the ribosome. After the first tRNA-amino acid complex has settled into the P site, a large ribosomal subunit fits into place and a complete ribosome has been formed. This ends the initiation phase of translation.

The same sort of amino acid selection quickly happens at the A site, with the second mRNA codon acting as the director (Fig. 12.7b). This begins what

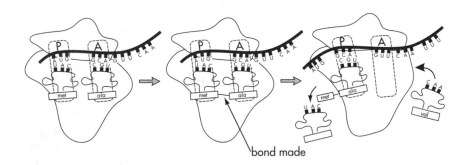

FIGURE 12.7b. *Elongation steps of translation. The amino acid alanine is brought to the A site, and is bonded to methionine. Then, the codon GUU moves to the A site; it directs the placement of a third amino acid, valine. These steps are repeated many times.*

is often called the *elongation* phase of the translation process. An enzyme built into the surface of the ribosome's larger subunit catalyzes the bonding of the two amino acids sitting on the surface, thus linking the first two of what will eventually be a long chain of amino acids.

In order for elongation to continue beyond the first two amino acids, more mRNA codons must come up to the surface of the ribosome. This is accomplished by a series of single-codon moves (Fig. 12.7b). It begins by the following sequence: the first codon moves beyond the P site, the second codon switches from the A site to the P site, and the third codon moves for the first time onto the ribosome surface, settling into the just-vacated A site. During the move, the tRNA that brought the first amino acid to the P site is released from both the first codon and the amino acid that it brought. This tRNA is free to become recharged with another copy of its amino acid for later translation activity. Since there is a yet-untranslated mRNA codon now available at the A site, its matching tRNA can be placed at this time. Quickly thereafter, the ribosome enzyme links the third amino to the second one. The rest of the elongation process is simply repeated actions of this type, with each mRNA codon taking its turn in the A site until the last codon arrives.

The final phase of translation is aptly named *termination* (Fig. 12.7c). It is triggered by the arrival at the A site of any one of three specific mRNA codons

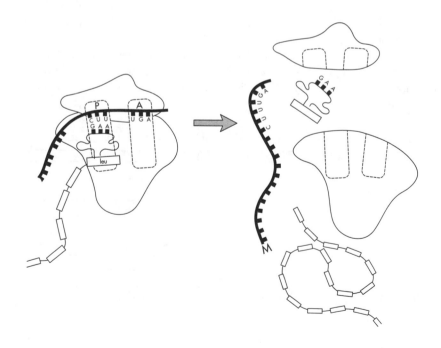

FIGURE 12.7c.   *Termination steps of translation. A termination codon (UGA) has arrived at the A site; no tRNA has an aniticodon to match this. Soon, all components fall away from each other, including the completed polypeptide.*

(see codons labeled termination in Table 12.1). No tRNA exists with an anti-codon that can bind to any of these codons; thus, no amino acid will be placed beside the one at the P site. With the help of an enzyme, the finished poly-peptide is released and floats free. It will spontaneously fold into its proper three-dimensional shape. The translation apparatus quickly becomes disman-tled, but all of the materials (mRNA, tRNA's, both ribosome subunits) can be used again. A typical mRNA of about 500 codons can direct the building of a 500-amino acid polypeptide in only 12 to 15 seconds. Errors sometime occur, but typically only one polypeptide in 20 will be found to have even a single amino acid misplaced.

Eukaryotic Organisms Show Variations Upon These Themes: Introns and Exons

All of the descriptions of both transcription and translation are most ac-curate for bacteria, in which the processes are best understood. Several dif-ferences are seen when one studies eukaryotic cells. However, for the most part the story is remarkably similar, providing a good example of the unity that underlies biological diversity. For instance, it was once believed that eukaryotic cells show an activity strikingly different from that of bacteria, at a point between the transcription and translation processes. Many eukaryotic genes include significantly large strings of codons that are transcribed but never translated. That is, mRNA is processed before it leaves the nucleus. One or more portions, those coded by DNA regions called *introns*, are en-zymatically snipped out, and the portions that remain (made under the di-rection of DNA regions called *exons*) are spliced together to reform a sin-gle but shorter strand of mRNA (Fig. 12.8). A typical exon is only 30 to 50

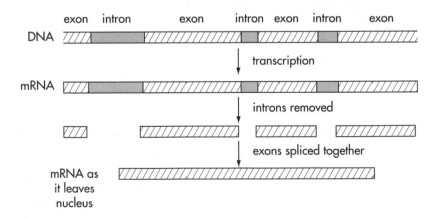

FIGURE 12.8.   *Processing of mRNA. Remaining regions are spliced together after portions that were made by DNA introns are removed.*

codons in length, so a processed mRNA might consist of 10 or more pieces spliced together. It is now known that this pre-translation processing activity also occurs with some bacterial (prokaryotic) mRNA's, and with the mRNA's made under the direction of the DNA residing in mitochondria and chloroplasts.

Is it useful to carry more DNA than is needed to build a polypeptide? There is still much to be learned about this, but one compelling hypothesis concerns a mechanism that would be valuable in evolution. A number of different genes contain exon regions that are very similar or even identical. It is believed that the production of new and potentially useful eukaryotic genes could occur by shuffling existing exons into new combinations. The introns of a gene would be something like the string that holds beads together in a necklace, without function other than keeping exons in proper order within the gene. An advantage of shuffling exons to make new genes is that this could happen much more quickly than if significant innovation had to occur by gradual accumulation of many individual mutations. It is something like building a new model of automobile by rearranging existing parts instead of building many entirely new parts.

THE GENETIC CODE DEMANDS A CLOSER LOOK

A codon is a string of three bases. From a series of ingenious experiments during the early 1960s, Nirenberg, Matthei, and colleagues were able to assign meaning to all of the 64 possible codons. Either DNA or RNA has only four types of bases and any three-base sequence can have meaning for the translation process. The resulting codon assignments for RNA are shown in Table 12.1. Obviously, one could convert this table into one for DNA by applying the base-pairing rules already described.

One Codon Acts as a Start Signal and Three Codons Are Termination Signals

Several striking facts become clear when one examines the table. First, with only one exception, any codon performs only a single function. For 61 of the codons, that function is to signify the placement of one particular amino acid during translation. In bacteria, from which this table was first devised, one of those 61 (signified as AUG) has a second function: to mark the position of the first codon that should be translated. AUG is acting not unlike the capital letter that indicates the beginning of a written sentence at the same time it is part of a word. Thus, a newly formed bacterial polypeptide should be expected to begin with the amino acid methionine, the one that is signified by AUG. Some polypeptides would not function properly with methionine at position one, so a post-translation removal of the first amino acid is performed in those cases.

TABLE 12.1. MESSENGER RNA* CODONS AND THEIR MEANINGS.

Amino Acid	Codon(s) Signifying It
Alanine	GCA, GCC, GCG, GCU
Arginine	AGA, AGG, CGA, CGC, CGG, CGU
Aspartic acid	GAC, GAU
Asparagine	AAC, AAU
Cysteine	UGC, UGU
Glutamic acid	GAA, GAG
Glutamine	CAA, CAG
Glycine	GGA, GGC, GGG, GGU
Histidine	CAC, CAU
Isoleucine	AUA, AUC, AUU
Leucine	UUA, UUG, CUA, CUC, CUG, CUU
Lysine	AAA, AAG
Methionine, initiation	AUG
Phenylalanine	UUC, UUU
Proline	CCA, CCC, CCG, CCU
Serine	AGC, AGU, UCA, UCC, UCG, UCU
Threonine	ACA, ACC, ACG, ACU
Tryptophan	UGG
Tyrosine	UAC, UAU
Valine	GUA, GUC, GUG, GUU
termination	UAA, UAG, UGA

*A similar table can be compiled for the three-base sets on DNA that produce these RNA codons, by observing the following conversion rules: C → G, G → C, U → A, A → T. Thus, GCU of RNA was produced because there was a GCA segment of DNA.

After assignment of an amino acid to each of 61 codons, workers found that each of the remaining three codons, indicated by the word termination in Table 12.1, acts to stop the translation process. This function, already mentioned in the description of translation, is akin to that performed by a period at the end of a written sentence.

Many Codons Are Synonymous

A second feature of the genetic code, also evident in Table 12.1, is that nearly all of the 20 amino acids used to make natural polypeptides can be signified by more than one codon. For example, look at the designations for the amino acid alanine: GCA, GCC, GCG, and GCU all call for alanine's placement. In a sense, this is the same as having synonyms in a written language. However, the redundancy found in the genetic code has a usefulness that is very unlikely with language synonyms: it provides some protection against the consequences of accidental changes. For the set of synonyms listed above, and for most of the other sets found in Table 12.1, the first two bases are identical. Thus, a mutation leading to substitution of a wrong base at the third position might not result in any change in the polypeptide. For example, altering a base in a gene so that the mRNA codon GCA was replaced by GCU would still allow placement of alanine in the proper place as translation proceeded. Any mutation whose effect is not shown in the polypeptide (and thus in the organism's phenotype) is called a *silent mutation*. The physical basis for this effect rests in the way a tRNA's anticodon pairs with an mRNA codon on the surface of a ribosome. Successful hydrogen bonding at the first two positions is often sufficient; the third base of a tRNA can remain unbonded and the wobble that ensues will not be sufficient movement to cause loss of contact. Another way of describing all of this is to say that a tRNA carrying its particular amino acid can bind to more than one mRNA codon, if the first two bases are the same.

Nearly All Organisms Use the Same Coding Scheme

A third feature of the genetic code is that it is nearly universally used among earth's organisms. With only a few minor exceptions, every tested organism has been found to be using the same designations. For instance, GCA signifies alanine placement for bacteria, corn, mushrooms, humans, etc. Obviously, this has great significance for evolutionary biologists, fitting the hypothesis that all organisms alive today have a common ancestry. The significance of the handful of aberrant codon assignments (mostly in mitochondrial genes) discovered since the late 1970s is still a matter of debate.

GENE CONTROL IS ESSENTIAL: METHODS IN BACTERIA

A typical bacterial cell is utilizing the information of only about 10 percent of its genes at any given time. Some of these genes provide housekeeping activity, which must occur constantly if the cell is to remain alive. But the information of others is much more specific, and is released only when the cell finds itself in specific environments. In only a few minutes after a bacterium's environment changes, the activity of some appropriate genes may increase (or decrease) by a thousand-fold. Such reversible activation, cued by the cell's needs, is evidence of gene control. Eukaryotic cells do this too, and some of their methods will be described later. However, the bacterial methods have been much studied and provide a clear model for our first understanding. Although control can be accomplished at several levels, our focus will be at the gene level, where transcription can be started or stopped.

In 1961, Francois Jacob and Jacques Monod published a landmark paper describing how the metabolism of the sugar lactose is controlled in the bacterium *Escherichia coli*. In order to understand what they found, one must be reminded of the concept of *negative feedback* (also called *feedback inhibition*), which was first discussed as a topic in physiology (Chapter 7). The essence of negative feedback is this: the quantity or activity of the product made by a system determines whether more of that product will continue to be made. When the quantity or activity is appropriately high, the system can sense this and shut down further production. Negative feedback provides appropriate control of a production activity, with the result that just the right amount of the product is present in the cell or organism.

An Inducible System of Control is Exemplified by the Lactose Operon

Jacob and Monod investigated a set of three side-by-side genes, which can produce a single mRNA. This, in turn, can be translated to make three enzymes, all of which are needed to cause the disaccharide lactose to enter the cell more easily and to be broken into two monosaccharides (galactose and glucose) that can then be metabolized to gain energy for the cell. It is appropriate to make all three of these enzymes available simultaneously when lactose is in the cell's environment, and it is equally appropriate to cease their production simultaneously when lactose is not available. The three genes making the mRNA are called *structural genes*, to distinguish them from other DNA regions that are involved in their control.

As described earlier in the chapter, the first step in transcription is the recognition of a promoter region by a molecule of RNA polymerase, and the temporary binding of this enzyme to the promoter. In bacteria, the promoter is very close to the first codon of a structural gene, but they are separated by a controlling DNA segment called the *operator* (Fig. 12.9). Jacob and Monod called the set of structural genes and the single operator that controls them an *operon*. If RNA polymerase successfully binds to the promoter, it can then easily move across the operator and begin its work of transcribing the first of the structural genes. However, the operator is sometimes covered by an attached protein called a *repressor*. This protein, made by a *regulator gene* located elsewhere on the chromosome, is shaped so that it can bind to the operator near this specific set of structural genes. When the repressor is in place on the operator, RNA polymerase can neither bind to the promoter nor move over to the first structural gene. Thus, transcription is turned off. This is the normal situation for the lactose operon when lactose is not present in the cell.

To understand how this off situation can be reversed, one must know two more things about the repressor protein. First, its binding to the operator is not permanent. It is shaped so that it forms weak bonds with the DNA; they are easily broken. Typically, a repressor stays on the operator for only a short while, after which it is normally replaced by another copy of the same protein. The second important point is that the repressor's shape can be induced to change so that it no longer fits on the operator. This change occurs if a molecule of lactose attaches to the repressor.

Now, imagine what happens if the bacterial cell moves to a place where much lactose surrounds it. Some lactose molecules diffuse into the cell, and bind to the copies of the repressor that are not currently attached to the op-

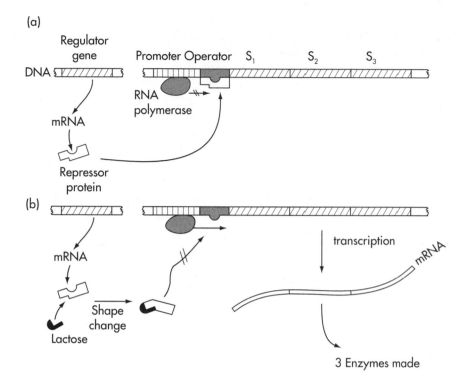

FIGURE 12.9.   *An inducible system: how the lactose operon works in E. coli. (a) Repressed condition. Three structural genes (S$_1$, S$_2$, S$_3$) are not transcribed because RNA polymerase cannot reach them. (b) Induced condition. Lactose changes the shape of the repressor; RNA polymerase can reach the set of structural genes.*

erator. They undergo the shape change described above. Soon thereafter, the single unmodified repressor happens to fall off of the operator, and it too undergoes binding by lactose and is also unable to cover the operator. In a matter of a few seconds, all available repressors have been altered so that they are no longer able to be effective in repressing; the lactose operon is now on, releasing its information. This is appropriate, since there are lactose molecules in the environment, available to be metabolized with the help of the three enzymes now being manufactured. Because the lactose operon is *induced* to release its information by a message arriving in the form of lactose, it is called an *inducible system* and the lactose is an *inducer*.

Are the lactose operon and other inducible systems good examples of negative feedback? The answer is yes, because of what happens when the cell eventually runs out of lactose. Consider the situation when all of the unbound lactose within the cell has been converted to galactose and glucose by the enzymes from the operon, and no more lactose is available from the environment. Soon, the lactose molecules attached to repressor proteins fall away and they too are metabolized. Repression of the operon can again occur, since the now-unbound repressors return to the shape that can attach to the operator. This is a case of negative feedback because the products of the system (the three enzymes) have, by their action, caused the system to return appropriately to an off condition.

Repressible Systems of Control Also Operate in Bacteria

Some other operons of *E. coli* are controlled in a fashion nearly opposite that of the lactose operon. Working in *repressible systems*, these operons feature structural genes that transcribe until there is a build-up of the end product that was made from their action. Accumulation of the molecule that is built when the operon's enzymes are working is the trigger for turning off a repressible operon; by contrast, with an inducible operon a *lack of the material that is to be changed* when the operon's enzymes are working is what causes it to turn off.

Here is the classical example of a repressible system. There is an operon that includes five structural genes whose five enzyme products work stepwise to manufacture the amino acid tryptophan. Usually, a cell has access to the precursor molecule that can be changed into tryptophan by these enzymes. However, it does not often have already made tryptophan in its environment, nor does it experience a build-up of self-manufactured tryptophan since the amino acid quickly becomes part of newly made proteins. Therefore, the normal situation is for this operon to be turned on. The regulator gene for the operon causes a repressor to be made, but its initial shape is not appropriate for attachment to the operator. Of course, this means that RNA polymerase is free to begin transcription; the operon is on. If, however, a molecule of tryptophan attaches to the inactive repressor, it acts as a *corepressor* by changing the repressor's shape into one that can bind to the operator; the operon is then off. A significant number of tryptophan repressors will be converted into their active shape only if there is much tryptophan in the cell, either from an outside source or from an over production by the enzymes of the cell's operon. Whatever the reason for repression, it occurs only when appropriate: when the cell would be wasting energy to manufacture unneeded tryptophan. Of course, this system is performing negative feedback, since a product (tryptophan) is announcing to a control unit that it exists in sufficient quantity and the control unit is responding by shutting down further production.

GENE CONTROL IS SOMEWHAT DIFFERENT IN EUKARYOTIC CELLS

Although the operon model developed for prokaryotes has provided many clues to understanding gene control in eukaryotes, our knowledge of the latter is still far from complete. One must remember that eukaryotic cells have far more genes, that these genes are located on chromosomes that have much more complicated structures, and that these cells usually must interact with neighboring cells by various chemical and physical signals. Thus, it is not surprising that eukaryotic gene control differs significantly from that of prokaryotic cells, and is more complex.

One major difference is that eukaryotic cells apparently do not cluster related genes in physically adjacent groups—they do not have operons. For example, the several genes for enzymes that are needed together to catalyze a multistep metabolic activity might be scattered over several chromosomes. Nevertheless, their expression is coordinated. Each gene is associated with a promoter region, where RNA polymerase begins its work. However, these promoters vary significantly in their structures, and one is sometimes located rather far from the gene that is to be transcribed. In spite of their variability, most eukaryotic promoters contain some nearly identical regions, as would be expected if they all act in a similar fashion.

Eukaryotic systems also include DNA regions called *enhancers*. Located near or even within a gene, an enhancer interacts with a regulatory molecule to start transcription and also to control the rate of transcription. The regulatory molecules that can attach to enhancers are proteins called *transcription factors*. Although the action of these factors and their enhancers is still not thoroughly understood, it is believed that they help to rearrange the three-dimensional form of the gene, placing it into a conformation that makes transcription more efficient.

Certain genes called *homeotic genes* are especially powerful in regulation. Found in many (if not all) eukaryotic organisms, these control the major embryonic steps that place body parts in their proper positions and determine such basic patterns as the anterior-posterior body axis. Mutation of a homeotic gene can lead to bizarre rearrangement of shape, such as replacing an antenna on a fruit fly's head with a fully developed leg. Obviously, a homeotic gene is a controller of many other genes, since a whole body region can be altered by changing one. All homeotic genes contain an identical (or nearly so) 180-base-pair region called the *homeobox*. This region codes for a part of a transcription factor protein. It is believed that a homeotic gene can influence many other genes during embryonic development by manufacturing this transcription factor at the appropriate time.

Control can also occur beyond the level of the gene. For instance, mRNA can be made in an unfertilized egg and stored until some time after the embryo begins to form. Embryonic cells can then selectively activate various stored mRNAs, causing them to begin the translation process. Control can occur even after translation. An example of this is that the hormone insulin (a protein) does not become active until after a significant portion of it is enzymatically removed. In other cases, proteins do not become active until carbohydrates are added to them. All of these mechanisms, and those occurring at the gene level, make a very complicated story of control in eukaryotic cells.

SOME GENES ARE ASSOCIATED WITH CANCER

One of the great puzzles of biology has been what causes a cell to be transformed from normal to cancerous. Part of the problem is that many quite different factors have been shown to be capable of triggering the changes that lead to malignancy. Some animal cancers, such as feline (cat) leukemia, are closely associated with infection by certain viruses. However, most cancers, and apparently all of those in humans, are not viral. Rather, they are either linked with exposure to chemical or physical insults or are of unknown origin. It is the wide variety of agents, with little or no apparent properties in common, that has made it difficult to determine how cancer gets started.

In the 1980s, cancer researchers began to understand that, while many things can trigger cancer development, all of them must ultimately act on the level of genes. When cancer-causing viruses were carefully examined, it was found that each carries a specific gene which becomes active in an infected cell and brings about all of the changes that make a previously normal cell cancerous. Since the study of cancer is termed oncology, this type of gene was named *oncogene*.

Armed with the knowledge of what viral oncogenes look like, workers began searching for them in a variety of cells. Surprisingly, oncogenes (or very similar genes) were found on chromosomes of virtually all human cells! A search for some benign function of oncogenes ensued and it soon became apparent that they are normal and essential genes, usually releasing their information only at very specific times during the life cycle. Most oncogenes are active during embryonic life, stimulating cells to reproduce at that time. Of course, reproduction is a normal and necessary activity of embryonic cells. With this knowledge, cancer researchers realized that cancer is usually a problem of faulty gene control. Oncogenes are master genes that regulate the functioning of many other genes associated with cell reproduction, and cancer ensues if one or more oncogenes are activated at an inappropriate time in the life cycle.

Normally oncogenes are irreversibly turned off after their function has been carried out, but they remain on the chromosomes and can be re-activated under unusual circumstances. One of those circumstances is alteration of their control elements by mutation. A wide variety of chemicals are known to be *mutagens* (causing changes in DNA), and many of these are also proven *carcinogens* (causing cancerous transformation). Additionally, certain forms of energy are capable of damaging DNA in ways that cause inactive oncogenes to come back to the on condition. Among these are (in increasing order of effectiveness) short-wave ultraviolet light, X-radiation, and gamma radiation. Another way that oncogenes can be activated is for them to move from one control region to another on the chromosomes. If an oncogene is moved so that it sits within a group of genes that are on, it too will become on, even if that is inappropriate for the cell. It is known that genes occasionally can be transferred to new positions within a set of chromosomes. This mechanism could also explain how virus-carried oncogenes cause cancer. Viral genes often become incorporated into the chromosomes of the host cell; if a viral oncogene were placed within a set of host genes that was releasing information, that oncogene would also do so and the cell would begin inappropriate reproduction. Yet another way in which oncogenes are believed to cause some forms of cancer involves an unusual replication of the gene until a cell has many copies of it. If only one or a few copies of the oncogene were releasing their information, the cell would remain normal, but if dozens or hundreds of copies were present and working simultaneously, the cell might respond to the accumulated effect and becomes cancerous. Some forms of human breast cancer are believed to be caused by such a gene amplification.

An exciting result of our being able to identify specific oncogenes is that physicians might be able to detect the mutated form in cells of a person who does not yet show any cancer symptoms. Since early detection is a key to successful treatment of cancer, a person known to carry the potentially activated gene could be carefully monitored and any tumors could be attacked while still very small. Of course, the ultimate goal of cancer research is to be able to prevent those initial changes in oncogenes, or at least to reverse them before they can cause damage. Our emerging understanding of oncogene activation is a giant step toward that goal.

GENETIC ENGINEERING IS ARTIFICIAL MANIPULATION OF GENES

One of the long-standing goals of geneticists has been to find ways of rearranging genes so that an organism has new and useful combinations. For most of the history of genetics, this was possible only by the cumbersome and time-consuming process of selective breeding—arranging matings, waiting for the maturation of the offspring, and hoping that they demonstrate new and valuable traits because of their combination of parents' alleles. Of course, this is limited to rearrangements of genes within a species; interspecific transfer of genes happens naturally only by relatively rare occurences among bacteria and viruses. However, in the early 1970s several discoveries were combined to make possible a new and vastly more flexible way to manipulate genes. This technique, the *recombinant DNA method*, led to a whole new field of genetics, often called *genetic engineering*. As described below, it is now possible in a short time to isolate a gene, to insert that gene into cells of a different species from the organism of its origin, and therefore to alter the recipient organism so that it does new and potentially useful things.

Restriction Enzymes Make Specific Cuts in Large DNA Molecules

The beginnings of this technology were in the 1960s, when *restriction enzymes* were discovered in bacteria. These are natural defenders against infective viruses because they recognize foreign DNA and slice it into fragments so that it cannot direct the reproduction of the virus within a bacterial cell. It was found that a whole family of these enzymes exists, and that each is restricted in how it cuts DNA. Each type finds a particular sequence of nitrogenous bases along the foreign DNA strand and cuts only at that particular place. For instance, the enzyme called EcoRI cuts DNA only where there is the base sequence GAATTC along one side of the molecule (Fig. 12.10), whereas the enzyme known as HindIII cuts only at the sequence AAGCTT. A feature of this cutting action that became very useful is that some of the restriction enzymes cut diagonally. That is, tails are produced on the two ends of the sliced DNA, rather than blunt ends (Fig. 12.10).

These tails of bases make the ends "sticky," enabling other DNA fragments with complementary tails to join them. This joining, enhanced by another enzyme called DNA ligase, can take place between DNA pieces from any source, and is the basis for the recombination that is so valuable to geneticists.

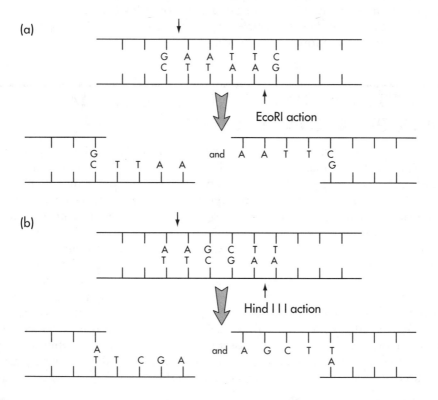

FIGURE 12.10.   *The cutting action of two restriction enzymes. Small arrows indicate the specific points at which the enzymes work. Note the tails that are produced.*

Plasmids Were the First Carriers of DNA for Genetic Engineering

In the early 1970s, a number of workers began to find ways to move foreign DNA into bacteria, using another naturally occurring material, the *plasmid*. Some bacteria carry DNA that is not part of their chromosome. This plasmid is a circular, double-stranded structure, composed of several genes' worth of DNA. Plasmids replicate when their host bacterium reproduces, and both of the daughter cells contain one or more plasmids identical to those in the parent cell.

Plasmids can be harvested from their hosts and manipulated. As shown in Figure 12.11, a restriction enzyme can be used to open up the circle of an isolated plasmid, with tails being created as the enzyme makes its cut. Then, if DNA from another source is added (having the same sort of tails because it had been prepared with use of the same restriction enzyme), and if DNA ligase is also provided, a plasmid can become circular again. But this circle is sometimes larger than that of the original plasmid because it contains the added DNA. If plasmids prepared in this way are placed in the vicinity of bacteria, occasionally one will enter a bacterial cell, carrying with it the added

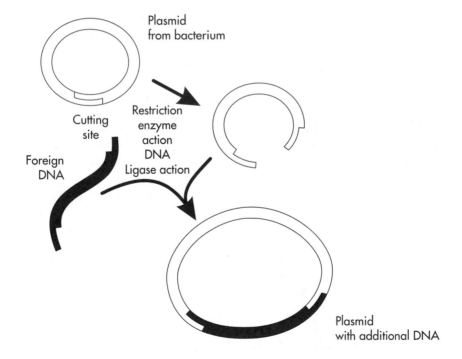

FIGURE 12.11. *A section of foreign DNA is added to a plasmid. Note how the complementary tails facilitate the reforming of a circle.*

DNA. A plasmid that does this is sometimes called a *vector* since it is the mode of transporting the DNA into a new environment (the bacterial cell).

Genetic Engineering Has Provided Valuable Products and Applications

If all goes well in the described procedures, a rapidly growing population of bacteria will form, each cell of which contains one or more copies of the added DNA. If that DNA includes a complete gene, and if that gene can be transcribed, the bacterial population can now manufacture the polypeptide encoded by the gene. There is no limit as to the origin of the added gene, and therefore the bacteria can be recombinant organisms—they can contain and use genetic information never naturally found in their species. Since this technique became well known in the mid-1970s, many applications have been found for it. The most common use is to provide a large amount of a valuable protein, for research, medicine, or industry. For example, bacteria have been engineered to provide human insulin, growth hormone, and blood clotting factors, all of which had been previously available only in small amounts and at great expense.

Plasmids inside bacterial hosts are also used as convenient repositories of foreign genes, since the genes can later be snipped out and moved elsewhere. It is possible to enzymatically chop up the DNA of an entire chromosome or set of chromosomes, then place each of the many fragments into a different bacterial colony for storage. A set of DNA segments stored within a series of plasmids and their bacteria is often called a *library.*

The entire genome of humans has been placed into a huge bacterial library, as a necessary step in the *human genome project.* This is an international effort to determine the base sequence of every gene in each of the 23 human chromosomes, as a valuable tool in understanding how the human body is formed under the direction of the approximately 100,000 functional human genes. With the cooperation of dozens of laboratories and the use of automated analytical instruments, it is hoped that this monumental project will be finished in the mid-1990s.

Although bacteria are still the hosts of choice for many genetic engineering events, other methods have been developed to allow nonbacterial cells to receive foreign genes. For instance, yeast cells are often the hosts of choice if the normal function of genes from eukaryotic organisms is to be studied. Cultured animal cells are necessarily the hosts if one wants to study antibody-producing genes, since only these kinds of cells can mediate the complex steps involved in making this product. Viruses are often used as vectors for carrying DNA into eukaryotic cells.

The Polymerase Chain Reaction Allows Analysis of Tiny Amounts of DNA

One of the most powerful tools in the current era of gene manipulation is the *polymerase chain reaction* (*PCR*), a method to manufacture billions of copies of a segment of DNA in a matter of a few hours. Although its details are beyond the scope of this book, it can be described as a sequence of steps that selects a particular portion of a large DNA region, then repeatedly amplifies this portion by enzymatically driven replication. Taking place in a totally artificial environment, the method duplicates each copy of the selected DNA approximately every five minutes. The growth is exponential (each newly formed copy becomes two in the next cycle), so a huge number of copies can be formed in a short period. Using PCR, an investigator can start with a very small amount of DNA (theoretically, only one copy) and quickly have enough for a very thorough analysis of its base sequence. It has been used in such diverse fields as criminology (identification of blood or semen specimens), evolutionary biology (comparison of DNA in fragments of extinct organisms with that of modern relatives), and medicine (analysis of single fetal cells for genetic diseases).

Gene Therapy May Be a Powerful Medical Tool

The term *gene therapy* has been invented to describe the addition of one or more genes to cells in order to replace the function of defective forms of the genes that had been inherited. It is best suited to attacking those diseases that are traceable to single-gene defects, such as cystic fibrosis and sickle-cell anemia. One of the problems that must be solved if this is to work is that of finding a suitable means of introducing the normal gene. The use of viruses as vectors and the direct injection of DNA into cells are both being tested. Another question is whether to treat the single cell that is the zygote (thus enabling all cells of the developing person to receive the normal gene), or to target the cells in a fully developed person that are being adversely affected by the presence of a defective allele. Initially, in the mid-1990s, the latter method is being clinically tested. At this writing cystic fibrosis (treating respiratory tract cells) and several blood cell defects (treating bone marrow stem cells) are being attacked by gene therapy. Injection of genes into just-fertilized eggs has been accomplished with several mammals and is contemplated for humans. This requires retrieval of eggs from the region of the ovary, artificial insemination while the eggs are cultivated *in vitro*, careful introduction of the DNA, and then replacement of the early embryos into the mother's uterus. The reward is great (all cells of the individual carry the normal allele of the gene), but the cost and medical risks are also great.

Transgenic Organisms Are Valuable in Both Basic and Applied Settings

The technique just described has also been used to produce mammals that are truly recombinant; that is, they receive genes from other species and incorporate them into their own genome. Known as *transgenic* animals, they are quite normal in most ways (a cow is still a cow), but they also have unique properties (the cow might produce large quantities of human growth hormone, which can be harvested from its milk). There is some contemplation of producing transgenic humans by introducing nonhuman genes into normal human zygotes in order to confer new abilities (such as resistance to pathogenic microorganisms), but the ethical considerations in doing this are very significant.

STUDY QUESTIONS

1. What are the four commonly used nitrogenous bases found in DNA? Which one of these is absent in RNA, replaced by uracil?

2. Describe hydrogen bonding between base pairs in DNA.

3. How does specific base-pairing work, and how does this make the exact replication of DNA happen? What is the role of base-pairing in the translation process?

4. Why is DNA replication described as being semiconservative?

5. What are the roles of the following enzymes?

    —DNA polymerase
    —RNA polymerase
    —restriction enzymes
    —DNA ligase

6. What are codons?

7. Describe how any particular amino acid is selected to be placed into a growing polypeptide on a ribosome's surface. What is the most important factor in selecting the correct amino acid from the available pool?

8. During translation, what mechanism is used to terminate the lengthening a a polypeptide chain at the appropriate point?

9. Describe what introns and exons are, in eukaryotic genes. Is there any evolutionary value in having these regions?

10. The genetic code includes a number of synonymous codons. What value might this have for an organism?

11. Weave the following terms into an accurate story of gene control in an inducible bacterial gene system: operon, operator, promoter, structural genes, regulator gene, inducer, RNA polymerase, repressor protein.

12. Modify the story you made for question 11, to describe the role of corepressors in a repressible bacterial gene system.

13. What are homeotic genes? Are they found only in eukaryotic cells?

14. Oncogenes have normal, useful function in cells, but can sometimes cause disastrous changes. What are these normal and abnormal activities of oncogenes?

15. What are mutagens? What are carcinogens?

16. Describe how plasmids can be modified to carry foreign genes into bacteria. What are some practical uses of doing this?

17. Describe some valuable applications of the polymerase chain reaction (PCR).

18. What are transgenic organisms? Why do people create them?

13
REPRODUCTION

What do we know about:
—*the advantages and disadvantages of sexual reproduction?*
—*practical uses of plant asexual reproduction in horticulture and agriculture?*
—*why mosses and ferns maintain permanent colonies only in moist environments?*
—*how plants can have both haploid and diploid portions?*
—*the anatomy of a flower?*
—*how some plants rely on insects for their sexual reproduction?*
—*animals that switch their sex?*
—*the sexual anatomy and behavior of humans?*
—*how contraceptive devices work?*
These and other intriguing topics will be pursued in this chapter.

REPRODUCTION IS A UNIVERSAL FEATURE OF ORGANISMS

Individuals live for only a limited period; species may live indefinitely. The failure of the individual is its inability to escape death; the success of the species is its ability to perpetuate itself. Consequently, reproduction is a process in the chain of life linking older individuals with newly created ones destined to take their places.

Reproduction is the creation of a new individual from one or two parents. Its simplest form, *asexual* reproduction, involves a single parent organism splitting, budding, or fragmenting to make one or more new individuals. For example, most members of Kingdoms Monera and Protista can reproduce by the asexual process called *binary fission*: a single cell divides into two of approximately equal size (Fig. 13.1 B).

In the highest forms of life, and in many lower forms as well, a third individual is produced from the union of parts (*sex cells* or *gametes*) of two individuals. This is, of course *sexual* reproduction. There are, however, certain processes sometimes considered reproductive where there are no new individuals

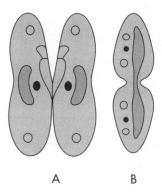

A   B

FIGURE 13.1.   *Paramecium. A: conjugation. B: fission.*

produced. An example is *conjugation* of the Protozoan known as *Paramecium* (Fig. 13.1 A), an organism that unites with a mate in a sex-like process. The two organisms separate without producing a third one, but for a while after separation both of them divide more actively. The primary function of conjugation is the interchange of genetic information, so that both individuals have the opportunity to do new things because they have a new combination of genes.

Upon reaching reproductive maturity, organisms have various adaptations to make reproduction likely. In higher animals where two parents are involved, there is often a set of mating behaviors that males and females display to signal their readiness to go into sexual activity. Similar physical unions cannot be achieved in stationary, insensible plant parents even though a union of their sex cells is just as important as in animals. For plants, the same results are obtained when water, wind, or insects act as agents in getting sperm cells or pollen to or near the eggs. In many aquatic plants and animals the sex cells are turned free in the water where it becomes a matter of chance as to whether reproduction can be initiated by union of male and female cells.

ASEXUAL PLANT REPRODUCTION IS MUCH USED IN AGRICULTURE AND HORTICULTURE

A form of asexual reproduction called *vegetative reproduction* has been much used for growing plants. It ensures that the offspring of any desirable plant will be exactly like the parent. The reason is that all cells of an organism (with the exception of gametes) have exactly the same kinds of genes. Therefore, any fragment of a plant will be subject to the same internal influences, whatever they may be. Horticulturists and nursery operators capitalize on this characteristic when they habitually propagate some plants by separating buds or cuttings from the parent and establishing them as individuals.

Peach, apple, and grape varieties, as well as many other fruit and ornamental plants, are propagated by removing buds or twigs and fusing them to another plant. The buds or twigs are known as *scions*, and the plant to which they are fused is the *stock*. The process itself is called *grafting*. Peaches are propagated by a form of grafting called budding, in which case a bud scion is attached to a stock. In any case, the growth layer (cambium) of scion and stock are placed together so they may become continuous and grow.

The scion contributes most of the above-ground parts, including the fruit-bearing structures, and the stock contributes the lower stem and roots. Each provides characteristics of its own parentage, and individual sets of genes in scion or stock ordinarily exert little influence beyond the place where the scion and stock fuse. There are some exceptional cases where chemicals diffuse from one to the other or where the stock may impose limitations on growth by regulating the absorption of minerals from the soil. Tomatoes grafted to jimson weed, a close relative, will accumulate the drug atropine produced by the stock. Dwarf fruit trees result when scions are grafted onto stocks that have smaller root systems than the plant from which the scions came. They cannot, therefore, supply water and minerals in sufficient quantities to support the most vigorous shoot growth.

There are several reasons for grafting. Many fruits such as peaches or apples do not breed true; that is, plants coming from seeds are not like their parents. The best way to propagate them is by vegetative means. Vegetative reproduction is also needed by some other plants like the bananas, which are sterile hybrids that cannot produce seeds. Another reason for using asexual methods in propagating commercial plants is to save time. As much as two or three years can be saved over the time required to grow them from seeds. Also, it is sometimes advantageous to graft some high-yielding but delicate scion to a hardier stock.

The more closely related the two plants are, the more successful is the grafting. Much grafting is practiced within the species or genus, but grafts have sometimes been successful between genera.

Some plants are propagated by putting buds or branches into the ground and letting roots and stems grow. Sections with buds (eyes) are cut from Irish potatoes and planted like seeds. Branches from rose bushes are stuck into the ground, where they become established as their basal ends sprout roots. Growth-regulating chemicals such as auxins (Chapter 10) may be applied to the cut end to speed root development. In still other cases, branches are pressed to the ground and covered a short distance behind the tip with a shovelful of soil. When roots sprout in the soil, the branch is cut from the original stock. Horticulturists call this practice layering.

In nature a large number of plants reproduce asexually. Many creeping plants have stems that run along the ground and take root. *Runners* are above-ground prostrate stems that root at most of their joints; *stolons* are similar except that they root only at the tips. These terms are sometimes used synonymously. Rooting in the sense used here actually means formation of a new plant. The new plant soon becomes established so that connection with the parent is no longer necessary.

Many other plants have underground stems that sprout branches. They usually have a considerable amount of food stored in them, a decided advantage in getting new plants started before they can synthesize adequate food for themselves. Such is the case of the banana tree, for the upright tree is a branch from a horizontal underground stem known as a *rhizome*. Large grassland areas throughout the world often catch fire and burn clean. Yet, the same plants that were burned are able to survive and grow again from surviving underground parts. An enlarged tip of a rhizome is a *tuber*. The Irish potato is an example. It is solid and bears a number of buds called eyes. As previously mentioned, the tuber can be fragmented and planted. Many garden plants are placed in the soil as *bulbs*. A closer look reveals that they are of two types. The true bulb is like the onion. In the heart of it is a bud surrounded by fleshy scales that are bases of leaves in which food is stored. The other kind of bulb is really a *corm*. It is somewhat flattened, is solid internally like a potato, has dry scales on the outside, and has one or more buds at the apex. Crocuses and gladioli come from corms.

Plants may also propagate from roots or leaves. In the spring of the year sweet potatoes, which are roots, are put in hotbeds. Many plants sprout from adventitious (i.e., occurring in unusual places) buds on them. The new plants, called *slips*, are pulled from the beds and planted. Dahlias and daylilies are also root propagated. A few plants are reproduced from leaves. Among them are African violet (*Saintpaulia*), life plant (Bryophyllum), and hens-and-chickens (Sempervivum).

BUDDING, FRAGMENTATION, AND SPORULATION ARE FORMS OF ASEXUAL REPRODUCTION

In addition to the form of asexual reproduction called binary fission (where the offspring are nearly equal in size because they form by the splitting in half of the parent), there are forms in which only a small portion of the parent becomes the offspring. Among these is *budding*, which is characteristic of a wide range of organisms (Fig. 13.2). It is the separation of a small reproductive fragment from a larger body. Multicellular buds are produced in the nonvascular plants, liverworts and mosses, where they are called *gemmae*. Fleshy buds called *bulbils* are produced in the seed plant tiger lily and others. Bulbils drop to the ground, where they grow into new plants. Budding is common in animals of Phylum Porifora (sponges) and Phylum Cnidaria (e.g., Hydra).

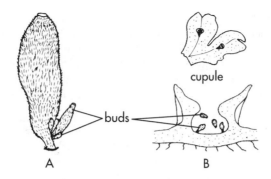

FIGURE 13.2. *Budding. A:* Grantia, *a sponge. B:* Marchantia, *a liverwort.*

There are some cases where an entire organism breaks into more than two parts. In other words, the parent is completely utilized in the process. This kind of asexual reproduction is *fragmentation.* For instance, inside red blood cells one stage of the malarial parasite breaks into numerous fragments called merozoites.

Sporulation is asexual reproduction by the production and release of *spores.* They reproduce without fertilization and are often encased in protective coverings that enable them to survive unfavorable periods. Spores are common in fungi and plants, where many of them are part of life cycles having alternating gametophytic and sporophytic generations (see below). Spores are the most important means of dispersal in plant groups lower than the seed plants.

BOTH ASEXUAL AND SEXUAL REPRODUCTION HAVE ADVANTAGES

Sexual reproduction is the most used method, a fact which could be interpreted to mean that it has advantages over the asexual method. The most-cited advantage is that every sexual union gives the opportunity for mixing the genetic materials of two somewhat different individuals: a *recombination* event. There are no guarantees that recombination brings about an individual with advantages over either parent; indeed, it could produce an inferior organism. However, most biologists in this field of study believe that, in the long run, a species improves by mixing genetic information and producing variation within the group. As will be discussed in Chapter 15, evolutionary change depends on such variation within a population.

However, sexual reproduction also requires extra use of energy on the part of participants. They must differentiate into males or females. They must find appropriate mates, or send their gametes off into the environment. Some animals participating in sex must use behavioral or chemical signals to overcome hostility of potential mates. All of these require the use of energy on the part of participants, in an endeavor that does not guarantee success for the species.

Asexual reproduction leads to the forming of a quite homogeneous population that might or might not be well suited for its current or future environments. Aside from the occasional mutational change, there is no way to create variation. However, asexual reproduction is certainly easier than sexual, since an isolated individual can achieve it. The success of bacteria is a testimony to the efficacy of asexual reproduction over very long time periods.

Actually, those organisms capable of performing both types of reproduction may be at the greatest advantage. They can perform the simpler asexual method if they are matching their environment well or if there are no appropriate mates in the area. They can then switch to the sexual style when the environment deteriorates. It is also possible that such organisms could periodically use the sexual phase to produce quite different body styles that would be appropriate for dispersal into new environments. For instance, the larva produced by sexual reproduction can be quite different in anatomy and abilities from the adults that produced it. Some animals that are sessile (nonmoving) use motile (moveable) larvae to disperse a population and avoid competition for resources with their offspring. An example is the sea anemone that reproduces asexually to make sessile adults like itself but can also make very motile larvae after sexual reproduction.

GAMETOGENESIS IS THE NECESSARY PRELUDE TO SEXUAL REPRODUCTION

As discussed above, there are some obvious advantages for species capable of producing gametes for sexual reproduction. Let us begin a discussion of this topic by examining the types of gametes that are made and where in the body they are produced. The overall view is that gametes must always be haploid by the time they contact each other to make a new individual. Chapter 4 describes the details of *meiosis*, the specialized movements of chromosomes that occur to bring cells from diploid to haploid. Remember that during the same time, there is a second event that helps sexual reproduction to be of value: crossing-over to increase the amount of genetic recombination. Because of the need to go through meiosis, not just any cell of the adult body can become a gamete. In animals, eligible cells are *primordial germ cells*, which look different from and are treated differently than other cells of the body from the embryonic period right on through the rest of life.

The process of maturing into gametes sometimes also includes changes in overall shape and cytoplasmic contents. Even if mature gametes look alike superficially, there is a physiological difference; for some of them are attracted to certain ones but not to others. The two types are labeled male and female or sometimes in lower organisms plus and minus strains. The common bread mold (a fungus) produces gametes whose sexes are indistinguishable unless they are tested by growing close to another whose sex (strain) is known. If zygotes form, the two strains are sexually different; if they do not form, they are alike. Usually sex cells distinctly differ from each other in both size and structure. Gametes visibly alike are called *isogametes*, whereas those that differ are called *heterogametes*.

Some Organisms Reproduce by Using Isogametes

Reproducing with isogametes is regarded as a primitive method, for the gametes show no signs of morphological specialization. Few organisms living today are isogamous, but this kind of reproduction is known in simple animals and is fairly common in algae and fungi. The freshwater filamentous alga *Ulothrix* is isogamous (Fig. 13.3 A). Some of the cells of its filament specialize as *gametangia* (gamete containers) and produce flagellated gametes. The gametes from one filament do not fertilize others from the same filament. Such lack of affinity indicates that there are two strains, visibly alike but sexually different. Some algae do not make and release isogametes that swim freely; rather, their cells temporarily fuse in a conjugating formation during which contents of the (+) strain pass into the (−) strain cell (Fig. 13.3 B). In this case, the sexually produced organism is the (−) cell which contains genetic information of both strains.

Anisogamy Is the More Prevalent Sexual Reproduction Method

Anisogamous gametes are visually distinguishable. Typically, the larger ones are female and the smaller ones male. Female gametes are *eggs* (*ova*). They contain stored food for the embryo that will be the result of fertilization. They generally have no methods of self-locomotion. Although they are the

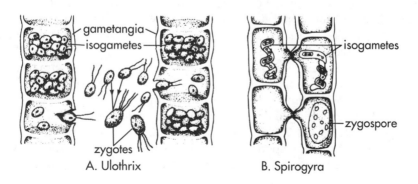

FIGURE 13.3.   *Isogamy in green algae. A:* Ulothrix *with swimming isogametes. B:* Spirogyra *in a sexual process with the conjugating cells functioning as isogametes.*

larger sex cells, they vary considerably in size among themselves. The difference in size is due primarily to the difference in amount of stored food (usually in the form of *yolk*). The eggs of flowering plants are invisible to the naked eye. In contrast, yolk of the hen's egg is about two centimeters in diameter. In cases such as the human, where the parent protects and nourishes the young internally from the very start, the eggs are small, about 0.15 millimeter in diameter. When the embryo develops outside the parent's body and when nourishment comes from the egg, more reserve food is necessary to ensure its survival. Generally speaking, the larger the egg the better the chance of survival rate, and the better the survival the fewer the eggs. Birds and some of the reptiles produce the largest eggs, which contain enough yolk to supply the growing young until they are well formed and hatched. In the chicken the incubation period is three weeks. By comparison to birds, the oyster, hookworm, tapeworm, fish, and frogs lay very small eggs. A single oyster is said to lay up to 60,000,000 eggs. A hookworm may lay up to 36,000,000 eggs. A tapeworm inside a cow has been estimated to lay 2,500,000,000 eggs over a period of ten years. The scarcity of reserve food in these small eggs and the hazards they encounter result in most of them dying before new individuals form. Reproductive success is possible only because eggs are produced in enormous quantities. In addition to the relationship between stored food and survival, there seems to be some relation between the number of eggs produced and the care given to them by the parents. When care is given, the number produced is often fewer.

The male gametes are *sperm*. They have little or no stored food and usually cannot survive for long. Of course, their small size enhances their ability to be mobile. They are produced in large quantities, and this compensates for their hazardous existence. It is thought that human sperm live in the female genital tract for only about 80 hours at most, perhaps being able to fertilize for only a portion of that time. An interesting contrast is the sperm of honey bees. No one knows exactly how they manage to survive so long, but they do remain alive and functional in the queen's body from the time she receives them until she dies, a period of about three to five years. Many other insect species also have long-lived sperm.

Most male gametes, especially in animals and lower plants, propel themselves with swimming appendages, flagellum-like structures often called tails. The tails of animal sperm cells move by rhythmic contractions of their microtubules, which are arranged in a configuration strikingly similar to those of flagella and cilia (Fig. 4.4) The human sperm has one tail, the moss sperm has two flagella, and the fern sperm has several flagella. Nonmotile sperm, which are typical of seed plants, are conducted through pollen tubes to the vicinity of the egg.

Gametes Are Produced in Specialized Regions

In animals, sperm cells are made in the organs called testes. A testis contains the germ cells that reproduce to (a) mature into sperm, or (b) maintain a constant cell line for making more sperm. In humans and many other animals, the germ cells are embedded in the walls of tiny tubes, called *seminiferous tubules*. As a human sperm cell matures, it gradually migrates toward the wall portion that borders the lumen (central cavity). Fully mature sperm fall into the lumen and are transported by fluid toward the outside of the testis. Many such tubules converge to become one tube, the *epididymis*, which is continuous with the tube called the *vas deferens*. The vas of each testis empties its contents into a single tube, the *urethra*, which opens to the outside of the body via the *penis*. Other animals may have quite different tubal arrangements for sperm release.

Surrounding the seminiferous tubules in a testis are *interstitial cells*. They are specialized to manufacture and release the male hormone, testosterone.

Eggs are produced in *ovaries*. The hormonally induced maturing of an animal egg within the ovary was described in Chapter 10. Many of the extra nutrients and cytoplasmic organelles that make an egg large are placed in it by the neighboring ovarian cells. In some animals, such as insects, these supporting cells are named *nurse cells* to denote their role.

The degree to which an egg completes the multiple steps of meiosis before being released from the ovary varies widely by species. In humans, the eggs complete Meiosis I (see Chapter 4) even before a female is born. However, they do not complete Meiosis II unless they are contacted by a sperm cell, after release from the ovary.

As with testes, ovaries produce sex hormones in addition to gametes. Chapter 10 describes the cycle of producing estrogen and progesterone by ovaries.

TRANSFER OF SPERM TO EGGS IS ALWAYS IN A LIQUID MEDIUM

A liquid medium is required if swimming sperm are to reach the eggs. There is no problem when they are released into bodies of water or within the female's body. The necessity for water, however, limits sexual reproduction of mosses and ferns to rainy periods, since sperm move to an egg over the surface of the plants. In animals, where sperm are deposited internally into a moist environment, the transfer is independent of environmental conditions and is more efficient.

By transferring *pollen grains*, seed plants overcome the dependency on water. The grains, carried chiefly by wind and insects, eventually produce two sperm apiece. At the flower, each grain produces a tube through which the sperm are carried to the vicinity of the egg. The tube tip ruptures, allowing a sperm to unite with the egg. This is described in greater detail later in the chapter.

The success of sperm transfer depends on the different methods used. In stationary animals (such as coral animals) and plants the only possibility is to release gametes, or at least the sperm, into the environment and depend on their swimming or being carried by some outside agent. A wasteful practice results when sperm are simply turned free in the water and only accidentally find the eggs. When animals can move around they are apt to be attracted together, thus closing the distance over which sperm must be transferred. The frog is thriftier than stationary animals in that the male attaches himself upon the female's back, and spreads sperm over the eggs as she lays them.

ANIMALS AND PLANTS TYPICALLY HAVE SPECIFIC MATING PERIODS

Frequently, mating behavior occurs only at certain seasons of the year that coincide with the production of gametes. The seasonal or periodic cycles of alternating fertile and sterile periods are associated with chemical changes within the body. In some animals the female passes through fertile periods called *estrus*. Only during this period will she allow copulation. The presence of a sexually ready female is sometimes advertised by her release of chemical signals called *pheromones*. This is particularly effective in many insects that are scattered over a wide range. For instance, the females of some moths emit pheromones that can be sensed by cells on the males' antennae. It is said that a single molecule of the proper pheromone attaching at an antennal receptor is sufficient to change the pattern of nerve activity in the male. The male moth is initally stimulated to fly upwind; then as it comes near the female it can sense the concentration gradient of pheromone, thus flying directly toward its source.

SEXUAL REPRODUCTION IN PLANTS IS KEYED TO ALTERNATION OF GENERATIONS

Mosses, ferns, and seed plants go through life cycles in which a sexual phase alternates with an asexual one. All have their special structures; but to illustrate the concept, a generalized life cycle, followed by a few representative examples, will be explained.

The Generalized Plant Life Cycle Includes Both Sexual and Asexual Periods

Unless a plant is propagated entirely by vegetative means, it will go through a cycle involving an alternation of sexual and asexual phases (Fig. 13.4). The

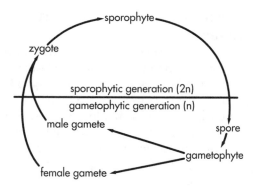

FIGURE 13.4.   *Generalized life cycle of plants. n represents haploid, 2n represents diploid. Gametes fuse to form a zygote in the process called fertilization. The sporophyte produces spores in the process called meiosis.*

sexual plant is called a *gametophyte* because it produces gametes, and the asexual plant is called a *sporophyte* because it produces spores. The chromosome number of the gametophyte is haploid and the sporophyte is diploid. Therefore any structure, whether whole plant or single reproductive cell, is assigned to the *gametophytic generation* if it is haploid and to the *sporophytic generation* if it is diploid.

The sexual individual (gametophyte), which is unlike most sexual animals in being haploid, produces gametes by mitosis. Two gametes unite and form a diploid *zygote*, the first cell of the sporophytic generation.

Each zygote grows into a multicellular plant called the sporophyte, which is also diploid. The sporophyte produces haploid spores (the gametophytic generation) by meiotic division. Spores grow into gametophytes, an event that brings the cycle again to the starting point.

It should be remembered that gametophytes do not produce gametophytes nor do sporophytes produce sporophytes. Rather, gametophytes produce sporophytes through the union of their gametes, and sporophytes produce gametophytes by meiosis. Also both gametes and spores are haploid and are of the gametophytic generation. The generalized life cycle in Fig. 13.4 is adaptable to most plants.

The plant of either generation may vary considerably in size from that of the other generation. Lower plants have more conspicuous gametophytes, and higher plants have more conspicuous sporophytes. In flowering vascular plants such as trees the sporophytic generation is what is seen as an independent organism. A moss plant is a gametophyte and cannot be compared with an oak tree. Instead it must be compared to the oak's gametophyte, which is a pollen grain or embryo sac (to be discussed later).

The Moss Life Cycle Features a Prominent Gametophyte Generation

The familiar moss plants are gametophytes that may have their sex organs in different plants or together in the same plant. Sex organs (gametangia) are produced at the tips of upright plants or branches (Fig. 13.5). Male structues are multicellular sperm-carrying bags called *antheridia* (singular: *antheridium*). Female structures are somewhat bottle-shaped structures called *archegonia* (singular: *archegonium*). Archegonia are multicellular and produce a single egg. The archegonium has an enlarged lower part which contains the egg and a hollow neck through which the sperm enters. Sperm can swim to the eggs only during wet periods.

After fertilization, the zygote grows into a sporophytic plant, which remains nutritionally dependent upon the female gametophytic plant. It consists of a leafless *stalk* with a *foot* on the bottom and *capsule* on top. The capsule produces haploid spores by meiosis. At this point there is a departure from the generalized plant life cycle given in Fig. 13.4. Instead of a spore's growing directly into the gametophytic plant proper, it germinates into a green algal-like

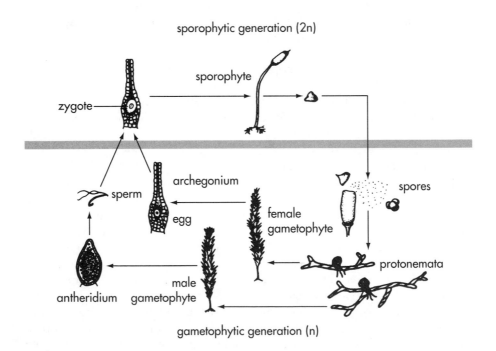

FIGURE 13.5. *Moss life cycle. Fertilization produces the zygote that becomes the sporophyte; meiosis produces the spores that become the gametophytes.*

filament, the *protonema*. This structure develops buds that grow into individual gametophytic plants.

The Fern Life Cycle Features a Prominent Sporophyte

Logically one begins a life cycle with the most conspicuous generation, which in the case of ferns is the sporophyte. Most ferns have cinnamon-colored, grainy-looking spots on the back of some of their leaflets or similar grainy material along the margins of leaflets (Fig. 13.6). The spots are known as *sori* (singular: *sorus*). Closer inspection reveals a sorus to be a clump of lens-shaped spore containers, the *sporangia* (singular: *sporangium*). When they mature and dry, the sporangia crack open and free the haploid spores produced within them.

A spore falling in a favorable place germinates into a gametophytic plant known as the *prothallus*. It is a simple, flat-growing, green organism somewhat heart-shaped. It is about a centimeter long and is attached to the ground by *rhizoids* (root-like structures) on the undersurface. A prothallus contains

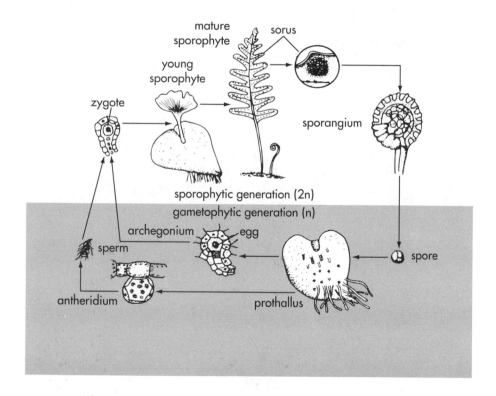

FIGURE 13.6.   *Fern life cycle. Sperm and egg unite in fertilization to produce the zygote that becomes the sporophyte; meiosis produces spores that become the organisms of the gametophytic generation.*

both sexes; both archegonia and antheridia are on the bottom surface. Archegonia are near the notch and antheridia near the point. On any particular prothallus, antheridia usually develop before the archegonia. The earlier maturation of antheridia is a barrier to self-fertilization. As in almost all lower plants, water is necessary for the motile sperm to get to the eggs of another prothallus. The zygote that follows fertilization grows into the familiar fern plant, the sporophyte.

THE FLOWERING SEED PLANT IS THE MOST HIGHLY DEVELOPED SPOROPHYTE

As will be shown in detail below, the flowering plants have several very specialized features, including the ability to protect the developing embryo by holding it deep within the sporophyte body. Its sporophyte form is the prominent portion of the life cycle; nevertheless, it follows alternation of generations as described above.

The Sporophyte Form Is a Prominent Flowering Plant

The flowering plant, like the fern, belongs to the sporophytic generation. It produces spores in the flower. Actually, the flowering plant produces two kinds of spores, male and female, either in a single flower or in separate flowers.

The flower of the plant is composed of a stem to which is attached one or more specialized parts (Fig. 13.7). The crown of the stem upon which floral parts are attached is the *receptacle*. If all possible floral parts are present, there will be four different groups, usually arranged in whorls. Beginning at the base, they are *sepals, petals, stamens,* and *carpels* (often a single carpel).

Sepals and petals are not essential for reproduction but are accessory parts

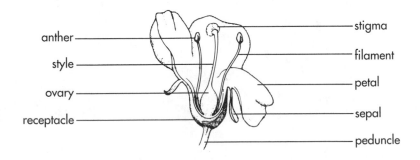

FIGURE 13.7.   *Structure of a flower.*

of a flower. Sepals are often small and green. They protect the upopened flower. Petals are the usual showy appendages that characteristically attract insects for pollination. As one might expect, wind-pollinated flowers usually do not have them. There are flowers like the tulip or lily in which both the sepals and petals are of a similar shape and color.

The essential parts of flowers are the stamens and carpels. Each stamen has a stalk called the *filament* and a spore-producing part called the anther. The anther produces small haploid spores, *microspores*, that develop into male gametophytes known as the *pollen grain.*

In its typical form a carpel has a swollen basal part called the *ovary*, an erect structure called the *style*, and a tip called the *stigma*. Some flowers have more than one carpel, fused together. There are several clues to look for to determine the number of carpels present; for example, multiple styles or lobes on the stigma, multiple lobes of the ovary, multiple cavities in the ovary, or rows of ovules. Cavities of the ovaries contain *ovules*, which are forerunners of seeds (Fig. 13.8). Each ovule in turn contains a sporangium that produces a single functional large spore, the *megaspore*. The megaspore undergoes cell reproduction to develop into a *megagametophyte* commonly known as the *embryo sac* (see details below). Fertilization occurs within this sac.

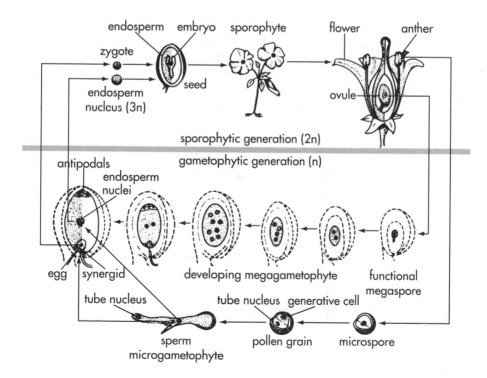

FIGURE 13.8. *Life cycle of a flowering seed plant.*

Flower Parts Are Used by Taxonomists for Classification

When flowering plants are identified, characteristics of their flowers are preferred for descriptions because they are not as variable as other parts, which are quite alterable by environmental influences. If one seriously intends to identify these plants, it is important to know certain terms that describe the flower as a whole. A *complete* flower is one having at least one member of all four sets of parts (sepals, petals, stamens, and carpels); an incomplete flower has at least one set absent. A *perfect* flower has at least one member of both sets of essential parts, stamens and carpels; an *imperfect* flower has one or both sets of the essential parts missing. Many specific terms are used to describe the various configurations that are taken by flower parts of various taxonomic groups, but these are beyond the scope of this book.

A flower is interpreted as a modified leafy twig in which the leaves are specialized as floral parts, and the stem to which they are attached is considered shortened through the reduction of the internodes. That this interpretation has merit can easily be seen in similarities to vegetative leaves. This is especially easy to see in sepals, many of which are leaf-like and green. Even petals may have leaf-like shapes; there is a green rose in which petals not only look like leaves but actually photosynthesize. Generally the stamens do not look leaf-like, but in a waterlily there is a continuous transition from petals, to petal-like structures bearing anthers, to typical stamens. The carpel seems quite like a leaf if one uses a little imagination.

A number of *fruits*, which are mature carpels, split into leaf-like parts. A good example is the pod of the garden pea. If it is split—as in shelling the peas—and the hull edges are spread apart, the pod resembles a leaf (Fig. 13.9 A). Such a resemblance is misleading in that the pod splits along the midvein. The pod actually develops with the seeds along the leaf margin (Fig. 13.9 B).

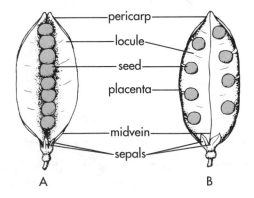

pericarp
locule
seed
placenta
midvein
sepals

A   B

FIGURE 13.9.   *Pea pod. A: pod split in usual way along midvein. B: pod split where margins of carpel are fused leaving midvein intact.*

In addition to the usual parts of a flower, other modified leaves may be attached below the sepals. They are called *bracts*. Bracts may be papery structures ensheathing a single flower as in the iris, or petaloid structures beneath a cluster of flowers as in the flowering dogwood. Generally they are incorrectly called petals in the dogwood. The same is true of poinsettias, in which the bracts are showy white or red leaves beneath relatively inconspicuous flowers. An entire family of plants, the *composites*, have compact heads of small flowers beneath which are bracts that look like sepals. Examples of this family are daisy, sunflower, dandelion, aster and dahlia. The so-called petals in this family are actually separate flowers.

The Female Gametophyte Portion of a Flower Must Develop Before Fertilization

The pattern of development of the female gametophyte as given here is but one of several common types. Development begins in the ovule, which contains a single *spore mother cell* (which is diploid). The mother cell undergoes meiosis, thus producing four haploid megaspores. One of them is large and functional, while the other three are small and vestigial. They are arranged in a column with the larger functional one most distant from the micropyle, the opening through which the pollen tube enters the ovule. The events that happen next to produce both the egg and supporting cells around it are rather complex.

The functional megaspore begins to enlarge, and simultaneously its nucleus divides by a series of three divisions to produce eight nuclei. Three of them move close to the micropyle, three to the opposite end, and two stay close together in the middle (Fig. 13.8). The six nuclei and the adjacent cytoplasm located at the ends of the embryo sac are organized into six small but distinct cells separated by thin membranes. The remaining central cell is much larger and contains two nuclei. One of the three cells at the micropyle end, usually the middle one, is an *egg*. The other two are called *synergids*. The three at the opposite end are the *antipodals*. The large cell in the middle contains two nuclei called *endosperm nuclei*. The entire unit of seven cells (with eight nuclei) is the female gametophyte (megagametophyte), which is also known as the *embryo sac*. Since all of its nuclei came from a single haploid nucleus, they are also haploid. The embryo sac is never seen except in the laboratory because it is contained within the ovule. When the embryo sac reaches the stage of development just described, the egg contained within it is ready for fertilization.

The Male Gametophyte Is Also Developing Before Fertilization

The point has been made that the anther of the stamen produces microspores. They come from diploid *spore mother cells*, which undergo meiosis. A single mother cell produces four haploid microspores that stick together in a *tetrad* when they are first formed. The microspores soon separate and develop into *male gametophytes* (microgametophytes) as described next.

The first step in the development of a male gametophyte (Fig. 13.8) is the division of the microspore nucleus into two nuclei. The two nuclei with their surrounding cytoplasm, sometimes separated by a membrane, are known as the *tube cell* and the *generative cell*. This two-celled unit is the pollen grain, which in reality is an immature male gametophyte. It is immature in the biological sense since it has not yet produced functional sperm. The pollen grain is transferred to the stigma of the carpel, where it continues to grow.

Pollination Leads to Union of Egg and Sperm

Pollination is the transfer of pollen (the immature microgametophyte) from the anther of one flower to the stigma of a second flower. Although pollination is obviously distinct from fertilization, it is a necessary prerequisite to it. Most flowers are insect- or wind-pollinated and often have devices that increase their chances for successful pollination. Insect-pollinated flowers have petals and nectar that attract insects, which seek nectar as a food source. The pollen tends to be sticky and clings more easily to insect bodies. Wind-pollinated flowers lack petals; or if present, they are relatively inconspicuous. They seldom produce nectar. Their pollen is light and dry and sometimes has membranous appendages; all of these characteristics make it easier to become wind-borne. Pollen is produced in large quantities, enhancing the opportunity for pollination, since much pollen is lost. The carpels may also be feathery, thus exposing a larger surface that catches the pollen.

Some flowers have unusual ways of effecting pollination. The mountain laurel (*Kalmia*) petals have folds into which the stamens are tucked and held under tension. When an insect alights on the petals, the stamens spring out of the folds and dust pollen over the body of the insect. One kind of mint has a bumper device on the stamen. The visiting insect, while seeking nectar, pushes against it, causing the anther to come down and rub pollen on its back. In some cases, flowers and the particular insects that pollinate them have evidently undergone co-evolution to make perfect matches. For instance, the carrion plant (*Stapelia*) has reddish blotches resembling blood on its petals and it smells like rotting meat. It is pollinated by flies looking for rotting meat to lay eggs in; when they mistake the flower for meat, they carry off its pollen on their hairs.

Cross-pollination Is the Ideal Event

In nature individuals do not commonly fertilize themselves, since this subverts the recombinational advantage of sex and may even lead to expression of more than the average number of recessive harmful mutations. It is true that most flowering plants bear both sexes in the same flower, and one must wonder how this does not often lead to self-fertilization. The answer is that there are often ways of preventing self-fertilization. The maturation of male and female parts at slightly different times, self-sterility, or mechanical barriers are the usual methods.

The Male Gametophyte Completes Its Maturation Only After Pollination

The development of the male gametophyte (pollen grain), which begins before pollination, is continued after pollen becomes attached to the stigma. A tube known as the *pollen tube* grows down through the style into the micropyle of the ovule to the vicinity of the egg. The development of an internal tube is an important advancement in that it eliminates the necessity of water for fertilization. Measured in terms of just hours, growth of the tube is usually at a rapid pace. As this growth occurs, the nucleus of the pollen's tube cell commonly moves along at the tip of the tube. Meanwhile, the generative cell divides to become two sperm cells (sometimes this occurs before shedding of pollen). The pollen tube penetrates the embryo sac, where it releases its contents.

Fertilization Brings Egg and Sperm Together and Also Produces Supporting Cells

One of the two sperm cells unites with the egg, thus giving rise to a zygote, which develops into a diploid embryonic sporophytic plant. The other sperm cell unites with the two endosperm nuclei of the embryo sac to form a cell with a triploid (3n) nucleus. The triploid cell proliferates into a food storage tissue (Fig. 13.8) called *endosperm*. This tissue comprises the bulk of many seeds like the grains, where it nourishes the young seedling plant. However, in other plants (such as the bean) the endosperm appears to be totally utilized by the developing embryo even before the seed goes into dormancy. Such an embryo commonly moves the endosperm nutrients into leaf-like storage places called *cotyledons*.

Thus, it can be seen that the flowering plants perform a unique *double fertilization*, with one sperm uniting with an egg to make a zygote and another sperm helping to produce a supporting tissue, the endosperm.

Seeds Are the Early Embryos Plus Supporting Tissues

Fertilization causes the *setting* of seeds, a process that is the maturing of ovules. Each seed contains an embryo and enough reserve food to get the new plant started. The seed is protected on the outside by a tough coat. Besides reproducing, seeds result in the dissemination of the plant, a necessity if sufficient numbers of them are to get to favorable sites for survival. They, or the fruits containing them, may have wings or other appendages that make them buoyant in the air, may have hooks or barbs that make them stick to animals as they pass, may be exploding fruits that expel seeds forcefully, may be collected and carried elsewhere by animals using them for food, or may have numerous other methods of dispersal. The rapid invasion of plants into denuded areas, such as newly plowed fields or newly constructed lakes, is a tribute to the effectiveness of the various methods. In lakes, seeds are often brought from distant places by migratory birds. Such seeds may be sticking to their bodies or may pass undigested through their intestines.

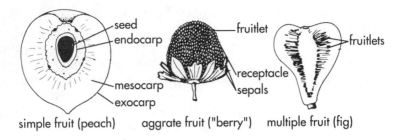

simple fruit (peach)   aggrate fruit ("berry")   multiple fruit (fig)

FIGURE 13.10.   *Fruit types.*

Fruits, Built Around Seeds, Provide Important Features

Flowering plants always produce seeds contained within fruits. Usually the fruit is an enlarged ovary, although it may sometimes include additional accessory tissues of the receptacle or some other part. It may be important in seed dispersal or may be a supplemental food source for germinating seeds.

Fruits are an extremely diverse lot (Fig. 13.10). They are classified as *simple* if they come from a single carpel of a single flower—for example, the peach. They are classified as *aggregate* if they come from several carpels of a single flower. Often aggregate fruits have accessory tissue that forms the bulk of the fruit. Examples are strawberries and blackberries. Fruits are classified as *multiple* if they are derived from carpels of several to many flowers clustered close together. Examples are figs, pineapples, mulberries, and corn.

Germination Is the Start of Embryo Growth
Leading to a Mature Sporophyte

A seed contains an early embryo that has become arrested in its development (awaiting favorable environmental conditions, such as moisture). The resumption of growth of an embryo plant is termed *germination*. In most cases the lapse of time between maturity and germination is a single season, but some seeds with hard coats or immature embryos may require longer. Some seeds have been known to germinate after more than a hundred years, but most of them remain viable for only a few years.

MOST ANIMALS CAN PERFORM SEXUAL REPRODUCTION

Even the simplest animals, such as sponges and jellyfish, rely heavily upon sexual reproduction (although many also perform budding and other asexual methods). Earlier portions of this chapter and sections of Chapters 4 and 10 have already covered several aspects of animal reproduction, such as meiotic production of gametes, endocrine control, and mating behavior. After description of some unusual variations upon the theme, the remainder of this chapter will focus on human reproduction.

Some Sexual Animals Bypass the Process

Parthenogenesis is the term applied when an animal that is anatomically capable of performing sexual reproduction instead goes to the asexual method. For instance, certain species of fish have both male and female individuals, which sometimes mate. However, females can produce haploid eggs that double their chromosomes and retain both sets to become diploid. These eggs then go into normal embryonic development just as if they had been fertilized by sperm.

A special case occurs with the honeybee. The queen receives sperm from a haploid male and stores it in a pouch. She can release sperm cells from this pouch to fertilize some of her eggs, which develop into females (workers or queens). Or she can release eggs without allowing them to contact sperm, in which case the embryos develop into males.

Some Animals Can Be Both Male and Female Simultaneously

Some animals are *hermaphroditic*, meaning that each body contains functional ovaries and testes. The common earthworm is monoecious. Two earthworms can mutually fertilize each other's eggs, but a single earthworm cannot fertilize its own. Some monoecious animals prevent self-fertilization by making the gonads of the two sexes functional at different times, even different years in the case of certain Molluscs.

Some Animals Are Capable of Switching Their Sex

In perhaps the most unusual cases, individuals of some fish species switch from being wholly one sex to being wholly the other, with both forms being functional. For instance, in one species of the fish commonly called the wrasse a male maintains a harem of females, with which he mates. But if that male dies, one of the females of the harem switches to being male and replaces him as the dominant fish.

HUMAN REPRODUCTION INVOLVES COMPLEX ANATOMY AND BEHAVIOR

Female Sexual Anatomy Allows Internal Fertilization and Nurture of a Baby

The organs producing eggs are the two *ovaries* located in the pelvic region (Fig. 13.11) and held in place by ligaments. They usually alternate in producing eggs. The eggs are produced by constantly maturing follicles (Chapter 10),

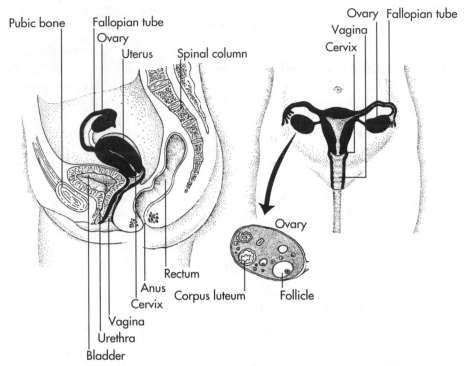

Pubic bone   Fallopian tube
              Ovary
                Uterus   Spinal column

Ovary   Fallopian tube
Vagina
Cervix

Ovary

Rectum
Anus   Corpus luteum   Follicle
Cervix
Vagina
Urethra
Bladder

FIGURE 13.11.   *Human female reproductive anatomy. (From John W. Kimball,* Biology, *6th ed. Copyright © 1994 Wm. C. Brown Communications, Inc., Dubuque, Iowa. All Rights Reserved. Reprinted by permission.)*

which rupture at the surface and free the eggs in the process known as *ovulation*. A woman periodically produces eggs, usually one at a time, throughout her fertile life, which typically extends from the early teen years into the late forties. Although the intervals between the production of eggs vary, a fair estimate of the average length of time would be about 28 days.

After ovulation, the egg enters the expanded free end of the nearby *fallopian tube* (also called *oviduct*) and moves toward the uterus, swept along by cilia of the tube's epithelial cells. Since the ovary and fallopian tube are not actually connected, it is possible (but rare) that an egg fails to enter the fallopian tube. It is even possible (but even rarer) that fertilization can occur just as such an egg slips into the abdominal cavity, where an unsuccessful pregnancy will occur.

The fallopian tubes lead to a muscular chamber, the *uterus*, in which embryonic development of a fertilized egg can occur. The uterus connects to the outside world via a tube, the *vagina*. The lower, narrow portion of the uterus that connects with the vagina is the *cervix*. The vagina serves two purposes: a place for deposition of sperm cells during intercourse and an exit for the baby at birth.

External female anatomy, collectively called the *vulva*, includes a pair of lip-like folds called the *labia minora*, surrounded by thicker folds, the *labia majora*. The vaginal opening lies within the labia, just dorsal to the exit of the urinary tract, the urethra. Ventral to the urethra is a small projection of flesh, the *clitoris*. Richly supplied with nerves and blood vessels, this is embryologically derived from the same tissues as the male penis and serves as the center of sexual sensations during intercourse.

Male Sexual Anatomy Provides Sperm and a Delivery Device

The organs that produce sperm are the two *testes* (Fig. 13.12) suspended in a sac called the *scrotum*, where the temperature is a little lower than in the body proper. This temperature difference is significant in maintaining their ability to produce viable sperm. As described earlier in the chapter, each testis contains a tangle of tiny tubes, the *seminiferous tubules*, in whose walls the sperm cells come to maturity. The route of sperm from these is in the following sequence. They go first to very small tubes called *vasa efferentia*, which collect into a single *epididymis*. As this tube leaves the testis it is called the *vas deferens*. The pair of vasa deferentia join at the urethra, which extends the length of the *penis*. At various times, therefore, the urethra serves to conduct either sperm cells or urine to the outside.

As sperm moves through the vas deferens, fluid is added to it by several glands: paired *seminal vesicles*, the *prostate*, and paired *Cowper's glands* (also called bulbourethral glands). These add volume to the sperm suspension and raise its pH. The material that enters the urethra, including sperm and added fluid, is called *semen*.

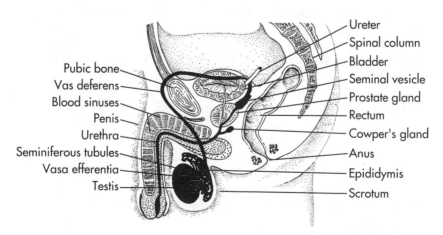

FIGURE 13.12. *Human male reproductive anatomy. (From John W. Kimball,* Biology *6th ed. Copyright © 1994 Wm. C. Brown Communications, Inc., Dubuque, Iowa. All Rights Reserved. Reprinted by permission.)*

The penis, like the clitoris, has many blood vessels and nerve endings. As sexual stimulation occurs, the arteries supplying the penis dilate, allowing more than the usual amount of blood to enter. This blood engorges *blood sinuses*, causing the penis to enlarge and stiffen. Veins of the penis are constricted by this shape change, slowing the flow of blood from the area. The *erection* that is thus formed enhances the ability of the penis to lie deeply within the vagina in preparation for depositing semen near the cervix of the uterus.

Human Sexual Activity Proceeds in Several Stages

Unlike most animals, humans do not go through cycles of sexual responsiveness interspersed between periods of nonreceptivity. Whenever both partners choose, a series of events leading to intercourse can begin. These begin with sexual *excitement*, during which muscles tense and both the male penis and the female clitoris become engorged with blood, as do the female's breasts. In the female, glands in the vaginal wall release lubricant materials and the vagina itself lengthens during the excitement phase. Blood pressure rises and pulse rate quickens.

The second phase is that of *orgasm*. In a male, this is the time when semen is released, an activity called *ejaculation*. Typically, a few milliliters of semen are ejected, containing over 50 million sperm cells. In the female, orgasm is induced by stimulation of the nerve endings of the clitoris. It is marked by rhythmic contractions of the pelvic muscles and the walls of the vagina, accompanied by greatly accelerated heart and breathing rates.

The final phase of the sexual response is called *resolution*. At this time, both partners experience a relaxed feeling of well being, accompanied by a return to prestimulatory conditions. Tissues lose their blood engorgement and heart and breathing rates return to normal. The male typically enters into a state of sexual nonresponsiveness, but a female may be responsive repeatedly in a relatively short time interval.

Successful Fertilization Occurs in the Fallopian Tube

Although fertilization of an egg by a sperm cell can occur anywhere from the surface of the ovary to the uterine chamber, if it is to lead to pregnancy it must happen somewhere in the upper third of the fallopian tube. This gives time for the embryo to develop enough so that it will be able to carry out the release of enzymes needed to *implant* itself into the uterine wall. Since this area of the reproductive tract is far from the place of sperm deposition, only a very small proportion of the sperm population reaches it. If the number of sperm in the ejaculate is significantly lower than 50 million, there may not be enough reaching the critical area of the fallopian tube to ensure fertilization.

Many Contraceptive Techniques Are Available

A couple wishing to engage in sexual activity without pregnancy has a variety of *contraceptive* methods available. The oldest of these is to avoid intercourse at times when ovulation is likely. However, many women have this event at unpredictable times, making this "rhythm method" an unreliable one. Some contraceptive methods rely upon placing a physical or chemical barrier between egg and sperm. The male *condom* is one of the most reliable methods in this category, since the barrier placed over the penis is almost always impenetrable by sperm. A *diaphragm, cervical cap,* or *sponge* worn within the female partner can also be effective, although less so than the condom. It is best to enhance the female's barrier devices with the simultaneous use of *spermicidal* (sperm-killing) chemicals introduced into the vagina.

Other contraceptive methods depend upon chemically mediated prevention of gamete release from gonads. Most forms of *birth control pills* or implants for women act at the pituitary-ovary level by fooling the endocrine system with artificial hormones. Similar chemical treatment for control of sperm release are being tested.

There are also methods relying upon preventing uterine implantation of the embryo. One of these is the placement into the uterus of a small mechanical object, the *intrauterine device (IUD)*. An IUD seems to work by irritating the uterine lining so that it will not be receptive to implantation. There are also chemical methods of making the uterus less receptive to implantation, or even to cause expulsion of an implanted embryo.

The permanent method of birth control is surgery upon the male or female reproductive tracts. In males, this a *vasectomy*; in females this is *tubal ligation*. Both procedures are very difficult to reverse.

Sexual Activity Can Be a Conduit for Certain Disease Agents

Since the variety of sexual activity among humans is internal fertilization, with a great deal of contact between tissues and fluids of two persons, it is not surprising that it can lead to infection by certain disease agents. Sexually transmitted bacterial diseases include syphilis, gonorrhea, and chlamydia infection. The latter is the most common sexually transmitted in the United States at this time. Viral diseases include genital herpes and acquired immunodeficiency syndrome (AIDS). Yeast infections and transmission of some Protozoan parasites can also accompany sexual activity. Sexually transmitted diseases are in epidemic proportions throughout the world.

STUDY QUESTIONS

1. In what respect is conjugation of paramecia like a sexual process?
2. What are some obvious advantages achieved by asexual reproduction? What does it lack that sexual reproduction provides?
3. How are dwarf fruit trees produced?
4. Name an animal that can reproduce by budding.

5. Give examples of plant propagation by roots, stems, and leaves.

6. What are isogametes? Describe an organism that uses them.

7. Compare and contrast the sperm of plants and animals.

8. How do flowering plants become independent of environmental water in their fertilization event?

9. How does plant fertilization differ from pollination? What is germination?

10. In a female animal's ovaries, what are nurse cells?

11. What relationship, if any, is there between the quantity of eggs produced and the care furnished by the parent?

12. When does meiosis occur in the life cycle of animals? Of most plants?

13. Describe the use of pheromones to aid animal reproduction.

14. In plants, what is the distinction between sporophytic and gametophytic generations?

15. What is meant by the term double fertilization in plant reproduction? In what group of plants does it occur?

16. What is the male gametophyte of seed plants?

17. What is an embryo sac?

18. What is endosperm? From what does it originate? What is its function?

19. In flowering plant reproduction, what is the function of a pollen tube?

20. Classify fruits and give an example of each group.

21. What is a monoecious animal? How does such an animal avoid self-fertilization?

22. What are the functions of male accessory glands such as the prostate?

23. In what anatomical area does fertilization occur in humans? Why cannot successful pregnancy be achieved if fertilization is elsewhere?

24. Discuss parthenogenesis in animals.

25. Why are the penis and clitoris considered to be quite similar in form and (some) function?

14
EMBRYONIC DEVELOPMENT

What do we know about:
—early ideas about how the adult form comes into being?
—the interaction between sperm and egg at fertilization?
—why gastrulation is said to be the most important period of one's life?
—whether gene induction and embryonic induction have anything in common?
—how shaping of organs is controlled?
—how a human mother is connected to her developing embryo?
—when a human first moves, pumps blood, and develops fingers and toes?
—the causes of multiple births?
—how plant embryos differ from animal embryos?
These and other intriguing topics will be pursued in this chapter.

DEVELOPMENT IS A CENTRAL THEME IN BIOLOGY

One of the most amazing things that one could watch is the sequence of changes occurring as a single cell becomes a complete organism with millions of cells in their proper places and specialized to carry out all of their complex activities. The events that occur to get this job done are said to be *developmental*, and the field of study is *developmental biology*. This is a central theme in the study of organisms, because none of the other activities of the body can be carried out unless embryonic development is done properly. It encompasses anatomy, genetics, cell structure and function, and many other areas of study. It even goes beyond the embryonic period to include later predictable changes such as growth, metamorphosis, and aging. However, as the title of this chapter advertises, most of our topics will focus on the changes that occur before either birth or hatching (in animals) or maturation (in plants). Recent advances in molecular biology have given biologists powerful new tools to study the genetic basis for developmental change. Many questions that have been asked for at least a hundred years are now coming closer to being answered. It is an exciting time for people working in this field.

THOUGHTS ABOUT EMBRYO DEVELOPMENT HAVE A RICH HISTORY

We are not far removed from the days when sperm, the smaller eggs, genes, and other such things were unknown and when correct explanations of fertilization and embryological events could not have been made. Ancients knew that copulation was a necessary reproductive event long before sperm and human eggs were discovered. That they came close to expressing reality is seen in such literature as the Bible, where the male component is called seed. Although they did not know about human or other mammalian eggs, they did know about large bird eggs and saw that young came from them.

Perhaps it was because they observed the hatching of birds that our ancestors got the idea that animals were preformed in the egg in miniature—but more or less like the adult—and that the miniature animal just increased in size up to the time of hatching or birth, an idea called preformation. After development of the microscope and the discovery of sperm, some imagined they could see miniature individuals in human sperm. Then, during the latter part of the seventeenth and early part of the eighteenth centuries, arguments arose between those who thought organisms were preformed in eggs and those who thought they were preformed in sperm. Taken to its logical conclusion, the first organism of each species would have had gonads containing gametes with preformed individuals. These in turn would have had miniature gonads and gametes also containing still smaller individuals and so on to the most recent member of the species.

Careful experiments in the late 1800s showed preformation to be unacceptable. It was replaced with the explanation now known as *epigenesis.* (Actually, this idea can be traced back to Aristotle, but was not experimentally supported until the nineteenth century.) According to the epigenesis hypothesis, gametes carry a *potential* for developing into individuals of specific form rather than carrying preformed individuals. The potential is in the form of genetic instructions at the chromosomes. After union of gametes, there is a complex chain of developmental events, during which the genetic instructions are used to put together complex structures from simpler materials. In other words, an adult has to be developed anew with each succeeding generation.

Providing accurate descriptions of the anatomical changes that occur during embryogenesis occupied much of the time of workers in the late 1800s and into the 1900s. Then, experimental embryology began to become the more important area, to attack the question of how as well as what. The first experiments were manipulations of embryos: taking away portions, transplanting pieces into unusual positions, even grafting two complete embryos together. As biochemistry and molecular genetics became mature fields from the middle of the twentieth century onward, their tools were made available to developmental biologists. Since the 1970s, a typical textbook in developmental biology has had nearly as much to say about genes as about embryonic anatomy.

DEVELOPMENTAL BIOLOGY IS ABOUT MORE THAN MOLECULES

In spite of recent successes, it is clear that understanding the molecular components of an embryo is not enough to grasp fully what the embryo is about. Development concerns change, and change occurs in the framework of time. Developmental changes also have to do with spatial relations: what is neighbor to what. So the new molecular view must be merged with classical embryo studies of anatomy and sequences of events if one is to gain full appreciation of what is occurring.

OTHER FIELDS OF BIOLOGY DEPEND ON STUDIES OF EMBRYONIC DEVELOPMENT

Just as embryology is indebted to other fields, others are indebted to it. One such area is that of taxonomy. When one wishes to place a newly discovered organism into its proper taxon, it is important to examine not only the adult anatomy but also the methods by which that anatomy develops in the embryo. Evolutionary studies have also benefited from embryology, since some embryonic structures and events give clues to ancestral relationships. For instance, the evolution of vertebrate animals can be studied by examining the sequential embryonic changes in the major blood vessel patterns of the living classes of the vertebrates.

There are also important medical advances derived from embryology. In recent years prenatal care as it affects both mother and child has come under such close scrutiny that many of the hazards formerly accompanying pregnancy have been greatly reduced. Nutritional and toxic influences on development have claimed considerable attention. It is now known that the viral disease rubella contracted by the mother during the first three months of pregnancy may cause certain defects in newborn babies. The drug thalidomide was used as a tranquilizer until it was discovered to cause serious limb deformities in babies when the mothers used it during pregnancy. It has recently been discovered that a set of physical and mental abnormalities called fetal alcohol syndrome can be attributed to the moderate to heavy use of alcoholic beverages during pregnancy. The incomplete development of the spine resulting in spina bifida is now known to be largely preventable by maternal ingestion of the vitamin folic acid. The entire area of infertility treatment (including, of course, *in vitro* fertilization) is indebted to embryology.

It has been estimated that about 20 percent of all pregnancies end in miscarriages (many before pregnancy is detected) and that about one child out of 165 is noticeably deformed. The alertness of doctors has made many of the physical and mental defects less mysterious, and many can be corrected. Inguinal (groin) and umbilical (navel) hernias in children may result from improper closure of tissue layers in the body wall and can be corrected surgically. One type of congenital heart defect results from the failure of the partition between the heart chambers to close; it can usually be corrected by open heart surgery shortly after birth. Some embryonic abnormalities can be detected by ultrasound well before birth, and a few of these have been corrected by surgery even before birth.

ANIMAL EMBRYOS DEVELOP IN A DISTINCT TIME-DEPENDENT SEQUENCE

Much happens from the time of fertilization until the individual takes on a body form constructed of tissues, organs, and systems. There is a great deal of variation in the specific shapes through which embryos of various animal groups progress, but the larger changes are about the same in all animals. In order as they appear, the following processes occur: fertilization, cleavage to produce many cells, blastulation, gastrulation, differentiation, and morphogenesis. These will be illustrated below.

Both Sperm and Egg Are Active Participants in Fertilization

Fertilization is the union of the two haploid gametes, the egg and sperm, to produce a single diploid cell, called a *zygote*. Both gametes are specialized and carry out important activities in the multistepped fertilization event.

The mature sperm that arrives at an egg is a cell highly specialized to move and to gain access to an egg surface (Fig. 14.1). Most of a normal cell's cytoplasm is missing in the sperm, giving it a streamlined shape that is easy to propel through a liquid medium. It has one to several mitochondria in its *midpiece*, to provide energy conversion for movement of its microtubule-containing *tail*. Its *head* contains mostly a nucleus, but that is capped with an important membranous bag called the *acrosome*. This structure, derived from the Golgi apparatus, is filled with enzymes that help the sperm as it nears the egg plasma membrane. A sperm cell is very much smaller than even the smallest egg.

Eggs of various animal groups present certain barriers that the sperm must traverse. Those of the sea urchin and amphibian are surrounded by a jelly-like

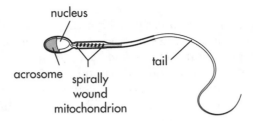

nucleus

acrosome

spirally
wound
mitochondrion

tail

FIGURE 14.1 *A human sperm cell. Acrosome and nucleus are in head region, mitochondrion is in midpiece.*

material. When a sperm arrives at this layer, its acrosomal membrane opens to release an enzyme that can dissolve the jelly and provide a pathway to the egg's plasma membrane. The mammalian egg also has a thin jelly-like layer, the *zona pellucida*, adhering to its membrane, but outside that layer is a much thicker layer of smaller cells, the follicle cells that came with it at the time of ovulation. This follicle cell layer, called the *corona radiata*, is held together by a glue-like material (the *hyaline layer*) that each follicle cell has secreted. As a sperm cell arrives at the outermost cells of this layer, its acrosome releases an enzyme that destroys hyaline material and therefore loosens the follicle cells, enabling the sperm to wiggle past them toward the egg. Upon reaching the zona pellucida, the sperm releases other acrosomal enzymes that dissolve a path to the egg surface.

If the sperm and egg are of the same species, their plasma membranes interact by linking together specific surface molecules to form an attachment. The touch of sperm membrane triggers immediate reactions on the part of the egg. First, it changes its surface in ways that prevent successful attachment of any other sperm. That is, it acts to prevent *polyspermy*, a condition that could lead to the zygote having too many copies of each chromosome (being *polyploid*). Then, the egg's cytoplasm rises up and surrounds the sperm, drawing it into the egg. Soon, all organelles of the accepted sperm degenerate, except for the nucleus.

At this point, there is a single cell containing two nuclei. Each of them, called a *pronucleus*, moves centrally until they meet and fuse together. At this point, most embryologists consider the act of fertilization to be completed. Thus, fertilization is not a single event but a series of activities beginning with initial contact between sperm and the layer(s) surrounding the egg and ending with the fusing of pronuclei. The cell that results can be called a *zygote* at this point. It soon undergoes cell division.

Cleavage Produces Many Cells from One

The initial task of the embryo is to produce enough cells to be able to form tissues and begin specialization of parts. Thus, the zygote soon enters into repeated cycles of cell division. These are not different from the activity described in Chapter 4, but in an early embryo they are usually referred to as *cleavages*. Different animal species have early cleavage at different rates. The cells of an embryonic fruit fly, *Drosophila*, can undergo cleavages at the rate of once every 10 minutes to produce thousands of cells in a few hours. Amphibian embryos typically cleave every 45 minutes to an hour, depending on the temperature of their watery environment. Human embryos are much slower initially, with only a few dozen cells being produced in the first three days.

It is important to note that these early cleavages produce cells that are substantially alike in form and function. Specialization comes later in most animal species. To find whether early embryo cells are still unspecialized, a simple experiment can be performed. Gently remove one cell of, say, a four-cell frog embryo. The remaining cells compensate for its loss and go on to produce a normal animal, demonstrating that none of them had yet undergone irreversible changes in the availability of their genetic information. Furthermore, the single removed cell can act as if it were a zygote, undergoing all normal developmental activities to produce a complete frog. The word to describe the ability of a cell to do this is *totipotency*. All animals eventually reach the point where their cells become so specialized that they can no longer do this.

A Blastula Is an Embryo with a Hollow Space

After a certain number of cells has been formed, the animal embryo typically takes on the shape of a hollow ball, with all cells at the periphery. This is the *blastula* stage. The exact shape of a blastula depends largely upon the amount of yolk that was deposited in the egg. If little or none is present, the blastula is likely to be a symmetrical ball with a significantly large fluid-filled space (the *blastocoel*) in its center (Fig. 14.2). If an intermediate amount of yolk is present, it takes up some of the blastocoel space. In the case of birds and reptiles, there is so much yolk that the cells of the embryo cannot cut into it, so the embryo sits like a cap upon the yolk. In such an animal, the blastula is significantly flattened and the blastocoel is a narrow slit within the cell mass.

Mammals produce an early embryo quite unlike all others. A hollow ball quickly forms (Fig. 14.2), but this is not the blastocoel. The cells forming this space are called *trophoblast* cells, and the fate of most of them is to become accessory tissue to connect the embryo to the mother's uterus and to form a protective bag (the amnion) around the embryo. The portion of the early mammal embryo that actually will become body parts is the *inner cell mass*. It is within this rather flattened mass that a blastocoel forms as a narrow slit resembling that of a bird.

Gastrulation Movements Put Cells into Proper Positions

Perhaps the most critical event in the early embryo is a set of precise movements called *gastrulation*. During gastrulation some cells that were previously

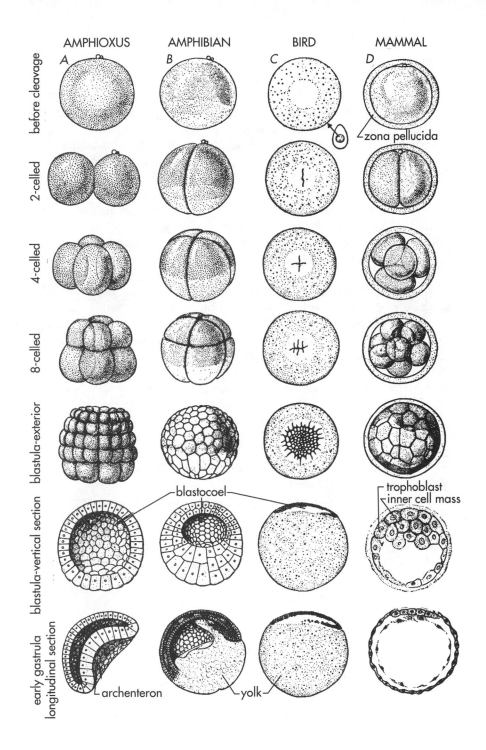

FIGURE 14.2. *Cleavage, blastulation, and early gastrulation of four Chordate animals. (Adapted from Storer, et al.,* General Zoology, *5th edition (1972), McGraw-Hill Book Co., New York.*

on the exterior of the blastula move to interior positions, enabling them to begin to build internal organs. This movement is far from random. Only certain blastula cells travel and they move in quite predictable pathways, probably guided by molecular markers on the surfaces of the cells over which they move. Movement begins with either a pushing inward of one side of the blastula (Amphioxus, amphibian, Fig. 14.2) or a movement of individual cells into the blastocoel (bird, mammal, Fig. 14.2).

The final form of the gastrula stage of an embryo depends somewhat upon the amount of yolk present, just as is the case with the blastula form. However, all gastrulas are similar in that they are composed of three layers of cells, each of which has a different fate in terms of what kinds of tissues and organs it will be responsible for making during later development. Figure 14.3 shows, in a very generalized way, those three layers, ectoderm, mesoderm, and endoderm. During gastrulation, a new internal space develops, the *archenteron*. It could be said that this is the first indication of an organ, since the archenteron and its surrounding endoderm cells will become the intestinal tract of the animal. Table 14.1 describes some of the structures that will be made from each of the three layers in a vertebrate animal.

Important changes occur within each of the cells as blastulation and gastrulation proceed. In many animal species, none of the cells has used its nuclear genes in the earlier stages of development, relying instead on information stored in the cytoplasm in the form of messenger RNA. This RNA had been made by the mother's genes, so it can be said that early stages of development are entirely under the control of the mother's genes. Transcription of some of the embryo's genes begins rather early, but it is at a minimal level until the middle of blastulation.

By the time of gastrulation movements, some cells are internally quite different from each other because they have used different genes. During gastrulation, the ability of cells to be totipotent (able to become all cell types) drops to nearly zero, even in frogs, the animals that seem to retain this ability longest. (Of course, some cells of any animal remain totipotent—those that will become gametes, charged with the production of the next generation). The loss of totipotency does

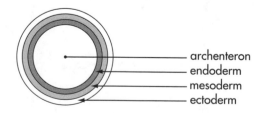

FIGURE 14.3. *Generalized gastrula, to show the relationships of the three layers of cells. The archenteron is an open space surrounded by endoderm.*

**TABLE 14.1. SOME TISSUES AND ORGANS
DERIVED FROM EACH EMBRYONIC LAYER IN A VERTEBRATE ANIMAL.
(SOME ORGANS ARE MIXTURES OF TWO TYPES; THE PREDOMINANT ONE IS LISTED.)**

Layer in Gastrula	Tissue or Organ
Ectoderm	skin, portions of inner ear, spinal cord, some peripheral neurons, portions of eyes, pigment cells
Mesoderm	kidneys, muscles, heart, connective tissues, bones, blood vessels and blood, gonads
Endoderm	intestinal tract, liver, pancreas, trachea and bronchi, lungs

not seem to involve removal of genes that are not going to be used for specialized activities, but rather the permanent inactivation of these genes.

Differentiation of Cells Is Induced by their Neighbors

If gastrulation has occurred successfully, some cells that were once in different regions on the blastula's surface are now next-door neighbors in the interior. The first cells that begin to *differentiate* (i.e., specialize toward their final anatomy and function) are a small patch of endoderm cells, at the area called the *roof of the archenteron*. Among the new things that these cells can do is the sending of a chemical message to overlying cells of the mesoderm. This is an *induction* event, defined as the sending of instructions from one tissue to another followed by the response of the receiving tissue by its becoming differentiated. In vertebrate animals, the mesoderm cells are induced to become a rod-like structure, the *notochord*. The notochord, in the process of becoming differentiated, begins to manufacture and release an inducer of its own, which diffuses into overlying ectoderm. This *secondary induction* leads to movements of the ectoderm to produce *neural folds* on the embryo surface (Fig. 14.4), the first outward sign that differentiation is occurring. Movement of this material continues and it drops under the surface to become the hollow *neural tube*. This will further differentiate into the spinal cord of the animal. These first inductions, in turn, lead to a cascade of others, each producing gene-level changes that start the induced cells toward becoming differentiated.

The proof of one tissue's inductive power over another is gained if the inducing cells are artificially removed to unnatural locations before they have exerted their influence. If this is done, the areas that were to be induced remain undifferentiated and (at least in some instances) the new neighbors of the inducing cells become differentiated into something that they would not otherwise have become. For instance, removal of notochord material to the ventral region of a frog embryo causes the development of a neural tube in the belly of the embryo, a quite odd arrangement.

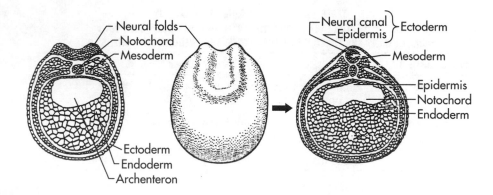

FIGURE 14.4. *Induction of neural folds and neural tube by underlying notochord, in a frog embryo. (Adapted from John W. Kimball,* Biology, *6th ed. Copyright © 1994 Wm. C. Brown Communications, Inc., Dubuque, Iowa. All Rights Reserved. Reprinted by permission.)*

The nature of the inducing chemical has been the subject of much research since the 1940s, and the source of much frustration. Any inducer is undoubtedly made in very tiny amounts and its isolation is therefore very difficult. Recent work (in the early 1990s) has led to some progress, to the point that certain identified proteins are likely natural inducers.

The term induction was used in Chapter 12, where the context was the initiation of gene transcription under the inducing influence of some molecule coming into the cell (remember the lactose operon of bacteria). Embryonic induction is probably similar. It involves the arrival of an inducing chemical that undoubtedly has at least a portion of its influence at the gene level. Animal embryos are, of course, eukaryotic, so it is unlikely that all of the steps of bacterial gene induction are applicable to this story. However, it really is the same story in the broadest sense. Embryologists with training in molecular genetics are hard at work to elucidate the steps of changing gene settings as the embryo goes through this and other stages.

Masses of Differentiated Cells Become Shaped into Organs

Soon after regions become differentiated, they grow and take the proper shape of the organs that they are becoming. This is called *morphogenesis*. The entire embryo also grows and takes on a shape that gradually becomes more like the free-living animal that it will become. Morphogenesis involves recognition on the part of each newly produced cell of (a) where it is in the organ so that it will differentiate into the proper type for that area, and (b) whether it should divide to enlarge the organ further. The latter question must be answered in order for cessation of organ growth to occur at the proper time. This problem extends into the post-embryonic period and includes the proportioning of the entire body. For instance, it would be disastrous for a person's left arm to continue elongating after the right arm had stopped. The fact that our right and left arms are remarkably similar in length is proof that cells in both of them have somehow recognized when it was time to stop reproducing.

Not much is known about how these two jobs are accomplished. Current hypotheses revolve around the idea that each cell in the developing organ has a particular set of membrane surface molecular characteristics, which provide it with an address or identification tag for other cells to read and respond to appropriately.

Some shaping involves cell death. For instance, the toes of many vertebrate animals are not initially separated from each other, since the tip of the appendage is formed in a paddle-like shape. Then, some cells are instructed to die, causing the characteristic separation. Other areas of the body are also shaped by such programmed cell death.

HUMAN EMBRYONIC DEVELOPMENT OCCURS WITHIN THE MOTHER

As a mammal, the human mother harbors her embryo within the uterus, where it obtains nutrients, protection, and warmth and has the opportunity to rid itself of toxic wastes. This marvelous position presents a problem for embryologists, however. Simply put, the human embryo is difficult to watch and study. However, in recent years several advances have given us greater opportunity for this. *In vitro* fertilization, taking place in a culture vessel, has provided an unobstructed view of the union of egg and sperm and the first several cleavage events. The use of viewing tools such as the fiber optic laparoscope (which can be inserted into the uterus with no threat to the embryo) have given close-up looks at the baby that develops after implantation. Combined with classical studies of aborted embryos and comparisons with nonhuman mammalian embryos, these new techniques are helping us to put together a rather complete story of human embryology. The following sections provide some of that story.

The Early Embryo's First Task Is to Implant

The first few days of a human are spent in its mother's fallopian tube, moving toward the uterus. During that time, it goes from a single cell to the multicellular configuration shown in Figure 14.5 A. As this happens, the uterine wall becomes prepared to nourish the embryo, by becoming thicker and more richly endowed with blood vessels. About six or seven days after fertilization, the embryo begins to attach to the internal wall of the uterus with the use of its own enzymes that cause a small pit to develop at the point where the em-

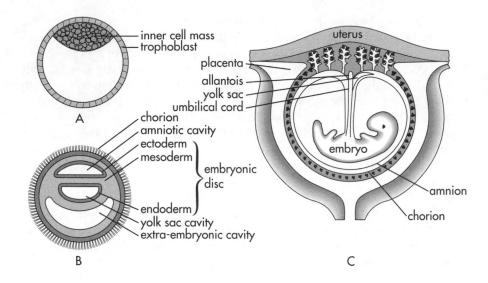

FIGURE 14.5.   *Early development of the human. A: before implantation. B: a few days after implantation (diagrammatic view). C: at about 5 weeks (diagrammatic view).*

bryo touches the wall. Shortly thereafter, the embryo sinks from sight into the tissue of the wall.

Extraembryonic Membranes Quickly Form for Nutrition and Protection

After implantation, while gastrulation is proceeding to make the three layers out of the cells of the inner cell mass (Fig. 14.5 B), the trophoblast cells go into rapid division. The trophoblast grows into a large mass of tissue that makes intimate contact with the uterine wall. The combination of this embryonically derived mass and the connected uterine cells will later become the *placenta* (Fig. 14.5 C). But for the first two months, the embryonic food and oxygen source is more direct, with many finger-like projections reaching into the uterine tissue from the outer wall of the embryo, the *chorion* (Fig. 14.5 B).

In the first three to four weeks, two cavities (*amniotic sac* and *yolk sac*) appear, derived from cells of the inner cell mass. One of these, the amniotic sac, comes to surround the developing embryo (Fig. 14.6) and fill with *amniotic fluid*. This sac and its fluid provide protection by cushioning the embryo that floats within. Of course, its volume must increase as the embryo grows.

The *yolk sac*, contrary to its name, contains no yolk. However, the name is logical because it connects to the primitive intestine of the embryo just as the yolk-filled sac of birds and reptiles does. If there were yolk present, it could

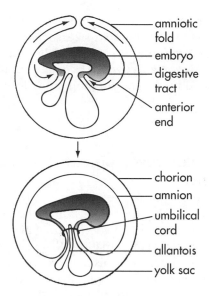

amniotic fold

embryo

digestive tract

anterior end

chorion

amnion

umbilical cord

allantois

yolk sac

FIGURE 14.6. *Growth of extraembryonic membranes to form amnion, yolk sac, and allantois.*

be brought into the embryo just as in birds. In the human and other mammals, the cells of the yolk sac become two very important things. Some are the *primordial germ cells*, which later migrate to the gonads to become the ancestors of all gametes. Other yolk sac cells become the first blood cells for the embryo.

One other sac-like structure develops from the body of the embryo, the *allantois* (Fig. 14.6). Extending from the intestinal region, it eventually becomes embedded in the *umbilical cord*. The cord is the connection between the embryo and the placenta, acting as a conduit for major arteries and veins. The walls of the allantois are the source of these blood vessels.

The Placenta Eventually Becomes the Primary Connection with the Mother

As the embryo grows, it eventually moves out of the uterine wall and into the cavity of the uterus itself. It must still retain connection with the blood supply of the wall, and it does this via the umbilical cord and the placenta. The close relationship of embryonic and maternal blood vessels in the placenta is necessary for proper exchange of materials between mother and developing embryo. The normal exchange of oxygen, carbon dioxide, foods, and metabolic wastes is accomplished by diffusion across membranes, since the blood vessels of the embryo remain intact (vessels of the mother break down and release whole blood into the region of the embryo's vessels). In addition, other materials with molecules of absorbable size such as alcohol, nicotine, and some drugs pass through it as they do into the cells of the mother. Viruses can also pass across the placenta, a fact which accounts for the 25 percent chance of a baby being infected by the human immunodeficiency virus (HIV) if the mother is infected.

The Embryo's Organ Development and Growth Occur at a Predictable Rate

Since this is not an embryology text, no attempt will be made to follow all of the steps of normal development. The human embryo goes through the same sequence of blastulation, gastrulation, differentiation, and morphogenesis as described earlier, but with the variations that mark it as a mammal. Some of the terms used to describe the organism are confusing. Throughout the period before birth, it can be called an embryo. However, the term *fetus* is often used for the period from nine weeks through birth. Nearly all internal organs are formed before the fetal period begins. The term baby is not one of precise scientific meaning and therefore may be used by individuals for any time period that they wish.

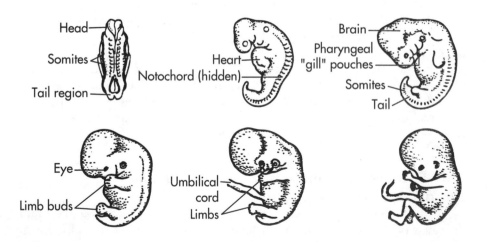

FIGURE 14.7.   *Typical human embryo shapes. In order, these represent embryos of weeks 3, 4, 5, 6, 6½, and 7 weeks. (Adapted from* Biology: The World of Life, *5th Edition by Robert A. Wallace. Copyright © 1990, 1987, 1981 Scott, Foresman and Company. Reprinted by permission of HarperCollins Publishers, Inc.)*

The following list indicates the approximate timing of some of the more important developmental changes that occur in the nine months of a baby's development.

Day 0	Fertilization
Day 6–7	Implantation
Day 22	Neural tube formed, rudimentary heart beating
Week 4	Rudimentary eyes and most internal organs present
Week 5	Distinct limb buds visible, distinct eye lens present
Week 6	Primitive reflex nerve activity possible, primordial germ cells arrive at gonads
Week 7	Individual fingers and toes are separated, first spontaneous body movement detectable by ultrasound
Month 3.5	Tooth buds present in jaws
Month 5	Regular periods of bodily motion, detectable by mother
Month 7	Earliest birth time when baby is expected to be able to breathe without mechanical aid
Month 9	Normal birth expected

MULTIPLE BIRTHS ARE POSSIBLE

Dizygotic twins come from separate zygotes. Either the two ovaries produce eggs at about the same time, or one ovary produces two follicles at about the same time. Usually the two ovaries alternate in producing a single egg in each menstrual cycle. Since these twins come from separate eggs and sperm, their inheritance is no more alike than among other brothers or sisters. It follows that they can be of the same or of the opposite sex. There is evidence that a tendency for this type of twinning is inherited through the mother. About two thirds of twins are dizygotic.

Monozygotic (identical) twins come from the same zygote. Most are believed to arise from a subdivision of the inner cell mass, thus presenting evidence that the cells of this stage are not yet irreversibly fated to be one part of the body or another. They share the same placenta and are of the same sex. The rarer type of identical twinning comes about by separation of cells at the two-cell stage. These are identified by being identical genetically but not sharing a common placenta since each of the two cells is the ancestor of a complete inner cell mass and trophoblast. About a third of the twins born in the United States are monozygotic. This type of twinning does not seem to be hereditary.

Births of more than two at a time may or may not be monozygotic. Before the development of *in vitro* fertilization techniques, it was found that one out of 90 births was twins; one out of 8,000 was triplets; one out of 600,000 was quadruplets; and one out of 52,000,000 was quintuplets. However, significantly more of these now occur in technologically advanced countries where *in vitro* fertilization is practiced. This is because the technique commonly involves inducing the ovulation of several eggs, their subsequent fertilization outside the body and placement into the uterus. Although the usual result is that only one embryo successfully implants and goes through development, there is some risk that multiple births will result.

METAMORPHOSIS IS AN EXAMPLE OF POST-EMBRYONIC DEVELOPMENT

Animals of some species experience rather spectacular but predictable changes in both body form and activity long after they have become free-living organisms. This phenomenon is *metamorphosis*. Well known examples are the change from crawling caterpillar to flying moth and from aquatic tadpole to terrestrial toad. We will examine insects as our examples of this developmental activity.

Aside from a very few that have no significant transformations between the embryo and the adult, insect species follow one of two patterns of post-embryonic development (Fig. 14.8). When metamorphosis is *complete*, they pass through a larval and pupal stage between the egg (actually the embryo developing within a protective egg case) and adult. All four stages are very different from each other. The *larva* is often worm-like. It may inhabit an entirely different habitat and have an entirely different diet from that of the adult that will come afterward. For example, the larva of the common house mosquito (*Culex pipiens*) is aquatic and feeds on microscopically small algae, while the adult is a flying animal that feeds on either plant juices (male) or bird and mammal blood (female).

The larval stage of an insect is always equipped to eat, but some adult forms do not even have functional mouth parts. The adults of these species live only long enough to reproduce, an activity that is almost universally reserved for the adult stage.

When complete metamorphosis is the style, the stage between larva and adult is the *pupa*. Although this stage is nonlocomotive and nonfeeding, it is far from quiet inside. Typically, the materials within the pupal case undergo a great deal of cell death (larval body parts) and cell reproduction (to create adult body parts). When the animal breaks out of its pupal exoskeleton, it is essentially a functioning adult.

The cells that grow into adult structures during the pupal stage were actually present and programmed for adult differentiation all through the larval life. Within a larva can be found a number of very small packages of quiescent cells, surrounded by the larva's working cells. Each of these packages, called *imaginal discs*, contains cells that are genetically programmed to become a particular adult body part. For instance, there are two wing discs in

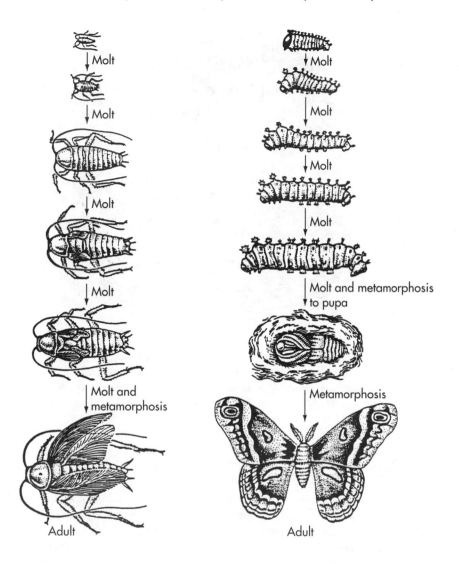

FIGURE 14.8   *Metamorphosis of insects. A: Incomplete metamorphosis, as in the cockroach. B: Complete metamorphosis, as in the moth. (From* Developmental Biology, *4th edition, Scott F. Gilbert, Sinauer Associates, Inc., Sunderland, MA, 1994.)*

a fly larva (whose adult form will have only two wings), two discs for building a pair of eye and antenna complexes, and a single disc that will become the genital apparatus. These discs reside in roughly the same positions in the larva as their tissues will take in the adult. The cells of the discs do essentially nothing in the larva except survive, but when the pupal stage is reached

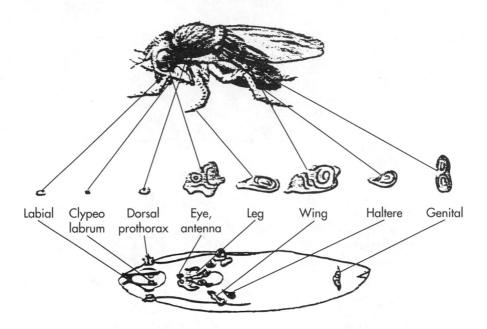

FIGURE 14.9. *Imaginal discs in* Drosophila, *the fruit fly, and the structures that they become in the adult. (Figure from* Biology, *3rd ed. by Eldra Pearl Solomon, Linda Berg, Diana Martin, and Claude Villee. Copyright © 1993 by Saunders College Publishing. Reproduced by permission of the publisher.)*

Labial   Clypeo labrum   Dorsal prothorax   Eye, antenna   Leg   Wing   Haltere   Genital

they reproduce and differentiate while all around them the larval cells are dying.

The other variety of insect metamorphosis is labeled *incomplete* (Fig. 14.8), to signify that it includes fewer distinct life stages. The grasshopper and its Orthopteran relatives are good examples. The animal that hatches from the egg case is a nearly perfect miniature of the mature adult. Called a *nymph*, it has nearly all of the behavioral abilities of the adult, lacking only size and the abilities to fly and to reproduce. Nevertheless, the period between hatching and becoming an adult does include distinct changes, so it is truly metamorphic. Those changes are a series of *molts*, during which the animal breaks out of its old exoskeleton, grows rapidly to a larger outside dimension, and then secretes a new chitinous exoskeleton that rapidly hardens. Some molts produce animals that are only larger, but others include more substantial change. With the grasshopper, each succeeding molt produces an animal with larger wings, until eventually they are large enough to sustain flight. The last molt of the grasshopper yields a sexually mature animal.

Insects with complete metamorphosis also go through a series of molts. The larva may pass through four or more molts as it grows, and the production of a hard pupal case is a molting event. Likewise, the breaking out of an adult from the pupa may be considered the last molting event, although it involves at least some production of the new exoskeleton even before the pupal case breaks open. Molting is controlled by a set of hormones. One of these, called *juvenile hormone*, must be released if the larval body is to be produced. When that hormone's production is significantly lowered, the next molt will form the pupa. When the production of juvenile hormone is completely inhibited, the next stage will be the adult.

PLANT DEVELOPMENT INVOLVES VERY DIFFERENT CHANGES

The embryonic development of higher plants bears little resemblance to that of animals, at least in terms of the anatomical changes that occur. However, before describing these differences, it should be pointed out that plants and animals likely use very similar devices at the molecular and genetic levels. They both must exert selective gene control in very specific sequences, and this is nearly identical among all eukaryotes.

The embryos of seed plants (Fig. 14.10) develop from zygotes contained in ovules (immature seeds). When the seed is mature the embryonic plant has a *hypocotyl* (stem), one or more *cotyledons* (leaves), and an *epicotyl* (bud).

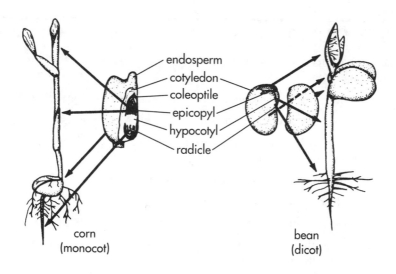

endosperm
cotyledon
coleoptile
epicopyl
hypocotyl
radicle

corn
(monocot)

bean
(dicot)

FIGURE 14.10.   *Seeds and seedlings of corn and bean. The embryo is contained within the seed and food is stored in endosperm or cotyledons.*

The cotyledons are the first leaves of a plant and are often strikingly different from later leaves. In some plants like the bean, they contain large quantities of stored food that get the embryo off to a good start until it can perform significant photosynthesis and bring minerals from the soil. Cotyledons usually persist for only a short time after germination. Before bean cotyledons drop off they turn green and wither as stored food is consumed. In many cases food is stored in a special tissue surrounding the embryo, and the cotyledons themselves are small. In corn a single cotyledon lies against the storage tissue and acts as an absorptive organ. Gymnosperms (such as pines) have irregular numbers of cotyledons, dicots (such as beans) usually have two, and monocots (such as corn) usually have one. Indeed, the words dicot and monocot refer to the number of cotyledons.

The hypocotyl bears the cotyledons and epicotyl at its upper end. At germination it grows a primary root from the lower tip (*radicle*). If the hypocotyl elongates, the cotyledons are lifted above the ground as in cotton, morning glories, and beans. If it does not elongate, the cotyledons remain in the seed coat below ground as in corn and peas.

It is virtually impossible to draw sharp distinctions among phases of plant development (growth, differentiation, and morphogenesis). Their interrelations are obvious, especially in the way differentiation contributes to morphogenetic changes. Much of the shaping of the post-embryonic plant body is under the control of growth regulators (Chapter 10) and environmental conditions.

STUDY QUESTIONS

1. Contrast preformation and epigenesis as explanations for the origin of individuals.

2. How are the fields of molecular biology and embryology related?

3. What are several medical problems related to embryology?

4. What are the roles of sperm and egg in fertilization? When is fertilization usually considered to be completed?

5. How does a sperm cell get past a mammalian egg's corona radiata?

6. Define totipotency. How can it be tested?

7. How does a blastula differ from a gastrula?

8. What are the three layers made by gastrulation? Name at least two structures eventually formed from each layer.

9. How does the amount of egg yolk influence blastulation and gastrulation?

10. What does the archenteron eventually become?

11. Define embryonic induction. Does it have any relationship with bacterial gene induction?

12. How could one test for inductive power of a given region of an embryo?

13. Show how the central nervous system originates from ectoderm.

14. Name the extraembryonic membranes (sacs). How do they originate? Of what use are they?

15. What is the relationship of the embryonic and maternal blood systems in the placenta? What kinds of materials can cross the placenta?

16. Why must a human embryo be several days old before it can successfully implant?

17. Distinguish between monozygotic and dizygotic twins. Which type is more prevalent? Which type would be more likely to allow successful skin transplants between its individuals?

18. Which stage or stages are missing if an insect is the type that performs incomplete metamorphosis?

19. What are imaginal discs?

20. What are the parts of the embryo of a seed plant?

21. From what two sources may plant embryos get food before they become independent?

15
EVOLUTIONARY CHANGE

What do we know about:
—*how we know that life has changed on earth?*
—*how fossils are formed?*
—*the age of the earth and of its oldest fossils?*
—*Darwin's predecessors and their ideas?*
—*the mechanism of natural selection?*
—*why evolution can happen more rapidly in small populations?*
—*how new species arise?*
These and other intriguing topics will be pursued in this chapter.

ARE BIOLOGICAL DIVERSITY AND UNITY JUST COINCIDENTAL?

Throughout the previous chapters of this book there have been some threads of continuity, among which are two that are very strong. First, there is certainly seen to be a great deal of *diversity* when one looks at the organisms of the world (see also Chapter 16). Six strikingly different kingdoms of organisms, a million or more identified species, special structures and behaviors to enable life to thrive in extreme environments, multiple ways to solve the same problems—all of these testify to the bewildering array of living things surrounding us. But simultaneously we see the theme of *unity* wherever we look. Consider the fact that most animals are bilaterally symmetrical (having nearly mirror-image left and right sides), or that all plants have cell walls but no animals do, or that the linked processes of glycolysis and the Krebs cycle can be performed by nearly all organisms, or that every organism from virus to human uses nucleic acids to hold genetic information and reads the same genetic language of codons.

As we review these facts, we are confronted with the question, "Are these twin features of diversity and unity just a coincidence or are they explainable by some underlying principles?" Biologists have struggled with this question for a long time. As they did, they added some more equally difficult questions. For instance, was there a time when no life existed on earth? If so, how did life first appear, and what did it look like? Has the diversity that we see now always existed? If not, how did the world get to be so populated with different kinds of organisms? Were there once organisms that no longer exist? If so, what happened to them? Are we all related in a biological sense that goes beyond anatomical similarity?

As in the case of all other unsolved questions, scientists have made it their business to seek the answers to these puzzles. Obviously they cannot retrace step by step the events of the past, but they have found some clues that can be fitted into a logical set of answers to at least some of these questions. It is the business of this chapter to examine the evidence that life provides and indicate what our current ideas are concerning the development of life on earth from the earliest times. The primary theme will be the concept that organisms have changed and continue to change—living things *evolve* by methods that are both natural and understandable.

Some fear that plugging the panorama of life's history into a set of rules similar to those that govern energy and molecules takes away our awe of the topic. An intention of this chapter is to demonstrate that it is satisfying to be able to grasp some of the underlying principles governing life and that this understanding takes away nothing from the astonishment of seeing how life operates.

THERE IS MUCH EVIDENCE THAT ORGANISMS HAVE CHANGED OVER TIME

Let us first examine the clues leading us to the conclusion that life has not always existed on earth and that it has changed since it first appeared. The evidence we have to work with does not easily lend itself to experimentation, that powerful tool of science. Remember from Chapter 1 that experiments require the manipulation of materials under controlled conditions. Since much of what is discussed in the context of the present chapter is historical (i.e., it happened in the past), and since we can never duplicate the past, observation becomes the primary tool rather than experimentation. Nevertheless, there are many observations to be made and they point to some common conclusions.

Fossil Remains of Organisms Provide a Rich Stock of Clues

It was inevitable that life would leave some evidence of its passing. As several pioneers in this discipline have stated, the layers of sedimentary rock are "written pages in nature's history book." From each page can be read an indisputable, though fragmentary, record of time that is untainted by human prejudices. The words of nature's history book are the fossils it contains.

Fossils are evidence of ancient life preserved in the earth's crust. They are found in sediments where they are either consolidated into rock or mixed with unconsolidated particles. Most sedimentary accumulations were originally deposited on the floors of large bodies of water. Since then some of these areas have emerged to become dry land, by upheaval of the earth's crust. Wind-blown material, avalanches, and volcanic explosions have likewise played their part in the burial of ancient life. Some other forms of life, not covered by transported sediments, were trapped and covered by exuding resins (insects in amber) or mired in mud or tar pits.

Out of the teeming abundance of life only a few organisms fall under conditions favorable for preservation. That most environmental circumstances favor the destruction of bodies is obvious from noting present-day events. Upon dying, organisms almost never lie undisturbed for long but are eaten by scavengers or broken down by decomposers. The conditions preventing their destruction, thus favoring their preservation, are immediate burial and possession of hard parts. Immediate burial excludes free oxygen, which is essential to most decay-causing organisms. Hard parts like bone or shell have little organic matter that can decay and are resistant to weathering processes. Nevertheless, even bones and shells do not remain for long when exposed to the weather.

In spite of how remote the chance is for any particular individual to be preserved, fossils are numerous. In some areas fossil marine shells are so abundant that they are scooped up by power shovels and used like gravel on road beds. In other places they are scarce and widely scattered. Often they are non-existent because of the nature of the rocks. Fossils beneath the surface are mainly inaccessible. Paleontologists (scientists specializing in the study of fossils) recover them from tunnels, mines, walls of gorges, road and railroad cuts, and material brought up from drillings or ditchings. The vast majority still remain concealed from human inspection.

The *actual remains* of an organism, either entire or in part, are sometimes preserved as fossils. In the La Brea tar pits in California many animals sank beneath the surface, where their entire bodies are nicely preserved. Mammoths, extinct for many years, have been discovered in the frozen desert of northern Siberia in such a good state of preservation that the flesh is still red and the DNA is nearly intact. Entire organisms are rare, however; more frequently only the hard parts persist.

Some fossils result from *altered parts* of bodies. For instance, spaces within bone or wood may be filled with mineral deposits. Some of the most faithfully altered fossils are petrified wood. Pieces of the fossil wood can be ground into thin sections to show the perfectly preserved cellular construction. Some organisms are carbonized; that is, the conspicuous material left from the alteration of their bodies is carbon. Coal is an example of carbonized plant material. A combination of heat, pressure, and water loss gradually changes plants into peat, then to lignite, to bituminous coal, to anthracite coal, and in some cases to graphite. Thin structures, like leaves or small animals, often remain as carbon prints on rocks.

A third type of fossil is an *indication* of the former presence of a plant or an animal. Some objects indicating the former presence of an earlier form of life are tracks, worm tubes, bird's nests, molds, casts, or dung (copro-lites). A mold is the external form preserved in rock or other material after the organism is gone. The people of the ancient Roman city of Pompeii were preserved as molds when volcanic ash solidified around their bodies. Casts are the result of inorganic matter filling molds or some other cavity and then solidifying. Even some flimsy jellyfish have been preserved in this manner.

A reasonable assumption is that under normal circumstances deeper layers of sediments with their included fossils are older than those in layers above them. From this, paleontologists and geologists have been able to reconstruct events of the past in correct sequence. At no place, however, can one find a continuous series of deposits from the beginning of the earth to the present. The earth's crust has undergone too many tumultuous changes for such a com-plete record to exist. Also, the break between one type of deposit and another lying next to it represents a discontinuity of some length of time. Regardless of these difficulties, paleontologists have made good progress in reconstruct-ing the record, although it probably can never be known in minute detail.

One revelation of the record is the story of the earth's age. It and related events are summarized in the *geologic sequence* in Table 15.1. The units of time given in this table represent periods characterized by certain conditions. The larger units are called *eras*, which are subdivided into periods, and the periods are subdivided into *epochs*. Each unit is assigned a span of time based on certain calculations.

Several methods have been used to calculate the ages of the rock strata, but the radioactive decay method is by far the most accurate. One element used for this purpose is uranium. Regardless of surrounding conditions, ura-nium changes (by many steps) to the element lead at a fixed rate that is set by some fundamental principles of atoms. The proportion of uranium that has decayed to lead in the various layers can be determined and its age fixed fairly accurately. This dating method requires that the decay rate has not changed over the time period in question. However, to assert that such a change has happened would require a completely different set of rules for the nature of atoms and their particles from that which the evidence of physics gives us. The methods used for dating geological strata is reliable. They indicate that the earth was formed about 4.6 billion years ago.

A useful way of calculating more recent time is the *radiocarbon method*, in which the amount of a radioactive isotope of carbon is measured. This is reasonably accurate, as proved by testing against human-fashioned objects of known age. The degree of error runs from 5 to 10 percent in young samples (1,000 years) and increases in older samples. The method has enough accu-racy to be usable to somewhat more than 25,000 years before present. It is used to date organic remains, which of course contain carbon. The measure-ment of the radioactive decay of an isotope of potassium is used for older fossils, providing reliable values for materials up to several hundred million years old.

Before the general acceptance of evolution as the means of change, some scientists noticed different sets of fossils in different layers of sediments and concluded that several violent catastrophes had occurred to wipe out all life on the earth at those times. They supposed that a whole new set was created in one broad sweep after each catastrophe. It is true that the discontinuities between the eras represent widespread geologic revolutions, but they did not destroy all life. A currently popular hypothesis is that at least some of these mass extinctions came soon after cometary or meteoric impacts on the earth and that the two events may have a cause-and-effect relationship.

Different kinds of organisms have left records through various lengths of time. Sponges and Protistan organisms extend from the Proterozoic era to the present; dinosaurs are confined to the Mesozoic era; mammals extend from the Mesozoic era to the present; grasses in large enough numbers to form grasslands extend from the middle Cenozoic era to the present; and humans left fossils only in the recent Cenozoic era. Judging from the quantity of fossils remaining, some kinds of organisms dominated certain periods of time. Trilobites, somewhat like modern crustaceans, dominated the Cambrian period; coal-producing plants (primarily ferns and their allies), the Carboniferous period; and dinosaurs and their relatives, the Mesozoic era.

One obvious fact is that the more recent layers of rock have a greater variety of organism types. In general, simple types of organisms are found in lower layers, and they become progressively more complex in each succeeding layer. The oldest fossils that have been found are those of bacteria, in rocks dated at about 3.5 billion years before present. However, fossils of some primitive eukaryotes are nearly as old. Where one type—for example, the sponges —extends through every layer from their beginning, the constituent species are mostly different in each layer.

Some periods of the earth's history have apparently been more amenable to the increase of biological diversity than others. The time during which the fossil record shows a greater increase than any other appears to be at the Cambrian period, beginning over 560 million years ago. Before that time there were bacteria, Protista, Fungi, and some invertebrate animals. In the relatively short period of about 10 million years, many new forms of organisms appeared, including all of the present phyla (the largest taxonomic grouping below kingdoms) with the possible exception of the chordate animals. In the same period were some organisms that have counterparts in no modern phylum. These lived a long time and must be considered successful, but did not send representatives into the present.

Occasionally it is possible to learn something of the habits of extinct animals by their structures or by other fossils associated with them. The nature of teeth and claws tells whether their owners were herbivorous or carnivorous. Fossilized material inside body cavities and fossilized excrement reveal the exact kinds of food an animal consumed.

TABLE 15.1. THE GEOLOGIC SEQUENCE, SHOWING MAJOR BIOLOGICAL EVENTS AND THEIR APPROXIMATE TIMES.

Eras	Periods	Epochs	Time*	Physical Events	Life
CENOZOIC "age of mammals, birds, and angiosperms"	Neogene	Pleistocene	1	glaciation and recession of ice	rise and dominance of humans, last of great mammals, relative increase in herbaceous plants over woody plants
		Pliocene	10 (11)	cool climate, coast ranges rising, deserts of Southwest develop	increase in herbaceous plants
		Miocene	15 (26)	cool climate, coast ranges rising, extensive lava flows	modern birds and trees, first grasses
	Paleogene	Oligocene	10 (36)	mild climate	first elephants
		Eocene	20 (56)	mild climate	first apes
		Paleocene	5 (61)	warm and humid climate	hardwood forests predominate, first placental animals
MESOZOIC "age of reptiles"	Cretaceous		65 (126)	warm climate, Rocky Mts. formed, inland submergence	early angiosperms and decline of gymnosperms, last of dinosaurs
	Jurassic		30 (156)	arid climate	first frogs, birds, and mammals; reptiles diversified, abundant gymnosperms
	Triassic		40 (196)	desert climate, volcanic activity	first dinosaurs, abundant gymnosperms

Era	Period	Duration (time)	Geology and climate	Life
PALEOZOIC	Permian	45 (241)	Appalachians elevated, widespread glaciation and aridity	first conifers, last of trilobites
	Pennsylvanian (Carboniferous) "coal age"	35 (276)	warm, moist climate; coal-forming swamps	first reptiles, swamp floras of giant ferns and seed ferns
	Mississippian (Carboniferous) "coal age"	25 (301)	warm, moist climate; considerable submergence	first insects, abundance of sharks, dominant plants lycopods and horsetails
	Devonian "age of fishes"	50 (361)	aridity, emergence	many fishes, first amphibians, first forests
	Silurian	30 (390)	extensive submergence	first land plants and animals, first freshwater fishes
	Ordovician	70 (461)	mild climate, extensive submergence	first vertebrates (fishes)
	Cambrian "age of invertebrates and thallophytes"	100 (561)	mild climate	all life marine, trilobites and brachiopods dominant
PROTEROZOIC		1000 (1561)	extensive igneous activity, first evidence of glaciation	marine algae, marine sponges, marine worms
ARCHEOZOIC		2000 (2561)	extensive igneous activity	bacteria and algae

Time in millions of years. First number is duration; second number in parentheses is time before the present.

Here is a series of statements summarizing the geological and fossil evidence.

1. The earth is very old, approximately 4.6 billion years.
2. Life was not always present on earth, but it appeared quite long ago, at least 3.5 billion years.
3. The first life was simple, unicellular.
4. Both the degree of variation and the complexity of organisms increased as time went by.
5. The major groups of modern organisms are represented in very old rocks, with most of them at least 500 million years old.
6. Not all groups of organisms that once existed have survived to the present.

Modern and Ancient Geographic Distributions Provide Clues

Geographical distribution of organisms is useful in showing their variance and in revealing their places of origin. Any particular kind of organism will spread in all directions from its place of origin until it reaches a barrier. That barrier could be an ocean, a mountain range, a desert, or any other topographic feature extensive enough to deter migration across it. Climatic barriers like temperature or humidity may be equally effective in stopping migration.

It is a great temptation to guess at the relative age of a species based on the distance it has spread. The longleaf pine is a tree of the coastal and sandhill region of the southeastern United States, and the Venus's-flytrap is endemic only to a small coastal area of the Carolinas. If everything else were equal, it would be logical to conclude that the flytrap population is younger, not having had time to spread as far as the pine. On the other hand, could it be that the flytrap is so restricted to a particular set of environmental conditions that it has not become dispersed to other places where those conditions are not exactly right, or has it not been disseminated across some geographic barrier? Or could it be old, without reproductive vigor, and dying, perhaps persisting as a relic in a smaller area than it once occupied?

That each isolated land mass has its own native fauna and flora is obvious. Where domesticated animals and cultivated plants started, however, is not always so obvious, because human activity has literally taken them all over the world. With much research, quite a few organisms that now exist in many areas have been traced to their place of origin.

In 1840 Asa Gray published some of his findings that showed striking similarities between life in eastern America and eastern Asia. Both areas share many of the same plant and animal genera. From the animals, the best-known example is the alligator, which is indigenous in no place on earth except the southeastern United States and the Yangtze River region in China. From the plants, the best-known examples are tulip poplar, magnolia, and sassafras trees. Interestingly, the plants are predominantly simple-leaved woody plants which, generally speaking, are thought to be primitive in the phylogenetic scale (compared to compound-leaved or herbaceous plants). Their antiquity is indicated further by fossil records. The largest association of these genera in Asia is in central China along the Yangtze Valley, where the climate presumably approximates what it was in ancient times. These two widely separated areas seem to be the relics of a forest that once extended from eastern America through a land bridge in the Bering Strait region to eastern Asia. The climate was uniform throughout and supported the same kinds of plants. At the same time, eastern America was isolated from western America by a large body of water. Later the Pleistocene ice sheet creeping down from the north separated eastern America from eastern Asia and wiped out most life in the area between them.

Significantly, the longer two land masses have been separated, the more different are their organisms, even though each has large areas suitable to the life of the other. During the lengthy separation, life developed into the present-day forms. Isolation of eastern America and eastern Asia, although recent geologically speaking, has been sufficiently long to permit the evolution of different species but not long enough to erase the old genera. As another example, geologic evidence points to the isolation of Australia and the neighboring islands from the Asiatic mainland during the Mesozoic era. Since then their animal and plant populations developed peculiarities quite unlike those found elsewhere.

The major lessons to be learned from biogeographical studies are that species of organisms have not necessarily been in the same places for their entire history and that distinct species have come into being where their ancestors were geographically isolated from others in the population. The latter is especially clear when one looks at related species on a group of islands and finds that each island might have its own species but that all of the species of the group bear similarity to an ancestral species that originally found its way to the whole island group. Furthermore, the longer each species is isolated from its relative species, the greater the difference grows between them.

Embryos Provide Clues

Embryos go through stages and have structures considerably different from their adult forms. What is the reason? No doubt some of the structures are adaptations to the environment in which they develop. The tail and gill pouches in the human embryo, however, appear to have nothing to do with adaptations during development. One thing is certain: embryonic tails and gill pouches indicate a kinship to other vertebrates, for they are all remarkably similar in the early embryonic stages (Fig. 15.1).

The close similarity suggests that some ancestral features are repeated as each individual develops; or to say it differently, the individual relives some of its ancestral history. It is tempting to say that the entire set of changes that an embryo undergoes accurately reflects the sequence of changes that its ancestors experienced, but such a sweeping generalization (popular in the late

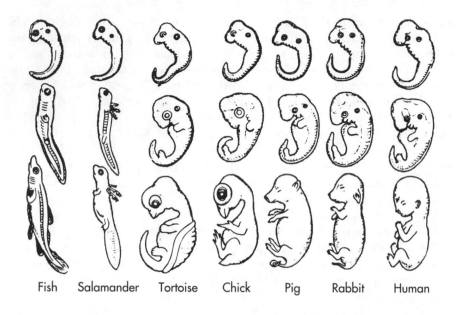

Fish   Salamander   Tortoise   Chick   Pig   Rabbit   Human

FIGURE 15.1. *Embryological development of vertebrates. Some retain the tail and gill regions to be used in the adult stage; others lose them before adulthood.*

1800s) does not bear scrutiny. Nevertheless, a number of the anatomical structures of embryos provides clues to ancestry and relatedness.

Comparison of the Adult Anatomy of Living Organisms Gives Many Clues

A study of comparative structures is one good way to observe natural affinities of organisms as well as to reveal evidence of changes. The closer that organisms are related by ancestry, the more similar are their body plans. Mammals, for instance, always have seven neck bones (manatees excepted). Similarly, the wing of a bat, the arm of a person, the flipper of a whale, and the front leg of a dog have identical embryonic origins; and the pattern of their bones (Fig. 15.2), musculature, blood vessels, and nerves are closely similar despite differences in proportions. The same similarity holds for other structures throughout their bodies. In plants comparative structures of the vascular system, flowers, life cycles, and pigmentation are useful in determining relationships.

An interesting observation is that some organisms, though remotely related, come to resemble one another closely in a superficial way. Living under the same aquatic conditions, whales and porpoises, which are lung-breathing mammals, are shaped like fish and have appendages comparable to fins. Insects, flying fish, birds, and bats are all in different taxa (classification groups)

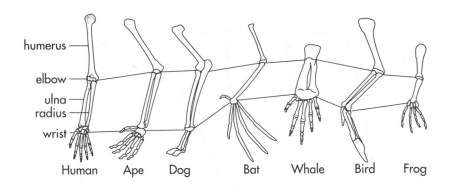

humerus

elbow

ulna
radius

wrist

Human   Ape   Dog   Bat   Whale   Bird   Frog

FIGURE 15.2.   *Comparison of bones in forelimbs of vertebrates.*

but have wings of one kind or another. The front appendages of moles and mole crickets are similar in that they are strong, short, paddle-like tunneling organs. The hummingbird moth reminds one of the hummingbird as it flits from flower to flower in search of nectar. The evolutionary pattern that results in different organisms resembling one another closely when they live in similar environments is *convergence.*

Another kind of evolutionary modification is seen in closely related individuals that are adapted to different environments. This is known as *divergence (adaptive radiation).* Mammals are a natural group that have diverged into forms that are adapted for life on land, in the sea, or in the air. One should recognize that mammals that are used as an example of divergence are at the same time illustrative of convergence in relation to animals of some more distantly related groups (Fig. 15.3).

Organs may be compared on the basis of common origin or common function. Structures are said to be *homologous* when they are of similar embryological origins regardless of any later modifications that may occur. The sucking mouth parts and chewing mouth parts of various insects, for example, are modified from patterns of identical embryonic origin. The same is true of the swim bladder of fish and lungs of other animals. Structures are said to be *analogous* when they have different embryological origins but similar uses. The gills of fish and lungs of air-breathing vertebrates are analogous but morphologically different. The wings of insects and the wings of birds are analogous and morphologically similar.

Another line of anatomical evidence of change comes from *vestigial structures* that are not essential or are even absolutely useless in an organism. It seems that the best explanation to offer for their presence is that they are remnants of ancestral structures previously functional when living conditions were different. It would be hard to justify a special (direct) creation in which useless or harmful structures such as these were made into the body. The human

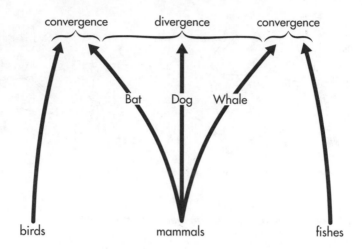

FIGURE 15.3.   *Convergence and divergence.*

is said to have over one hundred vestiges. There are vestigial pelvic and posterior limb bones in the whale, although the limbs themselves are absent. The
snake called a python has two projecting claws representing vestiges of hind
limbs. Moles (mammals that burrow in soil during nearly all of their life) have
nonfunctional eyes, as do some cave fish. Flightless birds have useless wing
bones. Flies and mosquitoes can no longer use their second pair of wings for
flight, for they are reduced to stalked, knob-like structures called halteres (Fig.
15.4), used in balancing. This example demonstrates that some structures are
vestigial with relation to their original use, but of value in a new way. Plants
likewise have vestiges. Rhizomes may have useless underground leaves (Fig.
15.4), and asparagus leaves are reduced to scales.

Physiology and Molecular Structures Indicate Relatedness of Some Organisms

Physiological processes frequently verify many of the relationships that
have already been observed from structures and may open new avenues for
investigation. Although not as obvious to the eye as structures, they are just
as real. One of the best lines of evidence comes from parasites. They are
rather specific in their requirements and do not find the right combination of
environmental factors to survive except in closely related host species. The
fact that polio viruses can infect only humans and certain apes indicates a
closer relationship between them than between the human and a cow, which
does not have the disease. Similarly, apple rust may infect apple, crab apple,
and hawthorn but not a pine or hickory tree. To be sure, some parasites may
pass through cycles involving several very different kinds of hosts, but any

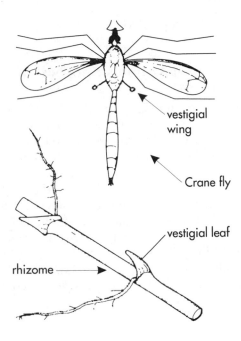

FIGURE 15.4.   *Examples of vestiges. See text.*

particular stage in a cycle is often confined to plants or animals that are closely related.

As the field of biochemistry has come of age, the structures of certain molecules (particularly proteins and nucleic acids) have been studied as further aids in identifying degrees of relatedness among organisms. For instance, human blood groups (determined by carbohydrates on red blood cell membranes) are identical with groups in other primates—groups *O* and *A* in chimpanzees and *A*, *B*, and *AB* in orangutans. Likewise, the Rh blood factor is common to humans and the rhesus monkey (Rh comes from the word rhesus).

The sequence of amino acids in a protein (or the sequence of codons of the DNA that determines amino acid sequence) can be compared to similar proteins made in other species and inferences of relatedness made. For instance, many vertebrate animals make hemoglobin as their oxygen-carrying protein. Comparison of their amino acid sequences may be as valuable an exercise as examining the shape and embryological origins of their body parts. What is usually found is that the molecular evidence rather closely matches the anatomical evidence and infers common origins of the animals. Furthermore, the approximate times when any two modern species shared a common ancestor can be estimated by assuming that (a) the differences arose by mutations in the hemoglobin gene and (b) the mutation rate of genetic material is fairly constant. Rather detailed trees showing divergence of related species can be devised on the evidence of modern molecular structures.

Breeding of Plants and Animals Shows that Change Can Occur

Fortunately for all, we do not find life a fixed creation. If it were so, our only sources of food would be wild plants and animals. People have been improving organisms for agricultural purposes for thousands of years, increasing yields and other features by selective breeding. The number of types of dogs seen at a kennel show testifies to the ability of a species to have many genetically determined variations. In such cases, humans do not do anything to provide the diversity, but simply choose among the varieties that have come about through mutation (however, see instant speciation, below). But we have learned much about how to *select* organisms with certain traits, allow only them to reproduce, and thereby see a gradual increase in the frequency of these traits as generations go by. This *artificial selection* to change a population is a strong hint that some sort of *natural selection* could change populations without human intervention.

CHARLES DARWIN USED THESE CLUES TO DEVELOP A MODEL FOR CHANGE

The period of the early 1800s was one of much observation of the organisms of the world. Exploring expeditions, by land and sea, often included one or more persons trained to take notes on the exotic plants and animals and to collect and bring them back to natural history museums. The French anatomist Georges Cuvier was developing the formal study of fossils. Geologists such as Charles Lyell were learning about how the earth had changed in prehistoric times and deciding that it was probably very old. Embryologists were comparing the prenatal anatomies of organisms.

Many of the clues described above had become available by this time, and some scientists began to conclude that there was strong evidence for a changing world, both geological and biological. A number of ideas were put forth to explain how organisms could evolve as the fossil and other evidence seemed to show they had. In 1809, Jean Baptiste de Lamarck proposed that evolutionary change comes about by the striving by organisms for some ideal. Central to this idea was the mechanism by which a striving was to occur. According to Lamarck, an organism that tries harder to get a job done and succeeds can genetically pass along to its descendants the ability that it has attained. In other words, acquired traits can be passed on. As each generation does a little better by trying harder, there is a cumulative change toward the goal of perfection. For instance, if you try to succeed at improving your running speed, your children should benefit from your gains and be able to improve beyond you. Lamark's idea is to be commended because it sounds logical and could work. However, it depends upon the ability of an organism's actions in life to change its genetic instructions, something that does not occur.

Charles Darwin, born in the same year that Lamarck published his hypothesis, used the evidence that was available to Lamarck but he added further observations that he acquired as the naturalist on board a ship that sailed around the world (in 1831–32). Darwin thought about how the clues fit together and developed an alternative hypothesis that he called *descent with modification*; the modern term is *organic evolution*. He had put together the major arguments by 1844, but for various reasons did not release them to the scientific community until shortly before he published a book, *The Origin of Species*, in 1859.

Darwin's Theory Is in Two Parts

Darwin proposed two main hypotheses: first, that *all life has evolved from a single origin*, and second, that *natural selection is the method that has led to the great diversity of species*. The first assertion explains why we see so many things in common among the presently living organisms, from shared body plans, to shared physiological strategies, to shared molecular structure and biochemical reactions. The second assertion provides a workable method for creating all of the diversity seen. It explains why we have more species now than in the earliest days of earth and how the diverse species fit fairly well into the habitats in which they are found.

Natural Selection Depends Upon Interaction Between Organisms and Environment

As Darwin set them out, there are several things that occur leading to evolutionary change by natural selection. First, a population of organisms (all of the same species, as the term population implies) is likely to have the ability to reproduce very well; that is, the potential to make more offspring than there were parents. Unless something stops it, the population will continuously grow in numbers. Second, as one observes the population over the generations, its total size is likely to remain relatively constant. This implies more organisms are produced than can be sustained by the resources in their environment, and that this counteracts the reproductive ability described as the first point. There is likely a competition among the organisms of the population to gain limited food or shelter, to avoid being eaten, and so on. Third, the population is likely to have variations among its individuals, in their appearances and abilities, and at least some (perhaps most) of these variations are inherited from generation to generation. This variability is now known to be inevitable, since genetic mutations happen within all populations. It is magnified in sexually reproducing populations, where genetic recombination also occurs.

If all of these conditions are present, the consequence is that those organisms whose combinations of phenotypes and abilities are most likely to let them compete successfully for the limited resources are the ones that will be more likely to pass their genes to the next generation. They have the advantage because they remain healthy long enough to reproduce and they reproduce with success. In other words, an appropriate match-up of their heritable characteristics with the environment leads to their being selected by the environment to send their genes to the next generation. (Or one could look at it from the opposite direction: those with less favorable matchups are *selected against.*)

The inevitable consequence of selection during one generation is that the population of the next generation has more organisms of the favored sort and fewer of those that were not able to reproduce as well. It is said that the population is *adapting* to its environment. The characteristics that have helped the successful organisms are said to be *adaptations.* Built into this consequence is the implication that such a population will become more homogeneous as generations go by and each generation experiences a weeding out of those individuals that are less fit for survival and reproduction. One might ask why all populations have not become nearly perfectly adapted to their environment, since that is the direction toward which selection leads. The best answer to this is that the environmental factors doing the selection are prone to change. Only populations that live in very constant environments (such as the depths of the ocean) are likely to be homogeneous; all others encounter occasional changes in temperature, water availability and quality, food sources, predators upon them, and so on. When the environment changes, the selection factors are different; it is a whole new ball game with new winning or losing combinations of genes.

In many cases, the selection process is quite subtle, with little difference in abilities among organisms and consequently little change in their frequencies over a single generation. Perhaps hundreds or even thousands of generations would have to pass before much noticeable change in the frequency of the various forms would have accumulated. However, there are a number of examples of natural selection observable over a shorter time. For example, a pasture of a Maryland farm was planted in clover and grass. One part was grazed and another part was kept fenced off from cattle. At the end of three years plants were taken from both parts and allowed to grow in an ungrazed area. A large percentage of plants removed from the grazed area tended to have a dwarfed, rambling shape, while the others were taller and erect. The conclusion is that the shorter, more rambling variations could survive the selective pressure of grazing better than the taller, erect ones.

Another interesting case involved moths. It was first observed by examination of insect collections that had been made near Manchester, England, beginning in 1850 and continuing for a number of years. A dark form of a normally light moth (*Biston betularia*) was originally seen in low frequency. It had the advantage of blending into the background of soot-covered tree trunks in this industrialized area, since the dark color helped conceal it from possible predators, an ability not possessed by the dominant light form of the moth. As years passed, the dark form became the more prevalent one. The human-polluted environment had resulted in evolution referred to as *industrial melanism*. Interestingly, after England incorporated antipollution laws and the trees gradually lost their extremely dark coatings, the population of the moth began to shift back toward the lighter form. This second change in frequency was possible because, although the dark form had become prevalent, there were still animals with the proper alleles to make the lighter color of wings. It points out an important concept: it is very advantageous for any population to maintain diversity, even if some of the forms have no current advantage, since they might be the ones to save the population in some future changed environment.

Have Darwin's ideas of evolution and natural selection held up under scrutiny? In large part, yes. The conceptual foundations are strong since they are logical and based on known facts about organisms. Darwin proposed his ideas before Mendel's work was widely disseminated, but everything that has since been found about genes fits into his model and even reinforces it. The mechanism of natural selection is now considered to be the major method by which evolutionary change occurs. The only serious modifications that have had to be made in well over 100 years are (a) natural selection has been joined by other methods for evolutionary change, (b) sometimes evolution proceeds much faster than Darwin envisioned, and (c) new species must be reproductively isolated from each other. These points will be discussed next.

NATURAL SELECTION IS NOT THE ONLY METHOD FOR EVOLUTION

Evolution of a species can occur by several mechanisms, only one of which is the natural selection discussed above. At least four other activities or conditions sometimes occur to cause change. Strikingly, most of these are not expected to be adaptive; that is, they can cause random change that may not lead to the population's becoming more matched to its environment. Only natural selection inevitably leads to useful adaptation.

At this point, it is appropriate to introduce some new terminology that Darwin could not have used. A population can be considered a *gene pool*. This rather myopic view is one of just the genes that are carried by the organisms. Since a typical population is reproducing by sexual activity, this set of genes can be considered as a pool, with much mixing as one generation produces the next. From this viewpoint, evolution is any change in the frequencies of alleles within the gene pool as the generations go by.

Let us now consider some evolutionary agents other than natural selection.

Genetic Drift Is a Form of Evolution
Within Small Populations

If a population is isolated from others of the same species, and if it consists of only a small number of individuals, then random chance can play a larger part in changing its gene pool than would occur in a large population. Any significant change by random chance in a small population is called *genetic drift*. Consider this example. There are twenty mountain goats in an isolated area. Eighteen of them have the genotype *AA* for a particular gene that governs coat color, and the other two have genotype *aa*. Suppose that one of the *aa* individuals slips and falls to its death before it has had the chance to reproduce. In a single accident having nothing to do with the gene's phenotype, there has been a loss of half of the *a* alleles from the gene pool. Now, of course, a very large population of goats would experience occasional random loss also, but each accident would have much less impact on the allele frequencies of that gene pool. It is the small size of a population that magnifies the impact of random events. How small must a population be, if one is to expect genetic drift to be important? There is no magic number, but evolutionary biologists consider that a population of 100 individuals or fewer certainly qualifies, and one of as many as 1,000 might show some evidence of genetic drift.

Random loss is not the only way that genetic drift can occur. For instance, there is also the *founder effect*, in which a small part of a large population becomes separated and founds a new population. Consider a large population of a certain bird species on a mainland area. It probably has much variation among its genes. Now suppose that over a period of several years windstorms carry just a few dozen of this population to an island that is normally far beyond the flying range of the species. The birds that arrive on the island were randomly chosen by the wind. It is likely that the frequencies of their various alleles do not exactly match those of the large mainland population; indeed, they might be quite different. In this case, the evolutionary change is not within the original population, but is a difference between that population and the newly founded population.

Of course, each of the two populations is likely to evolve differently after the second one has been established, for at least three reasons. First, they have different gene frequencies, as explained above. Second, the new population is very small, so it is susceptible to the sort of genetic drift described above. Third, it is likely that the two populations are under different selection pressures, being in different environments, so they will undergo natural selection in different directions.

Nonrandom Mating Causes Evolution of a Gene Pool

Another way for evolutionary change to occur involves mate selection. Individuals are most likely to mate with those organisms that are their closest neighbors, especially if they are in a large population and are not very mobile. A result of this *nonrandom mating* is *inbreeding*; that is, repeated reproduction between close relatives. Highly inbred portions of a population tend to have unusual frequencies of alleles. For instance, one inbred portion of the population might have many individuals with the *AA* genotype, and another portion might have a high frequency of *aa* individuals.

Another version of nonrandom mating is called *assortative mating*. Here, individuals in a population tend to choose mates that resemble themselves in some observable way. A human example of this is a tendency of short people to marry each other and tall people to marry each other. Assortative mating over many generations leads to a higher frequency of organisms with the homozygous genotype.

Mutation Inevitably Changes Allele Frequencies

Mutation, a change in the gene that produces a phenotype, is a most obvious way to change the gene pool. It is self-evident that an individual whose gamete has had a mutation to change allele *A* to allele *a* is going to influence the relative frequencies of these alleles in the next generation. However, this is not considered to be a major evolutionary change agent, since mutation at any single gene is a very rare event (perhaps one mutation per gene per million gametes). Making this rate even of less effect is the fact that *back mutation* can also happen, to cause allele *a* to become the *A* form. It should be noted, however, that a mutation that actually does happen within a *small* population will have much greater impact than if it happens in a large population.

Although mutation by itself cannot much influence a large population, it has major long-term impact upon the evolutionary history of a species. Remember that natural selection is a potent evolutionary mechanism, and that its occurrence depends upon there being variations in the population. Mutation is the method by which these variations have come into being.

Migrations Are Evolutionary Activities

If evolution is the changing of allele frequencies in a population, one of the most obvious ways it can happen is for mobile organisms to walk (or crawl or swim or fly or be carried) away from a population. As they leave, they take their genes with them and the population is different because of it. Of course, in-migration causes change of the gene pool, too. Migration has played a major role in altering human populations, especially in recent decades when transportation methods have drastically improved. It is becoming increasingly more difficulty to find isolated human populations whose gene pools have not been altered by sexual contact with others.

CAN NEW SPECIES ARISE FROM EVOLUTION?

All of the discussion so far has been concerned with change within a population but not to the extent that it becomes one or more new species. Remember that *species* is defined as a group of organisms that can successfully reproduce among themselves but not with other species (Chapter 2). Although most related species are different from each other in a number of their features, this need not be. If two groups differ in as little as a single feature of anatomy, mating behavior, gamete compatibility, etc., and if that feature inhibits successful reproduction, then they are separate species. Mutation and natural selection can lead to such a difference; no other mechanism need be invoked.

However, we must consider how those two groups could become separate. One such method is often called *allopatric speciation*, meaning that it involves the forming of species in two or more separated geographic localities. Suppose that a single population somehow gets split into two groups by some new feature of geography that isolates them from each other. For instance, a river shifts the course of its bed and splits a nonswimming animal population. Or, as a second example, a mountain range that had been barely passable by a species occupying both of its sides eventually rises (by internal geological forces) to the point where no animals of the species can any longer travel from one side to the other. Or perhaps there is a founding of a separate population of a species by random and extremely rare movement onto a distant island from the mainland (see founder effect, above).

In each of these cases, there are now two populations where there was once a single interbreeding one. There are also likely to be two different sets of selection pressures brought to bear on these populations, since they occupy different environments. Random mutations occur over time in both populations, and it is likely that some of them occur only in one of the two populations. The cumulative effect of all of these conditions, plus the possibility of any one or all of the other evolutionary activities described above, makes it likely that the two isolated populations will evolve to become different. This is not to say that they inevitably become different species; they might just become distinguishable subspecies. But, if there is gradual accumulation of differences, it is likely that, given enough time being isolated from each other, they will evolve to become different enough that they can no longer successfully reproduce even if given the chance to do so. Two species (not in contact with each other) will have been formed.

A second way that speciation might occur is called *sympatric speciation*. Here, there is not a need for geographical barriers to separate a group into two; the necessary changes occur within a single region. It is rare, but possible, for organisms of two related species to mate and produce a *hybrid* organism. Of course, the definition of species includes this possibility, but declares that such an offspring is expected to be sterile. Occasionally, especially in plant species, a hybrid organism, containing the genes of two species, can undergo certain chromosomal manipulations that lead to its becoming fertile (able to produce viable gametes). If two such hybrids mate (or if a hybrid self-pollinates), the resulting offspring are (a) different in characteristics from organisms of either of the parent species, and (b) able to carry on the line by reproduction. It has been an instant speciation event not requiring either geographical isolation or any of the gradual changes needed for allopatric speciation. A number of agriculturally important plant species have apparently arisen by sympatric speciation, as evidenced by examination of their chromosomes. For instance, modern wheat has the combined chromosomes of an early cultivated wheat and of a wild grass. Geneticists have learned how to induce the appropriate chromosomal changes needed to obtain successful mating between plant species and have done so to improve several crop organisms.

It is at least theoretically possible for animals to undergo sympatric speciation, but not by hybridization between species. For instance, if each of two forms within a population carries on very rigid assortative mating to exclude all reproduction with the other form, they have become as isolated from each other in a reproductive sense as if they had come to be separated geographically. Then, the mechanisms described for allopatric speciation can occur in each group.

This section began with a question, "Can new species arise from evolution?" The answer on theoretical grounds is yes. Is there fossil evidence for new species having arisen over the ages? Definitely yes. A typical species exists in the fossil record for about one to ten million years; then it is likely to disappear, perhaps because shifting environmental factors have driven it to extinction. Of course, when dealing with fossils, one must rely purely upon anatomical features to identify distinct species.

In recent years, two rather different mechanisms have been discussed for the sort of speciation that does not involve the instant hybridization of plants. One of these, often called gradualism, is the more traditional view, in which a gradual accumulation of differences comes into being until two groups can be called separate species. In 1972, Niles Eldridge and Stephen Jay Gould proposed an alternative mechanism that is often called *punctuated equilibrium*. They interpreted the fossil record to show that new species often come into being in a rather abrupt fashion (over a few hundred or thousand generations) with few or no intermediate subspecies fossils being found. They also assert that the fossil record shows the concurrent existence of the original species after the new one has come into existence. The gradualistic viewpoint would postulate many small steps between the start and end of a speciation event, and the existence of many intermediate fossils over perhaps a million years of records. Evolutionary biologists are still arguing about these two versions of speciation and the fossil records are fragmentary enough that both camps claim support from them. Gradualists say that most speciation events did not leave enough fossils to show the intermediate forms, and that those species in which a more complete series is available do show a gradual change.

A major question that the punctuated equilibrium advocates must answer concerns the mechanism for relatively rapid speciation. Is there a genetic basis for such a mechanism? Recent understanding of gene control may provide one. If a set of genes is operated by a single control gene (as is now known to be a common arrangement), then alteration of the control area has impact upon the operation of not one but several genes. Some of those genes, in turn, might have control over other sets. A single mutation (or alteration in the control conditions) might therefore initiate an immediate cascade of changes that could produce multiple phenotypic differences.

Of course, it is entirely possible that both the gradual and punctuated equilibrium methods happen concurrently in various populations. Perhaps further discoveries of fossils will resolve this question.

MACROEVOLUTION IS CHANGE ON THE GRAND SCALE

Evolutionary change on the level of the species or below has been called *microevolution*, to distinguish it from the larger changes of *macroevolution*. Although the fossil record shows that not all major groups of organisms (kingdoms, phyla, classes, etc.) came into existence simultaneously (see Table 15.1), much less is known about the details of these big changes. Perhaps no new evolutionary mechanisms need be invoked. Of course, to obtain big changes by the accumulation of smaller ones, much time is required. The age of the earth and the proportion of that time during which life has existed is sufficient to allow such accumulation.

No evolutionary biologist will claim that our understanding of the entire process will ever be complete. As more evidence is found, there may be large surprises, but the theory of evolution is becoming a stronger model rather than weakening as study continues. Recall the discussion of scientific theories in Chapter 1. The theory of evolution has all of the characteristics of a good theory—underpinnings in many thousand of observations, excellent predictive qualities, and the ability to be modified without seriously losing its structure. It is not without basis that evolution is considered to be the most important unifying idea in biology.

STUDY QUESTIONS

1. How are these types of fossils formed? (a) actual remains. (b) altered remains. (c) indications.

2. About how old is the earth? How do we know? What proportion of this time has seen life on earth? How do we know?

3. "Although all major groups of organisms have been present on earth since life began, the species within those groups have changed." According to the fossil record, what is wrong with this assertion?

4. What information is obtainable from fossil records besides the kinds of organisms present?

5. Why are animals and plants in Australia considerably different from those on the nearby Asiatic mainland?

6. Unrelated organisms that resemble one another are the result of what evolutionary process?

7. Give examples of homologous and analogous structures.

8. What kinds of evidence for evolution are seen by comparative embryology studies?

9. What are some examples of evolution observed over a short period of time?

10. What is a vestigial organ? Give an example.

11. How does evolution explain diversity among organisms? How does it explain similarities?

12. How did Lamarck's evolution model differ from Darwin's? Why is Lamarck's idea incorrect?

13. Define mutation and give some examples of it. How much influence does it have on evolution?

14. Outline the steps that together produce natural selection in a population. What does the selecting?

15. Has Darwin's theory been modified since first proposed? Elaborate on your answer.

16. Describe the founder effect. Why is it an example of genetic drift?

17. What is a gene pool?

18. Compare the gradualist and punctuated equilibrium hypotheses of speciation. What are fossil evidences for and against each idea?

16
THE DIVERSITY OF ORGANISMS

What do we know about:
—*why the classification of an organism is sometimes changed?*
—*why viruses cannot be fitted into any kingdom?*
—*the need to create a separate kingdom for the fungi?*
—*which group of organisms has the greatest number of species?*
—*why horsetails are not always parts of animals?*
—*whether the bilateral body plan has advantages over the radial body plan?*
—*engineering advantages provided to an animal that has a coelom?*
—*the difference between a cartilaginous fish and a bony fish?*
—*who are our closest nonhuman relatives?*
These and other intriguing topics will be pursued in this chapter.

THE ORGANISMS CAN BE ARRANGED LOGICALLY

The last chapter dealt with how we believe so many organisms have come into being. The present chapter provides the reader with a guide to the currently living organisms, presented in a way that indicates their taxonomic relationships with each other. It will be more than a catalogue, however, since many of the organisms will be described in terms of their habitats and adaptations to those habitats.

Chapter 1 introduced the reader to the field of systematics, whose practitioners attempt to group organisms (both living and extinct) into appropriate groups according to their evolutionary relationships. It will be recalled that this requires examination of adult and embryonic anatomy, fossil materials, and even the structure of the molecules made by the organisms. Refer to that chapter to be reminded of the many pitfalls to be avoided when one first examines a newly discovered organism and attempts to place it into its proper box of categories. One should also go back to that chapter to be reminded of the hierarchy of categories that is universally used, beginning with the smallest (species) and going progressively to more inclusive ones that culminate in the kingdom. In the present chapter, which does not attempt to be comprehensive, most of the organisms will be placed only into the larger categories of class, phylum, and kingdom.

One other thing must be clearly restated from Chapter 1 before we plunge into the task. Remember that classification is an inexact science, in the sense that the organisms do not always provide obvious and unambiguous clues to their relatedness to each other. The categories above that of species level are not directly testable, are (to some extent) reflective of the opinions of people working with them, and are therefore subject to change as more clues are discovered and the experts come to agreement. Undoubtedly, some of the placements of organisms used in this chapter will sooner or later become obsolete.

Even opinions on how many kingdoms exist have changed over the years. We will follow here the latest modification and place all of the organisms into six kingdoms. These were briefly described in Chapter 1. If one wishes to stick with the popular and quite useful system of five kingdoms, this can be done by simply lumping together the first two described below (Archaea and Bacteria), and calling that Kingdom Monera.

THE VIRUSES CANNOT BE PLACED INTO ANY KINGDOM

Since their discovery in the 1890s, viruses have puzzled biologists. They are unlike all other living things in that they are not cellular and cannot carry on any form of metabolism; instead, they use the metabolic processes of their hosts to gain energy. Some people even dispute their being called living organisms, although they are in the minority. Any virus consists of a single piece of nucleic acid and a protein coat surrounding it (a *capsid*). The variations in size and shape among different viruses are due to the capsid shapes.

Viruses can be classified according to several criteria. Viruses use either DNA or RNA as their molecules for storing information. Those that use RNA (including the infamous human immunodeficiency virus and the herpes viruses) are unique in this respect among all known living organisms. Those that use DNA can be further distinguished by whether the DNA is single-stranded or double-stranded. Another characteristic used for classification is the shape of each viral species. Viruses can also be distinguished according to whether they reproduce in prokaryotic cells (bacteria) or in eukaryotic cells. The former are called *bacteriophages* (or simply *phages*).

All viruses are parasitic upon cellular organisms and some can cause serious diseases leading to death of the hosts. Viruses affecting plants are usually transmitted by insects that are herbivores upon the plants. They often show their presence by pale blotching on leaves and flower petals. Viruses affecting humans are responsible for a long list of diseases, among which are smallpox, rubella, yellow fever, influenza, and mumps (in addition to those mentioned above). Nonhuman animals sometimes have viral diseases very similar to human ones (e.g., cowpox), but the usual rule is that each species has its own viruses and that viruses inhabiting one species are not very successful infecting cells of another species. Animal viruses are carried from host to host by a variety of methods, including air, water, biological fluids (blood, semen, etc.) and invertebrate vectors (mosquitoes, biting flies, etc.).

Because of their extreme simplicity and ease of obtaining large populations, viruses have been useful tools for geneticists attempting to understand the molecular nature of genes. They are also sometimes used to carry genes into eukaryotic cells in the artificial process of genetic engineering (Chapter 12).

KINGDOM ARCHAEA IS THE SMALLEST AND MOST RECENTLY NAMED KINGDOM

It was only a few decades ago that biologists began to consider that bacteria are different enough from all other organisms that they should be placed in a separate kingdom. Now, since the early 1990s, many are splitting that kingdom, Monera, into two. One group of bacteria (or bacteria-like organisms) is sometimes called the Archaebacteria and placed into its own kingdom, Archaea. (If a five-kingdom scheme is preferred, there will be two subkingdoms in Kingdom Monera: Archaebacteria and Eubacteria.) The archaebacteria have the following *prokaryotic* characteristics: being single-celled, lacking a nuclear membrane, and having a single circular chromosome that does not include permanent proteins.

The Archaean organisms have a cell wall that is molecularly different from that of any other organisms. They differ from other bacteria in a number of other features (of lipids, certain enzymes, ribosomal RNA, etc.) that together have persuaded many experts that they had a common ancestor but split very long ago. (Some even argue that the two bacteria-like groups arose from independent origins.)

Modern Archaeans are usually found in very extreme environments, where life is usually difficult to sustain. Some are *halophiles*, preferring very salty places such as salt ponds. Others are *thermophiles* (living in hot springs or hot vents deep in the ocean), or *methanogens* (living under strictly anaerobic conditions and producing the gas methane during their metabolism). Only a few genera of these organisms have been identified. They may be survivors from a period when the earth's atmosphere was very different from its present condition.

ORGANISMS OF KINGDOM BACTERIA ARE KNOWN AS THE TRUE BACTERIA

This newly split off kingdom comprises the more familiar Moneran organisms that are often called the true bacteria. Over 4,000 species of bacteria have been named, but there is good evidence (based on DNA studies) that this is only a very small fraction of those that exist. A problem in determining species of bacteria is that they do not usually reproduce sexually (or, at least not in classical sexual ways), and therefore cannot easily be classified into species by any ways except those of anatomy and biochemical activities.

For many reasons the true bacteria comprise an interesting group. Many bacteria are responsible for decay or fermentation. The spoiling of food, the decomposition of sewage in septic tanks and large disposal systems, the making of sauerkraut, the conversion of cider and wines into vinegar, the enriching of soil by fixing nitrogen, the depleting of soil by denitrification, and the decay of dead bodies all involve the metabolic activity of bacteria. Obviously the very same organism may be useful or harmful depending upon what it attacks. Decay bacteria are harmful if they destroy food or useful if they decay carcasses.

Parasitic bacteria, like other parasites, take foods and other essential substances away from the host. They can also release toxins that sicken the host. The effect of the damage they cause is disease. Many of the disease-causing bacteria are controlled by antibiotics, but they can rapidly adapt to antibiotics by undergoing mutations. A major current medical problem is that of antibiotic-resistant strains of bacteria that can harm humans.

Some nitrogen-fixing bacteria live in root cells, where they stimulate growth of nodules. This relationship between bacteria and host may be interpreted as mutualism, since both species benefit from the relationship.

A few bacteria live an independent existence by carrying on photosynthesis. These organisms, called Cyanobacteria, are sometimes found in ocean water but are more common in freshwater habitats. Another group of free-living bacteria is that of the *chemosynthesizers*. These are able to use inorganic molecules such as ammonia and hydrogen sulfide as energy sources. They are most often found in soil.

Bacteria vary considerably in size and shape (Fig. 16.1), features that are used in their classification. Shapes are round (*coccus*), elongate (*bacillus*), or twisted (*spirillum*). Bacteria often cling together in masses or strings.

In addition to this feature, another one used for classification is the staining quality of their cell walls. About 100 years ago, a man developed a staining procedure to help make microscopic identification of bacteria easier. Known as Gram stain for its inventor, it is taken up by some types of bacteria but not by others. Therefore, bacteria are classified as Gram-positive (staining) or Gram-negative (not staining). There is a third group of bacteria, the mycoplasmas, that lack a cell wall. These extremely small cells may be the

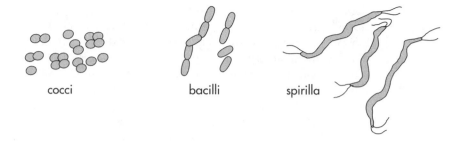

cocci bacilli spirilla

FIGURE 16.1. *Three types of bacterial shapes.*

most primitive cellular organisms. Some are implicated in certain human diseases, including forms of pneumonia.

Bacteria have been extremely useful as study organisms in genetics. They were used in the first investigations of the replication of DNA, the understanding of the genetic code, and the mechanisms of transcription, translation, and gene control. They were also the first organisms to be used in recombinant DNA studies that led to the field of genetic engineering.

KINGDOM PROTISTA INCLUDES THE UNICELLULAR EUKARYOTIC ORGANISMS

It will be remembered from Chapter 1 that all organisms other than viruses and bacteria are placed together in a group called the *eukaryotes*. They have a number of significant differences from the prokaryotes that lead evolutionary biologists to consider the possibility that they came into being independently.

Organisms of Kingdom Protista are the eukaryotes that are single-celled. However, some Protistans form extensive colonies that resemble multicellular organisms. Both asexual and sexual reproduction are found in the group, including many species that alternately use both methods. Protistans are usually mobile, using cilia, flagella, or a gliding (*amoeboid*) motion. The classification of Protistans is extremely difficult. Some experts place them into as many as 50 phyla. We will speak rather generally about their classification, using the loose term group as much as formal names.

Algae Are Photosynthetic Protistans

This is a diverse group that for many years was placed among the plants (and is still there, according to some biologists). They are often categorized on the basis of their colors, which are determined by their photosynthetic pigments. No attempt will be made to describe all photosynthetic Protistans.

Phylum Euglenophyta is a group that has both plant-like and animal-like characteristics. If they manufacture food, they have *pyrenoids,* which are granules having to do with the formation of paramylon, a storage polysaccharide. Pyrenoids are associated with the chloroplasts that the cell contains. Some of them lose their chloroplasts if cultivated in darkness and thereafter ingest organic materials, acting therefore as heterotrophs. It was this sort of mixed metabolic ability that led systematists to invent the category of Protista. Euglenoids move by flagella. Some euglenoids have a peculiar contracting and stretching action called euglenoid motion. *Euglena* (Fig. 16.2) is an example.

The *green algae,* phylum Chlorophyta, are an unusually diversified group with regard to both their shapes and habits. They vary in size from independent single cells to large colonies easily mistaken for multicellular plants (commonly called seaweed). They are common in fresh and salt water but are also found in soil, on bark of trees, and in lichens (see Fungi, below). They have chloroplasts that contain two forms of chlorophyll, a and b, plus carotenes and xanthophylls. They have pyrenoids around which polysaccharide is deposited. Their walls contain cellulose and pectin. Some members are flagellated. Most flagellated forms have two whip-like flagella. Many green algae perform sexual reproduction. Algae of lakes provide food for a host of small aquatic life such as worms and immature insects. Examples of green algae are *Protococcus* (Fig. 16.2), *Volvox,* and *Spirogyra* (Fig. 13.3). Because the congregations of some Chlorophyta are on the border of being multicellular organisms, some systematists prefer that this phylum be in Kingdom Plantae.

Like the green algae, the *golden algae* (phylum Chrysophyta) are a very diversified group. In addition to chlorophylls a, c, and e, they have carotenes

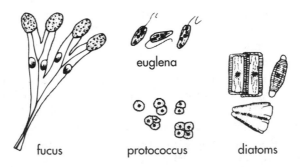

euglena

fucus   protococcus   diatoms

FIGURE 16.2.   *Some types of algae. The swollen tips of* Fucus *are receptacles containing reproductive organs; the bubble-like structures lower on the thallus are gas-filled bladders.*

and xanthophylls that usually mask the green, thereby giving them a golden-brown color. They do not store starch, but do store oils and a polysaccharide called leucosin. Their walls contain pectins, cellulose, and silica. Members are unicellular or colonial.

Organisms of phylum Bacillariophyta, the *diatoms* (Fig. 16.2), are cosmopolitan in distribution. Diatoms are single cells with box-like walls consisting of a high proportion of silica. Although diatoms are found in abundance in fresh water, they grow so prolifically in some parts of the oceans that their skeletons accumulate on the bottom as large deposits of ooze. Fossil diatomaceous deposits are mined and used for abrasives, insulation, and filters.

Brown algae, of phylum Phaeophyta, almost exclusively marine and more abundant in cooler oceans, are either in large colonies or form multicellular organisms, depending on which systematist is asked. If the latter, they should be in Kingdom Plantae. They are especially abundant in the intertidal zone or in shallow water. Besides containing chlorophylls a and c, they have carotenes and xanthophylls as pigments. Fucoxanthin, a xanthophyll, masks other pigments and gives the algae a brown color. Foods are stored as laminarin, a polysaccharide, and as mannitol, an alcohol. Their walls are composed of cellulose and algin, a gelatinous material used in foods and in several commercial processes.

Some brown algae have gas-filled bladders that make them buoyant in the water. The *kelps*, which are the largest algae, belong to this group. Some of them grow at least 50 meters long. *Fucus* (Fig. 16.2), or rockweed, is a well-known representative that lives along rocky shores. *Sargassum*, or gulfweed, is so abundant in a two million square mile area located near the middle of the north Atlantic Ocean (east of Florida) that the area is called the *Sargasso Sea*. Ordinarily Sargassum is attached, but when it breaks from its anchorage it floats. Many of the detached plants follow oceanic currents and accumulate in the Sargasso Sea.

Like the brown algae, *red algae* (phylum Rhodophyta) are mostly marine. They usually live in deeper and warmer waters than the brown algae and are often delicate in habit. They are not typical of aquatic plants in that they have no motile cells (sperm or spores). Rather obvious protoplasmic connections link the cells. They contain chlorophyll a and d, plus carotenes, xanthophylls, r-phycocyanin, and r-phycoerythrin as their pigments. They store a starch that is similar to glycogen. Walls contain cellulose and agar, the latter having many commercial uses. In fact, it is almost indispensable as a medium in which bacteria are grown in laboratories.

Some Protistan Are Like Fungi

There is a group of Protistan organisms that were once classified in the Kingdom Fungi. They apparently gained fungus-like characteristics by convergent evolution (Chapter 15) and were therefore foolers for a long time. They are now placed with other Protistan because their cell walls contain cellulose, a material not found in the walls of true fungi.

The major group here is the *cellular slime molds*. As these organisms feed, they are haploid individuals, resembling amoebae. But when food runs out or the habitat becomes dry, many of them congregate, form a mass (very close to being a multicellular organism) that produces spores for the next generation. Occasionally, two haploid cells join in sex to make a diploid zygote, which almost immediately goes into meiosis to make haploid cells again. Although slime molds are not observed by most people, they are much more common than believed. Most of the time they are concealed underneath or in organic litter. However, the spore-bearing stages are often seen but assumed to be some kind of fungi. The fungus-like Protistan are heterotrophic, having no photosynthetic apparatus.

Kingdom Protista Includes Completely Heterotrophic Animal-like Organisms

Many Protistan were formerly classed with animals because they (a) lack cell walls, and (b) are completely heterotrophic. Their general name is *protozoa*, signifying "first animal-like." Most protozoa are solitary, but some form colonies. Only some of their several phyla are described below.

The *flagellated protozoa* are in phylum Zoomastigophora (also called Zoomastigina or Mastigophora). Members of Zoomastigophora have a common characteristic in moving by means of one to many long flagella. Their reproduction is by simple fission, which is unique in being longitudinal. A well known parasitic member is *Trypanosoma*, which causes African sleeping sickness. It is transmitted by the tsetse fly. A mutualistic flagellate lives within the gut of the termite, providing the enzyme cellulase with which the insect digests its woody food.

A second major protozoan phylum is that of the *amoebas*, phylum Rhizopoda (sometimes called Sarcodina). Almost all of these protozoa move by means of cytoplasmic projections known as *pseudopodia* (Fig. 4.17 A). It is a characteristic motion called *amoeboid* (also performed by leukocytes of animals). Pseudopodia are often temporary and irregularly shaped. The familiar amoebas are asymmetrical, naked cells that are ceaselessly changing shapes. Some other members of phylum Rhizopoda have more definite shapes and are protected by shells. These shells are composed of calcium, silica, or chitin. One well-known parasitic member of class Rhizopoda is *Entamoeba histolytica*, which causes amoebic dysentery in humans.

Organisms of two phyla common in the plankton of the oceans leave shells when they die. They settle to the bottom where they accumulate as ooze. One phylum, Foraminifera, has calcium-based shells. The white cliffs of Dover (England) are chalky deposits of their remains that were uplifted by crustal movement of the earth. Foraminifera have a long geologic history, and distinctive members (index fossils) are useful to oil geologists, who can identify strata of rocks by the foraminiferal shells contained in them. The second such phylum, Actinopoda, includes organisms called *radiolarians*. These have shells of silica that accumulate as ooze on deeper oceanic floors.

The name *sporozoa* was given to organisms of phylum Apicomplexa because one stage in their life cycles involves the production of spores. Their life cycles are usually quite complicated. Mature sporozoans have no means of independent locomotion; all members are parasitic. The best-known example is *Plasmodium*, the malarial parasite that invades red blood cells and liver cells of humans. Some species of *Plasmodium* also parasitize lizards, birds, and other mammals. The intermediate host of the malarial parasite is the mosquito. *Eimeria* is a troublesome pest infecting intestinal cells of birds and mammals. Farmers have to fight it in chickens and cattle.

Members of phylum Ciliophora move by means of many short projections, cilia. *Ciliates* have two or more nuclei of two kinds. The larger type, the *macronucleus* (Fig. 13.1), controls release of RNA to direct day-to-day activities of the cell, while the smaller type (which may exist in multiples), the *micronucleus*, exchanges its genes with those of other strains during conjugation. Most ciliates are free-living inhabitants of water. The classic example used by biology classes is *Paramecium*.

KINGDOM FUNGI INCLUDES HETEROTROPHIC ABSORBERS OF FOOD

Although traditionally studied with the plants, the fungi must be considered as fundamentally different from them and be assigned their own kingdom. Indeed, as indicated in Chapter 1, recent molecular studies seem to place them closer to animals than plants. Fungi are multicellular heterotrophs, lacking the ability to perform photosynthesis. However, they do not actually eat organic materials; rather, they exude digestive enzymes onto an organic surface, then absorb the molecules that are released by the enzymes' activity. Most fungi reproduce by means of wind-blown spores produced by haploid bodies. Occasionally, there is sexual union of two organisms, but the resulting zygote quickly undergoes meiosis to bring the life cycle back to haploid condition. The fungal cells have walls that are chemically different from those of any other organisms. Most fungi are ecologically important because they are decomposers, releasing materials to be cycled through an ecosystem. The field that includes the study of fungi is called *mycology*. Since fungi were once considered to be plants, their largest taxonomic categories are called *divisions*, as are those of plants, rather than phyla.

Division *Ascomycota* is a large group (over 60,000 known species) of organisms that can reproduce by producing spores in a sac-like object called an ascus. *Yeasts* are unicellular members of this division (Fig. 16.3). Most other members are multicellular through most of the life cycle. Among these are the *truffles* prized by gourmet cooks, *powdery mildews*, most of the vari-colored *molds* that cause food to spoil, and many of the fungi that make symbiotic association with algae to form *lichens*. The genus called *Neurospora*

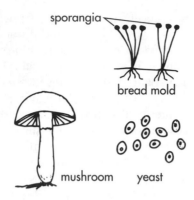

FIGURE 16.3.   *Some types of fungi. Bread mold is in division Zygomycota, the mushroom is in division Basidiomycota, and the yeast is in division Ascomycota.*

has been much used by geneticists to elucidate such mechanisms as crossing over.

Another fungal division is *Zygomycota*, whose best-known member is *black bread mold* (Fig. 16.3). Also in this group are fungi that form *mycorrhizae*, which establish mutualistic relationships between their organism and plant roots. The fungus provides some minerals to the plant root and helps extend its surface contact with soil, while the plant provides organic molecules to the fungus.

The *club fungi* are those of division *Basidiomycota*. They include some very familiar organisms, the *mushrooms, bracket* (or *shelf) fungi, puffballs, rusts,* and *smuts*. Their most distinctive characteristic is the manner in which they produce spores (*basidiospores*). The distinctive basidiospores (Fig. 16.4) are produced at the tips of minute stalks on the outside of characteristic *hyphae*

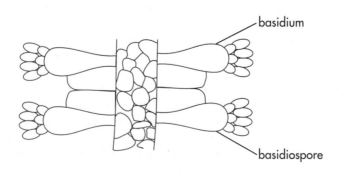

FIGURE 16.4.   *Reproductive portion of a mushroom, lying upon gills underneath the cap.*

(threadlike multicellular structures composing a fungal body). The hyphae bearing them have terminal cells that are often club-shaped: hence the name *basidia*. Basidia form from a *mycelium*, which in turn resulted from the fusion of two parental hyphae. Each basidium thus has a nucleus from each of the parent plants. They fuse and form a diploid zygotic nucleus, which then divides twice to produce four haploid spore nuclei. The spore nuclei migrate into the tips of the tiny projections from the basidium, where they become the nuclei of the basidiospores that soon detach from the basidium. The typical number of spores produced by a single basidium is four.

There are no rules that one can follow in separating poisonous and non-poisonous mushrooms; one just has to know them individually. If undisturbed, the mycelia of mushrooms spread in all directions from a central location. As the younger hyphae grow outward, the older parts toward the center die. This habit of growth results in a circular distribution of mycelia that is obvious at the time reproductive bodies (mushrooms) appear and form the familiar *fairy rings*.

Bracket fungi have reproductive bodies that grow like shelves on tree trunks. Some are parasitic on living trees, while others are decomposer organisms on dead trees. Rusts produce a kind of spore that is orange or rust-colored. They parasitize many plants—for instance, pines, blackberries, wheat, and apples. Many rusts have stages that parasitize two different hosts. The least economically important host is known as the *alternate host*. The barberry shrubs are alternate hosts of wheat rust, and junipers (red cedar) and related plants are alternate hosts of apple rust. Smuts produce a kind of spore that is distinctly smutty black. They do much damage to monocots such as corn, wheat, or barley. Other familiar basidiomycetes are puffballs, earthstars, bird's nest fungi, jelly fungi, and stinkhorns.

A number of fungi are so poorly known in regard to their sexual stages that they cannot be properly classified. They are put in an artificial catch-all division known commonly as the *imperfect fungi*, until their stages can be determined. The official division name is Deuteromycota and it comprises about 25,000 species.

KINGDOM PLANTAE IS THAT OF THE MULTICELLULAR PHOTOSYNTHESIZERS

We come now to the true plants, those organisms that have had such a profound impact upon the earth by their ability to capture the energy of light and provide it to both themselves and most of the rest of the organism types in the world. Technically, plants are multicellular autotrophs. Plants inhabit every conceivable habitat on earth and range from nearly microscopic to the size of giant sequoia trees. Although many are aquatic, their greatest success has been on land.

Plant reproduction is characterized by their ability to produce an embryo from the fertilized egg after sexual union of gametes. They have definite alternation of generations, between gametophyte generation and sporophyte generation (as discussed in detail in Chapter 13).

The plant kingdom can be divided into two major groups: those that cannot make seeds and those that can. Another way to divide them is by whether they have vascular systems (tubes for transport) or not. All told, there are approximately 250,000 identified plant species.

Bryophytes Are Nonvascular Plants

The members of division Bryophyta are the simplest of the plants. The best-known ones are the *mosses*. An absence of vascular tissue definitely limits the size they can achieve. Consequently they are all small plants. Some have flat, ribbon-like bodies that grow prostrate on the substratum; others are upright and leafy. Not having conducting tissues, they are not considered to have true roots, stems, or leaves. Their gametophytic generation is more conspicuous than the sporophytic generation (Fig. 13.5).

Bryophytes live in many kinds of habitats that range from extremely wet to extremely dry conditions. Some are submerged in water, while others live on bare rocks exposed to full sunlight. Many grow on the bark of trees, where growing conditions are favorable only intermittently, as when the weather is wet. Of course many species live in moist, shady places, which is probably the habitat most people associate with them

Mosses are erect or prostrate with erect branches. They are all leafy, and the leaves are spirally arranged. Gametophytic plants grow from buds produced on algae-like filaments known as protonema. In general, capsules of the sporophytes open by distinct lids. The moss life cycle is detailed as shown in Figure 13.5.

One group of Bryophyta, the *liverworts*, are either ribbon-shaped (Fig. 13.2 B) or leafy. In either case their bodies are anchored by *rhizoids* (structures that resemble root hairs). The leafy ones are superficially similar to a prostrate moss, but in most of them their ribless leaves are in two rows. Their sporophytes differ from those of mosses in having capsules that often split into two or more valves rather than opening by distinct lids. Many of the capsules split into four valves.

Ferns Are the Best Known Seedless Vascular Plants

Vascular plants are those that have conducting tissues (xylem and phloem), which means that they have true roots, stems, and leaves. Having vascular tissue is a sporophytic characteristic; conducting tissue is virtually never found in gametophytes. For the first time the sporophyte is the conspicuous plant, while the gametophyte is very abbreviated in size.

Four divisions comprise the vascular plants that are unable to make seeds. Three of the four are widespread but not commonly known plants; the fourth is that of the ferns.

Members of division Psilopsida are the most primitive of the vascular plants. It is composed of fewer than two dozen species worldwide. Psilotum (Fig. 16.5 A) is in tropical and subtropical America; *Tmesipteris* is in Australia, New Zealand, and New Caledonia. Members of both genera are terrestrial in habit and nearly always under twelve inches tall. For the most part their branches are dichotomous and arise from horizontal rhizomes. They have no roots, only unicellular rhizoids. There is no secondary growth. Leaves are either absent or extremely simple, and the vascular bundle of the stem has no leaf gaps.

Members of a second primitive vascular division, Lycophyta, are small plants that have persisted as relics of a group whose members were much larger and occupied a much more important position in ancient geologic times. Some of them retain the primitive characteristic of dichotomous branching (branches splitting to two, then each of these doing the same, as in Fig. 16.5 A). This group is represented by three genera—*Lycopodium* (Fig. 16.5 B), *Selaginella*, and *Isoetes*. Lycopodia are known as *club mosses* and include such members as ground pine and ground cedar. This genus is represented in temperate regions by terrestrial forms with stems commonly creeping along the ground and sending up branches less than a foot tall. They often grow close together in colonies.

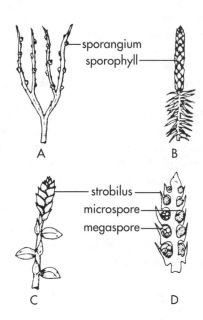

FIGURE 16.5   *Primitive vascular plants. A:* Psilotum, *dichotomously branched, with scale-like leaves, and with axillary sporangia. B:* Lycopodium *with terminal cone of sporophylls. C:* Selaginella *showing four rows of leaves and terminal cone. D:* Selaginella *cone containing megaspores and microspores.*

Selaginellas (Fig. 16.5 C and D) are known as *little club mosses* (some as resurrection plants). Their leaves are arranged spirally or in four rows. The spirally arranged leaves are aligned in four rows and are similar in shape to the leaves of higher plants. Selaginellas have one distinct advancement in having two kinds of spores, *microspores* and *megaspores*, that grow into separate gametophytes, *microgametophytes* and *megagametophytes* (see Chapter 13). Selaginellas in the temperate region are moss-like in general appearance and either creep along the surface or grow a few inches tall. Some of them grow in mats in very dry places as on exposed rocks, whereas others sprawl on wet soil, as along stream banks or margins of swamps.

All species of *Isoetes* live in very wet places. In general appearance they are somewhat grass-like. They have large sporangia at the base of their leaves and are like *Selaginela* in being *heterosporous* (with two kinds of spores). The megaspores are the largest of spores, easily visible to the naked eye.

Another less well known group of vascular seedless plants is division Sphenophyta. Its members are commonly called *horsetails* and are represented by the single genus, *Equisetum*. As a group, modern representatives compare feebly in numbers and size with the rich sphenopsid flora of the Carboniferous period. Modern horsetails are mostly under three feet tall and grow in somewhat open places such as meadows, roadsides, and stream banks. Horsetails are often called scouring rushes because they were used by American pioneers for scouring kitchen utensils. They are suitable for that purpose because of deposits of silica in their stems. Horsetails have underground perennial rhizomes from which erect stems grow. Some species have two kinds of erect branches—a branching vegetative one and a simple reproductive one with a spore-bearing cone on top. The stems are distinctly jointed and have longitudinal grooves and furrows.

By far the most widely recognized vascular seedless plants are the ferns, of division Pterophyta. They generally have broad leaves and leaf gaps. Leaf gaps are areas that are devoid of conducting tissue immediately above where the leaf is attached to the vascular bundle of the stem. The conducting tissue that would have occupied that space bends out into the leaf.

The range of habits and habitats of ferns is much more varied than one would expect from the impression sometimes gained from local situations. What one would perhaps call a typical fern (Fig. 13.6) has an underground rhizome but no stem above ground. The leaves, called *fronds*, have rusty-looking spots on the underside that are clusters of sporangia. When fronds emerge from the ground they unroll from base to tip. While in this emerging stage they are called *fiddleheads*. Some may be treelike or climbing vines; others may be delicate and filmy or coarse and weedy. Ferns often grow in shady damp places, but many of them grow under extreme conditions—for instance, on trees, on rocks, in water, or in dry open places. Fern gametophytes are small, flat structures known as prothalli (singular: prothallus—see Chapter 13). Their sperm are spiral, multiflagellated cells.

The Gymnosperms Are the Most Primitive Vascular Plants with Seeds

The advantages and methods of producing seeds were described in the chapter on reproduction (Chapter 13). Some divisions of seed plants protect their seeds within a fleshy fruit, but the more primitive ones do not. There are four divisions of the latter type, together called the *gymnosperms* (from the Greek: naked seed). The seeds of gymnosperms are produced on the outside of the female structure, which could be called a carpel or pistil if it were compared to a similar structure in a flower. In the cone-bearing gymnosperms (of division Coniferophyta), which are by far the largest group, these seed-bearing structures are the scales of the cone. The seeds are attached at the base of the upper surface.

This is the first group of plants to have pollen, male gametophytes that produce the sperm. Pollen makes fertilization without water possible, giving gymnosperms and other seed plants the distinct advantage of being able to reproduce in dry weather. Almost all of them have nonmotile sperm.

Commonly, gymnosperms are evergreens with needle-like or scale-like leaves. (A few, like the Ginkgo tree, have deciduous, broad leaves). They commonly have an excurrent shape; that is, they are shaped like a Christmas tree. Their wood is composed of tracheids; their lumber is known as softwood. Gymnosperms include cycads, the Ginkgo, conifers, *Ephedra* (Mormon tea), and *Welwitschia*. Conifers, the most familiar gymnosperms, include pines, junipers, cypresses, cedars, firs, spruces, arbor vitaes, and sequoias. *Ginkgo* has a long geologic history and came to this country from Chinese gardens. *Metasequoia* was known only as a fossil until it was discovered in inland China in 1946. Now it is extensively propagated in America. *Welwitschia* is an odd, sprawling, two-leafed gymnosperm of the deserts of southwest Africa. *Ephedra* is a shrub of the United States' southwestern deserts and elsewhere. Gymnosperms are by far more abundant in the northern hemisphere.

The Angiosperms Protect Their Seeds Inside Fruits

Angiosperms, including only the single division Anthophyta, are the dominant land plants today. They are contrasted with gymnosperms by the manner in which they produce seeds, for in this group seeds are enclosed in fruits (Figs. 13.8 and 13.10). Fruits come from flowers, also a distinctive characteristic of angiosperms. Flowers are not always showy and fragrant, but they are present even in weeds, grasses, and trees, where they are not always noticed. Angiosperms have pollen that produces nonmotile sperm. Their wood is composed chiefly of fibers and vessels (tracheae). Their lumber is known as hardwood. The life cycle of an angiosperm is given in Chapter 13. About 235,000 flowering plant species have been identified.

Two classes of division Anthophyta exist: class Magnoliopsida (commonly called the dicots) and class Liliopsida (commonly called the monocots). *Dicots* are plants that have embryos with two cotyledons (seed leaves). Cotyledons are leaves that are present on the embryo contained within the seed (Fig. 14.10). Often they are thickened with stored food. Usually cotyledons are ephemeral and have a different shape from leaves that come later. Of course cotyledons are not usually available, so one has to look for other characteristics to recognize plants as dicots. The leaves are usually net-veined, flower parts (not flowers) are usually in fours or fives or their multiples, vascular bundles are arranged in a cylinder (Fig. 8.9), and vascular bundles are open. Open vascular bundles are those that contain cambium and are open to further growth; that is, they have secondary growth. Dicots include such plants as rose, daisy, celery, buttercup, oak, and maple.

Monocots have only one cotyledon. They are thought by some to be derived from dicots, in which case one cotyledon was lost. The leaves are usually parallel-veined, flower parts are usually in threes or sixes, vascular bundles are scattered in a parenchymatous ground mass (Fig. 8.9). The vascular bundles are closed (closed to growth because they contain no cambium). Examples of monocots are sedge, lily, orchid, amaryllis, iris, corn, and grass. Grasses are of great economic importance to humans as the source of all grains, as pasturage for their cattle and sheep, as stabilizers of soil to prevent erosion, and as enhancers of the beauty of lawns and golf courses.

ANIMALIA IS THE LARGEST KINGDOM

The kingdom of animals contains more than 1 million identified living species, with many more undoubtedly to be discovered. Some estimate that as many as 20 million species currently exist. As variable as animals are, they have some characteristics common to most of them. They are, of course, the kingdom of multicellular heterotrophs that eat their food and then digest it internally.

For the most part animals are more responsive than plants. This characteristic is particularly related to their having a coordinating nervous system and muscles that can contract. Certainly a nervous system capable of perceiving stimuli, interpreting the environment, storing information, and directing responses plus muscular tissue capable of reacting to stimuli are milestones of achievement in the animal line of descent. Response through muscular action that allows the organism to seek another environment or alter the one in which it lives rather than submit to whatever may come improves its chances of survival. In the case of some sessile animals, which are aquatic, it provides the means to circulate the environment. Locomotion also disperses adults, a function that is performed by reproductive structures in sedentary animals and most plants. Locomotion is possible because of the flexibility of animal bodies. Flexibility is possible because of the absence of cell walls and the presence of joints in the skeletons of many higher animals. Motion is an energy-consuming activity that far exceeds the energy required by plant processes.

Animal cells do not have cell walls, chloroplasts, plant storage products, or large vacuoles. On the other hand, they do have cell centers that participate in the formation of asters during mitosis, features missing in plants.

As a consequence of not having cell walls that make their bodies rigid throughout, many animals have developed supporting skeletons. The skeletons may be external (*exoskeleton*) or internal (*endoskeleton*). If they are external, they may also serve an important function of protection. There is no question but that exoskeletons are characteristic of smaller animals like most of the invertebrates and that endoskeletons support the larger animals. By the very mechanics of construction, an exoskeleton does not support as well as an endoskeleton. Furthermore, the animal is limited to the confines of the skeleton unless it molts and rebuilds another larger one (Chapter 14). Endoskeletons can keep pace with the growth of other tissues and support large land animals. A much stronger skeleton is required to support weight in air than in water because water is buoyant. The mass of whales is manageable by their skeletons as long as they are in water, but they get hopelessly grounded on beaches during storms.

While not all animals have skeletons, the animals that do have them range in size from microscopic to the largest organisms ever occurring on earth (the whales). Sponges secrete *spongin fibers* (made of protein) or mineral needles (*spicules*) that give a supporting framework. Exoskeletons of some Molluscs are the familiar sea shells. They support and protect very weak and delicate bodies. They are enlarged by accretion as the animals grow. Sea shells are mainly calcareous. Insects, crustaceans, and other Arthropods have exoskeletons of a polysaccharide material called chitin. Echinoderms (starfish and others) have internal plates for support. The vertebrates have internal, jointed skeletons composed of cartilage, bone, or both.

That there is a relationship between form and function has been recognized for a long time. An obvious expression of it is seen in the difference in the general body plan of attached versus free-moving animals (Fig. 4.17 C and D). Animals that remain attached or move only very slowly from place to place are more likely to be radially symmetrical. Since this plan has body parts radiating out in all directions, its possessor can meet challenges or catch food coming from any area. A moving body, which is typical of most animals, should have a form that is easy to manage. This ability to control motion is best achieved in a bilaterally symmetrical body plan, which includes a head equipped with sensory apparatus to explore territory being approached.

With respect to locomotion, any adaptation for speed would be advantageous. Reducing friction by streamlining the body or by getting the body out of water or soil could be advantageous. Or the lifting of bodies off the ground by having legs underneath rather than extending from the sides would be more efficient. The forms that animals have taken have incorporated these features that have functional advantages.

With this general review behind us, let us now examine the major phyla of Kingdom Animalia. As with the other kingdoms, we will not try to be comprehensive, confining our study to the animals having the largest numbers or greatest impact on others.

Phylum Porifera Has Animals without Tissues

Poriferan organisms have some specialized cells, but are not at the level of placing them into efficient groups called tissues. All other major phyla of the kingdom are at least at the tissue level of organization. The *sponges*, comprising this phylum, are sedentary in their adult stages but achieve dispersal through their motile larvae. They are asymmetrical in body shape and often form large colonies. They are always aquatic, mainly marine.

The general body shape of sponges in their simplest form is somewhat like a vase, but sponges are often shapeless masses. If the body is vase-shaped (Fig. 13.2 A), the hollow central part is the *central cavity*. The body walls are perforated with pores through which water-borne organic materials enter the central cavity. Sometimes the pores are openings to tiny *canals* that go straight through to the central cavity, but in a vast majority of cases they open to a more complicated arrangement of canals. Water is circulated by flagellated cells called either *collar cells* or *choanocytes*, which are the most common cells of the inner lining of the central cavity and some of the canals. Every flagellated cell has a collar surrounding the base of the flagellum. This collar is a surface for the capture of water-borne particles that might be food. Thus, the sponge is a filter-feeder (Chapter 7); it digests intracellularly. After filtered water enters the central cavity, it flows out of the animal via the large top opening. This *osculum* is therefore not a mouth but an exitway. Each of the three classes of sponges has a distinctive kind of skeletal material, either calcareous, siliceous, or proteinaceous. The calcareous and silicious supporting material is in the form of simple or branched needles called *spicules*. The protein material is in the form of *fibers*.

Phylum Cnidaria Is Characterized by Possession of Stinging Cells

Cnidarians (formerly called Coelenterates) are radially symmetrical, diploblastic animals. The term *diploblastic* means that there are two distinct cellular layers in the body, an outer ectoderm and an inner endoderm (Fig. 16.6). Between these two layers is a third layer, the mesoglea, which is predominantly gelatinous, not cellular. The extreme thickness of the gelatinous mesoglea in *jellyfishes* is the characteristic that gives them their name. The digestive cavity, the *gastrovascular* cavity, has only one opening, which is referred to as the mouth. Food is obtained through the mouth and waste is expelled from it. A ring or fringe of *tentacles* surrounds the mouth. Upon their surfaces are numerous stinging cells called *cnidocytes*, for which the phylum is named. Upon being touched, a cnidocyte expels a harpoon-like structure called a *nematocyst*, which contains a paralyzing poison. Prey are digested mostly in the large gastrovascular cavity, although some intracellular digestion takes place in endoderm cells. The cavity also serves to provide oxygen-rich water to endoderm cells lining it, thus relieving the animal of having to provide a blood system.

A Cnidarian body can take one of two forms, *polyp* (Fig. 16.6 A) or *medusa* (Fig. 16.6 B). Some species have both body forms, which alternate with each other in their life cycles. A polyp is attached, has a thin mesoglea, and often reproduces asexually. A medusa, the shape of a jellyfish, is free-swimming, has a thick mesoglea, and reproduces sexually. All Cnidaria are aquatic and predominantly marine.

There are many interesting examples in this phylum. The freshwater *hydra*, which is a classic teaching laboratory example, is particularly well known. Others are *Portuguese men-of-war* (actually colonies of polyps), *jellyfish*, *sea anemones*, and *corals*. The latter are colonial polyps that secrete exoskeletons of stone (calcium carbonate). They build up large accumulations that take the form of fringing reefs, barrier reefs, or atolls.

Phylum Ctenophora Contains the Comb Jellies

The approximately 100 species of *comb jellies* resemble jellyfish in having much mesoglea in proportion to cellular tissues. Nearly spherical and only about a centimeter in diameter, they usually have eight rows of *comb plates* used for locomotion. They are advanced over the Cnidaria in that the digestive tract is complete, having both a mouth and anus. Comb jellies usually have two retractile *tentacles* with adhesive surfaces to capture prey. Reproduction is entirely sexual. They are entirely marine. Since they have both similarity to and difference from the Cnidaria, it is difficult to establish the evolutionary relationship of the two phyla.

Flatworms Are in Phylum Platyhelminthes

Flatworms are bilaterally symmetrical, triploblastic animals with flattened bodies. They are *triploblastic* because they have three complete cell layers;

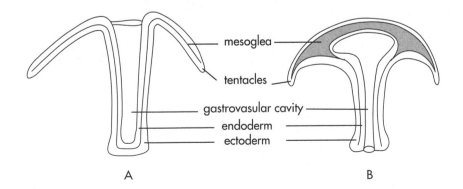

mesoglea

tentacles

gastrovasular cavity

endoderm

ectoderm

A   B

FIGURE 16.6.   *The Cnidarian body plan can take two forms. A: the polyp (more sessile) form. B: the medusa (more mobile) form. Note the difference in amount of mesoglea.*

the mesoglea of Cnidaria is replaced by the *mesoderm* (Figs. 14.3, 16.7 A). All other animals to follow are triploblastic and unless otherwise stated are also bilateral. Flatworms have soft bodies unsupported by skeletal material. They have only one opening to their digestive system, which is often a much branched cavity extending into remote parts of the body. This branching gives many more cells contact with the food that is digested partially within them (intracellular digestion). The cavity shape also allows distribution of food to all cells, a necessity since there is no blood system. There is no body cavity other than the digestive tract (that is, they have no coelom—see below).

There are free-living, commensal, and parasitic species. The freshwater *planarian* is a free-living one. *Flukes* and *tapeworms* (Fig. 7.3) are parasitic types. Some parasitic species have unusually complex life cycles during which some stages are subjected to precarious environmental conditions. Survival is assured because they reproduce in prodigious quantities, a circumstance that permits heavy losses without annihilation. Often a stage has degenerated into little more than a reproductive organ. Such is the case of the tapeworm, which has no digestive system or special sense organs.

Phylum Nematoda and Others Have the Pseudocoelomate Body Style

The flatworms have the advantage of three layers of cells, but they are relatively inflexible because of a tightly packed body (Fig. 16.7 A). A better way

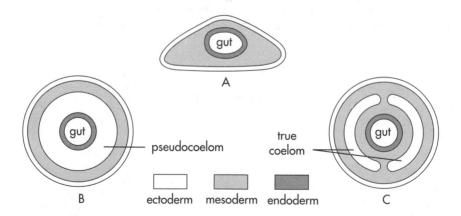

FIGURE 16.7. *Comparison of body plans for the triploblastic animals. All views are of cross sections through the body. A: phylum Platyhelminthes has no open space other than the lumen of the gut. B: phylum Nematoda has a pseudocoelom. C: several phyla have a true coelom, which is entirely formed within mesoderm.*

to arrange the three layers is to have some open space between layers. Such a space, called a *coelom*, allows the body to bend upon itself with greater ease (Fig. 16.7 B, C). If it is filled with fluid, it also becomes an avenue for moving materials; for instance, the digested materials leaving the endoderm of the gut and oxygen entering at the body surface. All animal phyla above the Platyhelminthes have such an opening.

Two versions of the coelomate body style occur. The more primitive form, featuring a *pseudocoelom*, is found in phylum Nematoda and several smaller phyla (Fig. 16.7 B). The nematodes are worms sometimes called *roundworms* to distinguish them from the flatworms of phylum Platyhelminthes. Note that the pseudocoelom has mesoderm as its outer margin and endoderm (of the gut) as its inner margin.

Roundworms are soft-bodied, with no skeletal support. They and members of all phyla that are to follow normally have complete digestive tracts with an entrance (mouth) and an exit (anus). Roundworms move around in a thrashing motion, for their locomotor muscles are all in longitudinal bands. Widely distributed and unusually abundant, they live in soil, water, and the bodies of plants and animals. Several invade the human body. Among these are Ascaris (an intestinal worm), hookworm, *Trichinella*, pinworm, Guinea worm, and filaria (causing elephantiasis).

A tiny roundworm living freely in soil, *Caenorhabditis elegans*, has been much studied by developmental biologists and geneticists. Every one of its approximately 1,000 cells has been mapped throughout embryonic development. By studying mutants that have various of these cells missing or duplicated, developmental biologists gain insight into the normal processes by which the organs are made.

Animals with a True Coelom Have Further Advantages

If a *true coelom* (Fig. 16.7 C) is the form used, the animal retains the advantages of the pseudocoelom and adds two more. Note that this type of opening is formed entirely within mesoderm. This means that the intestinal tract, made of endoderm, is bounded by a layer of mesoderm. Some of this mesoderm differentiates into circular muscles and these make possible the efficient peristaltic movement of food through the tract. The other advantage of a true coelom also comes with the internal mesoderm layer. It becomes a region of blood vessels, serving the intestinal tissue and connecting to the rest of the body. Thus, food can be more efficiently moved from the area of absorption and oxygen can be more efficiently supplied to that area deep within the body. Each of the phyla remaining to be discussed possesses a true coelom.

Phylum Mollusca Includes Quite Diverse Body Shapes

Molluscs are soft-bodied animals enclosed in folds of skin called a *mantle*. The body proper is divided into a *visceral mass* and a muscular *foot* ventrally located. The foot is modified in several ways and is a basis for separating individuals into classes. In most cases mantles secrete external *shells* that protect the delicate flesh beneath. Shells may be of two parts (valves) hinged together or of one part coiled. The circulatory system is of the *open type*, where blood is not always confined to vessels. Molluscs live on land and in water, but most of them are marine. One member of this group, a giant squid, is the largest living invertebrate.

Phylum Mollusca is a large group composed of many members that are economically important. Clams, oysters, scallops, snails, abalone, and squid are consumed for food by humans and many other animals. There are several members of the phylum that are negatively important. The shipworm burrows into pilings or other wooden structures and is considerably destructive. Snails and slugs sometimes heavily damage vegetation, and some snails are intermediate hosts of worm parasites that affect humans and other animals.

In addition to the examples already mentioned in connection with economic uses, there are chitons, tooth shells, sea hares, conchs (types of snails), mussels (clams), octopuses, and pearly (chambered) nautiluses.

Phylum Annelida Is That of the Segmented Worms

Annelids (Fig. 16.8) are worms that have bodies divided into segments. Segmentation is evident externally by circular constrictions around the body and internally by partitions called *septa*. Some structures are segmentally arranged; this means they are repeated in successive segments (as with the appendages of the clamworm, Fig. 16.8 A). Their bodies are soft and unsupported by skel-

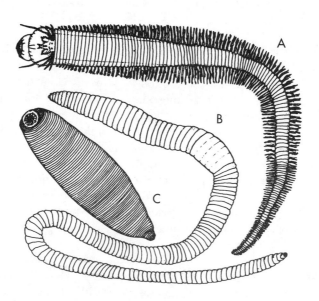

FIGURE 16.8. *Annelids. A: clamworm. B: earthworm. C: leech.*

etal material. They usually have tiny bristle-like setae that project from the body and aid in locomotion. The setae can be felt by running the finger over the body of an earthworm. If one tries to pull an earthworm out of its burrow, it will anchor itself in the ground by projecting the setae into the side of the burrow. Annelids have both circular and longitudinal muscles in their body walls, a structural feature that allows them to extend and draw up their bodies rather than thrash about like a roundworm. The nerve cord is double and ventrally located. The circulatory system is closed. The excretory system consists of segmentally arranged, paired tubes, the nephridia.

Annelids live mainly in the soil or in water. Earthworms are considered useful for cultivating soil because they eat large quantities of soil and eliminate the indigestible part of it at the surface of the ground. Their burrows make the circulation of air easier and make the soil more absorbent of rainwater. The leech is an external parasite most often found in fresh water (some are land leeches). They suck blood from almost any vertebrate animal available. In the days when bloodletting was practiced to cure diseases, the leech was a drugstore item. There are many curious-looking Annelids in the ocean. One of them, *Chaetopterus*, lives entrapped in a U-shaped tube of its own making. The palolo worms of southern Pacific coral reefs have posteriors filled with reproductive structures separable from the rest of the body. At specific times of the year, the last quarter of the October-November moon, they mature and become detached.

Phylum Arthropoda Is the Largest of All Phyla

The Arthropod group is by far the largest group of animals with respect to the number of species it contains. They share the characteristics of having externally segmented bodies, jointed appendages, and exoskeletons. Most segments of the body may be distinct from end to end as they are in centipedes or they may be nearly indistinguishable as in most spiders. More often, as in crustaceans and insects, segments in the front part of the body are more or less obliterated, whereas segments in the abdominal region remain distinct.

This is the only group of animals studied thus far that have jointed appendages. Jointing is a necessary corollary to the development of skeletons if animals are to move; otherwise a skeleton would give them the rigidity of a tree. Some Arthropods have a pair of appendages for each segment of the body; in others the appendages are limited to certain regions. The skeleton is composed of a firm polysaccharide material called chitin. If the animal grows, the skeleton has to be discarded and another one secreted (see molting, Chapter 14).

Usually the body has a recognizable *head*, *thorax*, and *abdomen*, but the first two of these may be fused into a *cephalothorax*. Arthropods usually have at least two kinds of special sense organs, *antennae* and *compound eyes*. The circulatory system is open. The dorsally located heart is a tubular structure that pumps blood out through the arteries. Blood passes through body spaces and reenters the heart through pores. The nerve cord is quite similar to that of the

Annelids in having a double construction and in being ventrally located.

Arthropods of class Malacostraca are primarily gill-breathing aquatics. Commonly called crustaceans, they are the crayfishes, lobsters, crabs, barnacles, shrimp, water fleas, and sow bugs (pill bugs). The latter is a land form commonly found under damp boards, bricks, and in similar places. Crustaceans have a pair of appendages on each segment of their bodies, and the first two pairs of them are antennae.

The members of class Insecta comprise the largest group of Arthropods, and indeed of any organism. Over one million species of insects have been named and most biologists believe that there are many times this number still to be identified, especially in the tropical rain forests. They have bodies that are distinctly divided into head, thorax, and abdomen. The latter is obviously segmented. The head is equipped with a pair of antennae. The thorax ordinarily bears three pairs of legs and two pairs of wings.

Breathing is accomplished by inhaling and exhaling air through segmentally arranged tubes called *tracheae* (Fig. 8.19), which ventilate all of the tissues. The excretory organs are *Malpighian tubules* (Chapter 9) that encircle and empty into the digestive tract at a constriction between the stomach and intestine. Metamorphosis is usually complete or incomplete, but a few insects have no metamorphosis at all. Many insect larvae are destructive to crops, for they are voracious feeders. Some adult insects are directly destructive; others are vectors of disease organisms. Some of them have economically useful purposes like the pollination of flowers or the production of honey and shellac. Undoubtedly, insects of many sorts have a profound impact upon the nature of any ecosystem.

The Arthropods of class Arachnida include spiders, harvestmen (daddy longlegs), ticks, chiggers, and scorpions. Perhaps there are only two seriously dangerous spiders in the United States. The best-known one is the black widow, but the one more capable of harming people is the rather rare brown recluse. The black widow is easily recognized by its slick, black body with red (orange-red) markings on its abdomen. Spiders cause only a handful of fatalities per year in the United States. Scorpions are poisonous but not all are dangerous. Arachnids have the reputation of being an obnoxious lot, but only a few of them are harmful. They have bodies usually divided into cephalothorax and abdomen and have four pairs of legs. They have no antennae or mandibles.

Centipedes, in class Chilopoda, have worm-like, segmented bodies that are flattened on top and bottom. Segments characteristically bear one pair of legs apiece. Centipedes are carnivorous, and the first segment of their bodies is equipped with a pair of *poison claws* that aids in capturing food and in maintaining defense. Although the name suggests that they have one hundred legs, the number is usually fewer but is sometimes more.

Millipedes, of class Diplopoda, have worm-like, segmented bodies that are round in cross section. Their segments, which are actually pairs of segments fused together, characteristically bear two pairs of legs. They are herbivorous but may be scavengers at times. They defend themselves by rolling up and lying quietly (playing dead) or by secreting an offensive-smelling chemical.

The name suggests that they have one thousand legs, but none have that many. They usually have closer to one hundred of them.

Other, minor classes of phylum Arthropoda exist but will not be discussed here. Some of them are known only from fossil records.

Phylum Echinodermata Is That of the Spiny-skinned Animals

All Echinoderms are *radially symmetrical* animals derived from *bilateral* larvae. Their radial bodies are pentamerous, which means that parts are in fives or multiples of five. A starfish with five radiating arms clearly illustrates that condition. Echinoderms are called spiny-skinned animals because many of them do have spines that project from the skin, sometimes very long ones. Others are spineless, however.

They have a unique *water vascular system*, an arrangement of tubes ending in many tubular feet. The system contains water, not blood, and is used both for distributing materials and for locomotion. Some starfish use the system's tube feet, which can act like many tiny suction cups, to pull open clams in order to eat their soft body parts.

There is an internal skeleton consisting of calcareous plates. Over their surfaces many, but not all, Echinoderms have myriads of tiny pincers, the *pedicellariae*, that are said to keep the surface free of debris. All Echinoderms live in saltwater.

Echinoderms are represented by starfish, brittle stars, sea urchins, sand dollars, sea cucumbers, and sea lilies. Brittle stars have serpentine arms distinctly marked off from a central disc. The sharp separation of arms and disk is quite unlike the condition of starfish. Sea urchins are spiny, hemispherical animals that have a unique set of jaw-like structures around the mouth, called *Aristotle's lantern*. Cleaned skeletons, which are sometimes found on beaches, have five distinct areas over the top. It looks as if the animal had five arms like a starfish, which were folded back over the top and fused together. Sand dollars are flat, like a coin. Their cleaned skeletons have five petal-like areas on the top that correspond to arms. Sea cucumbers are soft-bodied, elongated animals that may vaguely resemble a cucumber or a very thick worm. Their skeletal material is reduced to microscopic pieces. Their mouths are surrounded by a fringe of tentacles. Sea lilies (crinoids) are often attached by stalks, giving the animals a plant-like appearance. Their arms are feathery. During the Paleozoic era crinoids were unusually abundant, and they have been plentifully preserved as fossils. Some other attached forms belonging to two other classes are now extinct.

The Chordate Animals Include Ourselves

The members of phylum Chordata as a whole have reached what is generally considered the highest level of animal development. At some stage in their lives they have a dorsally located, gristle-like supporting rod called the *noto-*

chord. In primitive forms it persists throughout life, but in advanced forms it is replaced by bone (the spine). At some stages of development Chordates have *gill pouches* or even *gill slits.* Except in fishes, these do not usually persist into adulthood. The central nervous system, composed of a *hollow nerve cord* derived from ectoderm (Fig. 14.4), is dorsally located, a position opposite that found in other animal phyla. All Chordates have a *tail* extending beyond the anus (Fig. 15.1), although some resorb it into the posterior body before maturity. The Chordates, sea squirts excepted, have internally segmented bodies. Some segmental features may not survive beyond the embryonic stage, but others like the pairing of cranial and spinal nerves or the vertebrae of the spinal column persist throughout life.

The tunicates compose a small subphylum called Urochordata. They are marine animals so named because they secrete a tunic-like covering over their bodies. They are often called sea squirts because of a habit of squirting water from their *excurrent siphons* (openings). Their adult bodies are not at all like bodies one would expect in the Chordate group. Their general shape varies but is somewhat like a coffeepot with an entrance for water (*incurrent siphon*) in the lid position and an exit (excurrent siphon) in the spout position. Most of them are sessile although some are free-living. They vary in size from about the dimension of a pin head to over a foot in diameter.

The typical tunicate has a looped digestive tract. The pharynx is large and contains many slits (gill slits) in its wall. The pharynx acts as a sieve in collecting food, for water passes through the slits and through a cavity to the outside, while bits of food are left behind (a form of filter-feeding).

From the larval stage one is able to recognize tunicates as Chordates. The immature stage looks somewhat like a tadpole and has a distinct notochord and tubular nerve cord. Both of these characteristics are lost at maturity. During development larvae attach themselves and undergo a metamorphosis involving the resorption of the tail and other transformations.

A second small subphylum of phylum Chordate is Cephalochordata. It is best represented by the lancelet, or Amphioxus, a fish-like marine animal of small size, usually under 100 millimeters long. Ordinarily it remains burrowed in sand with only the mouth end projecting. It has a laterally flattened body with distinct muscular segments (*myotomes*). The dorsally located notochord and tubular nerve cord run essentially from end to end. The pharynx is like that of the tunicates in being a straining organ with numerous gill slits.

All preceding animals are invertebrates; all of the following ones are vertebrates, in the final Chordate subphylum, Vertebrata. One of the reasons for the separation of vertebrates from other animals is their achievement of a segmented spinal column (*vertebral column* or backbone), the main axis of an endoskeleton composed of cartilage, bone, or both. The backbone replaces the notochord early in development except in the most primitive members, where the notochord is reinforced by cartilage or bone. Skeletons are jointed, thereby giving flexibility to the body.

Vertebrates have several additional distinctive features other than the usual Chordate characteristics. Their bodies are segmented, although this distinction is not recognized externally. Their bodies are composed of three or four regions—head, neck (sometimes absent), trunk, and post-anal tail (sometimes absent in adults).The heads contain enlarged brains protected by a cranium and are provided with paired, special sense organs. They often have jaws and teeth for pulverizing food. The circulatory system is closed. Unlike that of the invertebrates, the heart is ventrally located. Vertebrates have paired excretory organs known as kidneys, composed of numerous nephrons. Bodies are externally protected by cornified epidermis, mucous and oily secretions, scales, plates, feathers, or hair. For perhaps the first time in the animal kingdom some members (birds and mammals) regulate their body temperature (Chapter 9).

The fishes comprise three classes (in the most-used classification scheme) whose members have similar body forms and adaptations for an aquatic existence. Chief among them is the presence of *gills* for breathing, although some invertebrates have these too. They have a two-chambered heart consisting of one atrium and one ventricle.

The most primitive fishes are the members of class Agnatha, the jawless fishes. Among them are lampreys and hagfishes. They lead a parasitic existence by attaching themselves by their mouths to other fish and sucking body juices from them. While they are attached to their hosts, they cannot take water into their mouths and pass it through the gills to the outside in the usual manner of breathing. Consequently, they have acquired the ability to draw water in and out of the gill slits in a breathing action similar to inhaling and exhaling of air by lung breathers. Gill slits, six to sixteen pairs, open separately to the outside. The notochord is persistent in the adult as a part of the supporting structure. The rest of the skeleton is composed of cartilage. Vertebrae, which are characteristic of the subphylum, are imperfectly formed. Agnathans have neither scales nor paired fins.

The cartilaginous fishes, of class Chondrichthyes, are a group that have paired fins, paired jaws, and a skeleton of cartilage. The notochord is rudimentary in adulthood. Vertebrae are distinctly formed. The mouth is ventrally placed and equipped with enameled teeth. The intestine has an unusual device called the *spiral valve* that slows the passage of food and increases the absorbing surface. The tail fins of cartilaginous fishes are usually of the asymmetrical type (*heterocercal*). Five to seven pairs of gill slits open separately to the outside. Most have *placoid scales*, which project from the surface. This type of scale is distinctive in having dentine and enamel like that in teeth. Unlike the bony fishes (see below), cartilaginous fishes have no air bladder for hydrostatic function. Fishes belonging to this class are sharks, rays, and skates.

The largest group of fishes, that of class Osteichthyes, has skeletons that are chiefly bone but are sometimes also cartilage. Bony fishes are the ones that most people have in mind when they talk about fishes. Their bodies are generally covered with flattened scales. Gills and gill slits are covered by a bony flap called the operculum. The tail is usually symmetrical (*homocercal*).

In addition, most bony fishes have an air (swim) bladder, whose varying amounts of contained air determine the density of the fish and thus whether it will sink or rise in water.

The remainder of the Chordate classes are *tetrapods*, animals with two pairs of appendages used for walking. Sometimes, of course, one pair is wings or arms. Some, like snakes or legless lizards, have no appendages, although vestiges may be present internally. Most tetrapods, some amphibians excepted, are lung breathers. Also, all of them with the exception of amphibians have internal fertilization.

Animals of class Amphibia are primitive tetrapods that are somewhat transitional in habits and structures between gill-breathing fishes and lung-breathing tetrapods. Amphibians usually pass through an aquatic, fish-like larval stage that depends on gills for breathing. A few even retain gills for adult use. As adults they are usually lung breathers that live on land or in and out of water (always fresh water). They have no diaphragms, so their intake of air is more like a swallowing action than an inhalation. Skin, soft and scaleless, often plays an important role in breathing when in water. Many lung-breathing amphibians stay under water for a long period of time and breathe entirely through their skin while submerged. For the first time in the animal kingdom, some members have *vocal cords*. Amphibians, however, do not have necks. They have a three-chambered heart consisting of one ventricle and two atria.

One group of amphibians, the salamanders, are lizard-like but are easily distinguishable from lizards because they have neither scales on their bodies nor claws on their toes. They do have tails. Some of them may be commonly referred to as mud puppies, water dogs, newts, or hellbenders. Another important amphibian group is the frogs and toads, both tailless when they become adults. Frogs are smooth and toads rough-skinned.

Members of class Reptilia are distinctly equipped for a terrestrial existence. Their bodies are covered by protective scales or scutes (horny plates), and they are lung-breathers. They have internal fertilization of eggs. Reptiles do not have vocal cords but may blow or hiss. Legs of modern reptiles spread to the side, so in effect they are belly-sprawlers. The toes, if present, have claws. In addition, these are the first animals with necks. They also have a nearly four-chambered heart, with the ventricle imperfectly divided in two.

Lizards and snakes belong in the same order. Lizards usually have legs. They also usually have movable eyelids. Some better known lizards are the fence swift, skink, chameleon, horned toad, and Gila monster. The latter is a large, poisonous lizard with black (or brown) and pinkish (or yellowish or orange) colors and a beaded exterior. It is the only poisonous lizard in the United States. Poison, however, is not injected by fangs but oozes from the base of the teeth.

More than almost any other group of animals, snakes arouse our interest and fear. The only poisonous ones in the United States are rattlesnakes, copperheads, cottonmouth moccasins, and coral snakes. One should learn to avoid these and enjoy the rest. The first three poisonous snakes listed are pit vipers, recognizable by a pit between the eyes and nostrils. This organ is extremely sensitive to infrared (heat) radiation, useful in recognizing the presence and direction of endothermic animals.

The turtles, terrapins, and tortoises have a box-like shell made of bony plates and covered with horny material. The shell is composed of an upper part, the carapace, and a lower part, the *plastron*. Some have the plastron hinged in such a way that the head and legs can be retracted and the shell completely closed. They have toothless beaks.

Alligators, crocodiles, and gavials are lizard-shaped but large. They are covered by tough, leathery skin marked off in sections called *scales*. Bony (dermal) plates are underneath the dorsal scales. They have four nearly separate heart chambers. The American alligator has a blunt snout; the crocodile, a pointed snout. Both have much stronger muscles to close the mouth than to open it. When humans wrestle with alligators they take advantage of that weakness and hold shut the reptile's mouth.

Reptiles were so abundant in the Mesozoic era that it is often referred to as the age of reptiles. They adapted to all kinds of habitats. Of these, the dinosaurs are the best known to the lay public.

Birds (class Aves) are animals with some reptilian characteristics and adaptations for flight. Among some of the reptile-like characteristics are the habit of laying large shelled eggs, scales on the body (legs and feet), horny beaks (like turtles), and certain bone structures. Adaptations useful in flight are strong, light bodies, a temperature-regulating mechanism (endothermic), an efficient four-chambered heart, and keen senses of sight and hearing. In addition they have forelimbs modified into wings, lungs connected to air sacs that extend among the internal organs, and a body covered with light feathers. Birds have necks and ventrally located legs. Almost all have toothless beaks; therefore, they have a part of the stomach modified into a muscular gizzard for grinding the food. The voice is well developed and originates in a *syrinx* at the base of the trachea (not the larynx used by mammals for sound production). The left ovary and oviduct are generally the only ones to develop.

That birds have spread into all habitats is revealed in adaptations they have achieved. Habits and habitats are especially reflected in the kinds of beaks, necks, feet, and legs they have. Some, like the ostrich, emu, rhea, penguin, and kiwi, are grounded by having wings too small for their bodies.

Nesting habits of birds are about as variable as the birds themselves. Many male birds claim *territories* for themselves and their mates, especially during the breeding season. A territory is vigorously defended from intruders, which are always more competitive if they are of the same species. The size of the territory may be as small as a room for a tiny bird or several square kilometers for a carnivorous eagle. Territories of different birds with different food requirements overlap. Some birds congregate in large numbers to nest in an area that has come to be called a *rookery*.

The highest level of development in terms of mental endowment is achieved in mammals (class Mammalia). They also reproduce more efficiently in that fertilization is internal and the developing embryo is protected and nourished inside the mother's body (although some other animals do both). Then after birth the young are nourished by milk secreted by *mammary glands*. It is from this characteristic that the class gets its name. Mammals are built for activity by having a complete separation of the pulmonary and systemic circulation (distinct four-chambered heart, no mixing of oxygen-rich with oxygen-poor blood), a diaphragm to aid in breathing, a mechanism for maintaining constant temperature (endothermic), hair to insulate against loss of heat, and legs supporting the body from below (in all land forms). Unlike those of the birds, vocal cords are in the *larynx* at the top of the trachea. Sound is directed into the ear by external flaps, the *pinnae*. Mammals have necks, most of which have seven vertebrae, and most mammals have long tails. Reproductive characteristics are the production of tiny eggs and a penis for copulation.

The most primitive mammals are the *monotremes*, the duckbilled platypus and the spiny anteaters. They lay shelled eggs and have a *cloaca* (single opening for digestive and nitrogenous wastes and gametes) like the birds and reptiles.

The *marsupials* are the pouched animals. The pouch (marsupium) is a brood pouch. The young, born in an extremely immature stage, are placed there by their mothers. They continue to feed (from mammary glands) and develop in the pouch. Examples of the marsupials are opossums, kangaroos, and koala bears.

The *placental* mammals are considered the highest of the mammals. The young develop in a uterus and are nourished through a *placenta* (Fig. 14.5). They include such groups as the carnivores, bats, hoofed mammals, rodents, elephants, whales, and primates, all of which are quite familiar. The primates are especially interesting because order Primates is the taxonomic group to which humans belong. Primates have brains with an unusually high degree of development, especially of the cerebrum. They also have thumbs for grasping, an important advancement for the manipulation of tools. Also, primates have stereoscopic vision, which permits a more accurate judging of distances.

STUDY QUESTIONS

1. Why are viruses not placed in any of the recognized kingdoms?

2. What exactly is a bacteriophage?

3. What are characteristics of Kingdom Protista?

4. Why are biologists considering a sixth kingdom? Which organisms are placed in it?

5. Algae are hard to place into a single kingdom. Why? In which two kingdoms have they sometimes been placed?

6. What are cellular slime molds?

7. From what seaweed does the Sargasso Sea get its name? To what division does it belong?

8. Name a medically significant Rhizopoda. Do the same for a sporozoan and a roundworm.

9. What is the most significant difference between mosses and ferns?

10. Classify the following: bread mold, yeast, mushroom, rust, and smut.

11. What is the composition of a lichen body?

12. Which generation is most conspicuous in vascular plants?

13. Of what significant advantages are seeds to plants that have them?

14. What is meant by having naked seeds, a characteristic of gymnosperms? Give several examples of gymnosperms.

15. What advantages do animals have over plants by having muscular and nervous systems? Enumerate other characteristics of animals.

16. What advantages does an endoskeleton have over an exoskeleton?

17. How is the bilaterally symmetrical body form of free-living organisms related to their behavior?

18. What is the phylum of sponges?

19. In which phyla are animals diploblastic?

20. What body forms do Cnidaria have? Which form has a thick mesoglea? Which form is free-swimming?

21. Define the following: cnidocytes, chitin, pedicellaria, notochord, and mantle.

22. Describe the digestive system of flatworms in general.

23. How does a true coelom differ anatomically from a pseudocoelom? Describe the advantages conferred by possession of each.

24. Where is the nerve cord located in annelid worms? In Arthropods? In vertebrates?

25. What are three characteristics of phylum Arthropoda? Which is the largest class in the phylum?

26. What animals are in class Arachnida of phylum Arthropoda? What characteristics do they have in common?

27. Is a spider a type of insect? How can you tell?

28. How do vertebrates differ from other animals? Vertebrates of which classes regulate their temperatures?

29. Which subphyla of the Chordates are invertebrates?

30. What are the best-known examples of amphibians? What is unique about their breathing? How can they stay submerged in water for long periods of time without breathing?

31. How are reptiles equipped for a terrestrial existence?
32. Name a member of class Chondrichthyes. Of which Chordate subphylum is it?
33. How are birds similar to reptiles?
34. Name a mammal that lays shelled eggs.

17
ORGANISMS AND THEIR ENVIRONMENTS

What do we know about:
—ways that species are adapted to their environments?
—how populations become dispersed into new places?
—the predictable sequence of changes undergone by any habi-
tat?
—the characteristics of the major biomes of the world?
—what the difference is between a forest and a jungle?
—what the difference is between a prairie and a plain?
—why all deserts are not hot?
—the current loss of biodiversity?
—why the loss of upper-atmosphere ozone is a problem?
These and other intriguing topics will be pursued in this chapter.

ORGANISMS CANNOT EXIST
IN ISOLATION

To remain alive, all individuals must obtain raw materials from their surroundings. Their degree of success in obtaining these materials is a measure
of their ability to survive. Among the more obvious and immediate needs are
oxygen, water, and foods. Of the three, oxygen is continuously necessary since
it cannot be stored. If it is not available as free oxygen, the organism must respire anaerobically. Water is second in terms of urgency of need, although a
few plants and animals, particularly those of the desert, can conserve or store
water produced by metabolism. Probably most living individuals can store
some foods. Regardless of special adaptations to lay away a reserve of needed
materials or of adaptations to conserve them—through inactivity, for instance—the supply becomes exhausted rapidly and must be replenished. Thus
no living individual can long be separated from its natural surroundings without dying.

ECOLOGY IS THE SCIENCE OF RELATIONSHIPS

The inseparability of organisms and the environments in which they live serves to emphasize the importance of learning about the relationships between them. The science that deals with these relationships is *ecology*. It deals with the manner in which environments influence organisms and organisms influence environments. It even deals with how organisms influence other organisms; after all, organisms are a part of the environment. It can be safely said that ecology is the most complex and wide-ranging of the fields of biology, since one might be dealing with every level from molecules through interspecies relations during the course of a single research project. It can also be safely said that this field has become one of the most important for all of us, scientists or nonscientists, because of the threats to the world's environments that have become obvious in recent decades.

SOME DEFINITIONS ARE NECESSARY

The physical home of an individual or population is known as its *habitat*, a composite of all the surrounding conditions with which the individual is in contact. It can be extensive, like that of some birds, or more limited like that of an intestinal bacterium.

The *niche* is a much more complex concept. It is the position occupied by any species, within which it can maintain a viable population. It includes the physical habitat, but also all of the biological activities and interactions that occur and are necessary to keep the organisms of the population in good health. Since no two kinds of organisms are identical in all biotic and environmental relationships, any particular kind of organism occupies a niche unique to itself. In other words, the niche it occupies cannot be occupied by any other kind of organism in the same location, at least not permanently. The niche differs from the habitat in that the niche is a position or status, not simply a place. It follows that many species could occupy the same habitat, each with its own niche.

THE ENVIRONMENT UNDERGOES CONSTANT CHANGE

Everyone is conscious of changes constantly transpiring, and geologic history has recorded changes from the beginning of life on earth. Land erodes, weather changes, volcanoes explode, land masses rise and fall, fires sweep through vegetation, and storms and floods scar the earth's surface.

Some changes in the environment have been accelerated by the presence and activities of humans, while others have apparently been seen on earth for the first time because of humans. As a species, we have stripped large areas of their natural cover of vegetation for lumber and cultivatable land, dug into the earth for valuable minerals, and built cultural features like roads, tunnels, dams, canals, and buildings. The result is a drastic change in many sections of the earth.

ORGANISMS WITHIN A POPULATION EXHIBIT VARIATIONS AND ADAPTIVITY

Most populations experience the production of more offspring than can possibly find favorable conditions to survive. Among the offspring of parents reproducing sexually, there are no two individuals exactly alike. If all of the offspring are in the same habitat, some will be more suited to it than others and in general will survive in competition with their less adapted variants. In time, the only survivors will be the offspring of individuals who by chance make changes harmonious with the environment. This is, of course, a description of natural selection at work (see Chapter 15 for details).

Variations are *physiological* or *structural*. Physiological variations are not so obvious but are just as important in survival. A physiological variation making a plant resistant to a disease gives it a tremendous advantage. Or a variation making a plant tolerant to wider temperature ranges or better adapted to conserve water vastly increases its geographical distribution.

The great diversity of life seen today is a tribute to the ability to adapt to a varied and changing world. In so far as survival is concerned, it makes no difference how adaptation occurs as long as it permits a species to live in its niche. True, some forms are so poorly adapted to current conditions that their future existence is in jeopardy, while others are so well adapted that the species is flourishing. The periodic appearance of large populations does not necessarily mean that the species is unusually well adapted, for some temporary upset in the balance of nature may give a certain species a temporary advantage not normally enjoyed.

Inability to Compete Leads to Extinction

Whenever the harmonious adjustment between an organism and its surroundings is no longer possible, extinction of the organism follows. Extinction is a failure on the species level. The sedimentary rocks are sprinkled with fossils showing the prevalence of extinction. Among them are the dinosaurs, which may have become extinct because the climate changed (perhaps because of the impact of some cometary or meteoritic body), because arising mammals ate their eggs, because they were stricken by disease, or because some unknown obstacle presented a situation to which they could not adapt.

The passenger pigeon became extinct in the twentieth century because it could not cope with the efficiency of guns and the shortsightedness of humans.

A High Degree of Specialization Can Be Hazardous

A high degree of specialization (physiological or structural) reduces the ability to survive if conditions change, while less specialization favors survival. The very fact that a fish is specialized for breathing in water narrows the conditions under which it can live to an aquatic habitat. If the same fish is adapted for life in salt water, the conditions under which it could live are narrowed still further. Some insects are so specialized that they will eat only one species of plant. Obviously, their chance of finding food is less than that of insects that are not so narrow in their host plant specificity. One type of moss restricts itself to recently burned places and another type to the knotholes of trees. Although there may sometimes be burned areas and knotholes available, those particular places are considerably restricted ones.

One of the most interesting types of specialization is parasitism. The continued existence of the parasite depends upon the availability of the host. Such a relationship constitutes a fine balance between the parasite thriving and its becoming so prevalent that it kills all host organisms and thus itself too.

Adaptations of parasites to a dependent existence sometimes results in degeneration or loss of their body parts as well as the acquisition of new features. Intestinal worms are degenerate in comparison to their free-living relatives. They have no specialized sense organs, and in the case of the tapeworm no digestive system. They have acquired the ability to resist the action of digestive enzymes, a feature that saves them from being digested with the food. The most hazardous part of a parasite's existence is the period during reproduction. As is frequently the case, this weakness is compensated for by their producing prodigious numbers of offspring out of which a few may, by chance, find conditions favorable enough to survive.

Humans Are Generalists

There are no totally generalized (unspecialized) individuals, but some plants and animals are much less specialized than others. When individuals are generalized, they are more apt to survive when conditions change. Humans are a rather generalized species. We cannot jump as well as a flea (if so, a high jump would be about 140 meters), nor hear like a bat (20–18,000 cycles per second compared to 50,000–100,000 cycles); neither do we have the strength of a gorilla. For food we eat both plants and animals. We have been able to adapt to a wider variety of environmental conditions than nearly any other organism and have a much wider distribution than most. The major credit for this adaptability and wide distribution cannot be credited to generalization of the body, however. It really should be ascribed to a specialized brain, which permits us to alter the environment by building shelter, making clothes, generating heat, and so on. It is a case of changing the environment to adapt to the animal rather than the animal's adapting itself to the environment. Some credit for our success should go to the possession of thumbs. These specialized structures make possible the ability to grasp and use tools. Another specialized characteristic, held in common with other mammals and the birds, is the ability to keep a constant internal temperature, giving greater independence over external temperature fluctuations.

Some Species Have Been Bred to be Specialists

Cultivated crops and domesticated animals are bred selectively to get specific results. They do not survive well in competition with plants and animals that are naturally selected in the wild. They are protected and pampered by human hands in an artificial environment of our own making. Cultivated corn nicely illustrates the point. The fruits (grains) are compacted on a cob and are not at all suited for dispersal. If left alone the ear falls to the ground, where it is eaten by animals or decomposed by fungi. When some of the grains do survive the winter, they germinate into plants that grow so close together that they can seldom if ever produce fruits. In addition, modern corn plants fare poorly in competition with other plants.

Animal Species Show Specific Adaptations to Their Niches

In addition to coming under the influence of physical and chemical forces in the environment, every individual interacts with a variety of other living organisms. The company is often threatening, for it seems that almost every living being has natural enemies ready to consume it for food or compete with it for the same food or space. It is not surprising then that numerous adaptations are useful for both offense and defense.

As a general rule animals are similar in color and patterns to their surroundings, which makes them more difficult to see. This is *mimicry*. A random collection of insects from any meadow will impress the observer with the high percentage of greenish and brownish species that blend into the background. The same is true for many amphibians, snakes, and birds. The latter sometimes lay speckled eggs with a good camouflage effect. (Exceptions among birds are those in which one sex, usually the male, wears bright plumage used in attracting the mate and distinguishing among related species.) The polar bear, varying hare, weasel (when on snow), and ghost crab (on beaches) are white like their backgrounds. It is even more striking that many animals—e.g., some frogs, the chameleon, the flounder, and the squid—can change their colors or color patterns rather quickly to match the background.

Some animals are disguised as other organisms. An insect called the walkingstick looks like a twig; a leafhopper looks like a brier; an edible butterfly looks like a foul-tasting one; another butterfly looks like a dead leaf. If a defenseless and edible animal's color and shape (and sometimes behavior) make it look like an obnoxious one, this is called *Batesian mimicry*. If a set of obnoxious species all have the same general patterns (to teach predators that this is the warning pattern of being obnoxious), this is a form called *Mullerian mimicry*. Sometimes biologists mistake one type of mimicry for the other until further research sets them straight.

Possibly the large spots on the thorax of a click beetle and on the backs of some insect larvae may be mistaken for large eyes and be sufficient to scare away a would-be enemy (the *startle effect*). "Eyespots" on the wings of many butterflies and moths either cause a startle effect or give a predator a false head to attack.

An assortment of some more obvious specializations (adaptations) are included here for illustration. Toads, millipedes, stink bugs, and skunks produce offensive odors or tastes that repel enemies. Of course, they might not be disagreeable enough to deter some animals that have poor senses of taste and smell. The turtles, snails, clams, barnacles, hermit crabs, bryozoans, corals, and oysters rely on hard coverings into which they can withdraw when danger threatens. The centipedes, scorpions, sting rays, some snakes, some spiders, some lizards (Gila monster), some insect larvae, and some Cnidaria (jellyfish and Portuguese man-of-war) produce poisons. Some animals depend on swiftness to flee or pursue. Roaches, crickets, and centipedes have somewhat flattened bodies, making it easy for them to slip under bark, logs, rocks, or debris. The flea is compressed in the left-right dimension, making easier its movement among closely spaced hairs of a dog or cat. Some reptiles, birds, and mammals have claws, beaks, teeth, or horns that cut flesh. Lice have appendages for holding to hair.

Plant Species Also Have Specific Adaptations to Their Niches

Unlike most animals, plants are stationary and neither attack not flee; they have to take conditions as they exist. Fortunately, the main needs of green plants are simple and often abundant. They also have numerous ways to get what they need and to conserve water. As for defense against animals, plants are mainly ineffective. A few have stinging hairs, thorns, sticky secretions, or bad tastes and odors that may repel some animals.

Plants can be separated into three groups based on the availability of water in the places where they grow. They are *hydrophytes*, *mesophytes*, and *xerophytes*. For the most part, specific plants are fitted for living under a narrow range of humidity variations.

Hydrophytes live with an abundance of water, some being entirely submerged. Their roots are poorly developed and nearly always lack root hairs. Conducting and supporting tissues are at a minimum, and air spaces are abundant throughout the plant. Their leaves are thin. If submerged, the leaves are frequently finely dissected or ribbon-like as well as uncutinized. They do not have stomata or palisade cells. If leaves float on water, they have stomata in the upper epidermis and well-developed palisade cells.

Mesophytes live in intermediate situations with respect to the supply of water. Roots, stems, and leaves are well developed and amply supplied with conducting and supporting tissues. Roots are extensive and provided with root hairs. Leaves are broad, thin, moderately cutinized, and abundantly ventilated by stomata, frequently located on both sides. If leaves are horizontal, palisade cells are well developed.

Xerophytes live in hot deserts, on bare rocks, in frozen places, and in salty areas where available water is scarce because of osmotic conditions. Root growth is extensive, thus making it possible for the plant to gather more water. Conducting tissues are well developed. Parts above ground, from which evaporation of water is possible, are abbreviated. Some stems are leafless and succulent (with water-storage tissue); they are green and perform the function of leaves. Frequently, stems bear spines or thorns that represent reduced leaves or branches. When leaves are present, they are usually small, leathery, and heavily cutinized. For further protection against evaporation, they may have several layers of epidermis, a minimum of space between the cells, well-developed palisade cells, and stomata that are sunk beneath the surface on the shaded lower side. Either the stems or leaves may be protected by surface hairs or scales. Some desert plants like the palo verde or ocotillo quickly shed leaves when the weather gets too dry and replace them when sufficient rains come.

Dispersal Is an Important Adaptive Ability

Many animals have some form of locomotion and can seek a more favorable habitat if the need arises. Plants themselves do not move around but have numerous ways of dispersing their seeds or spores, which, from the point of view of species survival serve the same purpose as locomotion in animals. Many lower plants produce large numbers of spores that may be carried over great distances by wind or water. Some are sure to fall where conditions are favorable. Seed plants may bear seeds or fruits with wings or fluffy fibers that make them airborne, seeds or fruits with hooks or barbs that attach them to moving animals, explosive fruits that throw the seeds, corky seeds that float on water, and so on.

Organisms invade new areas until they reach some barrier beyond which they cannot pass. If no barriers existed, many species would be much more widely distributed over the world than they are now. Some obstacles to the free movement of species are climatic, geographical, structural, and physiological. Therefore, one could not expect a tropical orchid to grow in Alaska, a swamp plant to grow on a desert, a fish to live on land, or an earthworm to live in trees.

Invasion of new locations results from (1) individuals moving into a place where they have never been before or into disturbed areas, (2) new variations of species that originate from time to time, and (3) accidental and deliberate *introductions* from one part of the world to another by humans. Some have compared the spreading of a species into a new area to the invasion of an army in which there is a constant battle between the invader and the defender (in this case the environment). In favorable times the species advances; in unfavorable times it retreats.

Sometimes an invading individual finds the new environment so hospitable that the invader flourishes beyond anything possible in its natural habitat. Under such circumstances the invader is apt to become a pest (from the human viewpoint). Rabbits in Australia, starlings in America, the agent of Dutch elm disease (a fungus) in America are but a few well-known cases.

The place from which a species originates and spreads is called the *center of origin* (or *center of dispersal*). If nothing complicates the picture, the larger the area of dispersal, the greater the age of that particular form of life. At times a form once widely dispersed dies out except in a few isolated spots. It is then referred to as a *relic*. Of course, there are cases where it may not be easy to tell whether a form is a relic or is recently evolved. There is a plant called *Shortia galacifolia*, which has a very limited distribution in the Southern Appalachian Mountains. Another species of *Shortia* is also found in eastern Asia. What was once a genus of wide distribution is now represented in the United States by only a relic.

COMMUNITIES ARE ASSOCIATED SPECIES INTERACTING TOGETHER

A particular set of environmental conditions supports a specific group of species. Without attempting to go into details of terminology, such a specific group, regardless of its size, shall be defined as a *community*. In most land areas plants seem to exert more influence than animals. Consequently it is often customary to name the community after the type of vegetation that is most conspicuous (and dominant). The oak-hickory, hemlock, or longleaf pine communities illustrate the point. The study of such groupings is sometimes called *community ecology*.

Succession Is the Change of Communities in a Region

The denuding of any part of the earth is followed by a *succession* of different communities of life. As community succeeds community the environment is altered, making possible the invasion of different organisms in an area where they could not have lived before. The invasion of new forms of life changes the environment enough to wipe out earlier inhabitants. The moving in of the new and the exclusion of the old continues, if no new disturbance occurs, until a *climax community* is reached. Then succession ceases. The *climax* is that particular point where the types of life present are in harmonious balance with the existing environmental conditions. When the climax is reached, the same kinds of organisms reproduce themselves continually until disturbing forces like fire or a change in climate upset the balance and start a new chain of succession.

A trained biologist frequently can travel along a highway and tell exactly how many years a formerly cultivated field has been abandoned by noting the kinds of plants growing there. As the years pass there is a succession of plants. Especially in the first few years after abandonment, plants of each successive year may be different from those in the preceding one.

Succession is forever at work filling lakes, covering clearings, and even covering bare rocks. When humans cultivate crops, they are energetically at war against succession in defense of the status quo. The maintenance of grassy lawns and golf greens is but another example of how humans come to grips with succession.

One of the best ways to observe the results of succession is in *lake zonation* (Fig. 17.1). Almost any lake with a little age and which has not been subject to extreme fluctuation in water level will show distinct evidence of succession. The deeper offshore zone may have *floating hydrophytes* (water plants) on the surface. Nearer the shore, where water is shallow enough to allow good light penetration, there may be a zone of *rooted hydrophytes* submerged beneath the water surface. Still closer to shore, there may be a zone of *rooted hydrophytes with floating leaves* (water lilies). The swampy

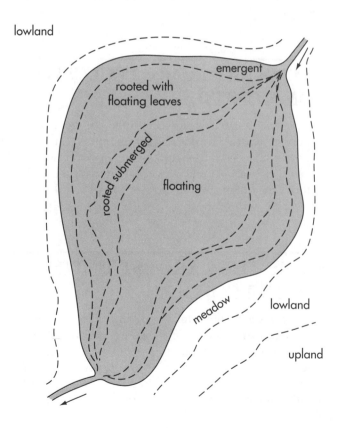

FIGURE 17.1.   *Zonation in a mature lake. Some zones may be missing.*

shore line is a zone of *emergent plants* like sedges, rushes, swamp shrubs, and trees (e.g., willows). Water usually stands in this zone. The next zone is the *meadow* (bog), which is drier. Plants are similar or identical to those in the previous zone. The zone outside the meadow is still drier and consists of *lowland* species typical of the area. Higher and still drier ground supports an *upland* community of plants, which may be climax for the region. Not all zones listed here will necessarily be found in any particular lake, but several may be.

The constant filling of lakes with mineral sediments and organic debris is a drying-up process that results in a gradual migration of the zones toward the center. The floating hydrophytic zone is replaced by submerged plants that become rooted. The area they formerly occupied is replaced by the zone immediately outside it. All of the other zones move inward likewise. If uninterrupted, the process theoretically continues until conditions of the entire area are brought to an intermediate position with respect to available water; that is, it is neither extremely wet nor extremely dry. Starting from either extreme of wet or dry, succession moves in the direction of an intermediate position.

ECOSYSTEMS ARE MORE INCLUSIVE THAN COMMUNITIES

An *ecosystem* is the sum of a community and the geographical region in which the community exists. Thus, it is a larger entity than either population (one species) or community (various interacting species). Actually, an ecologist often ignores these distinctions, since it is nearly impossible to study a community without including examination of the physical world in which it exists, and vice-versa.

BIOMES ARE THE LARGEST UNITS OF STUDY IN ECOLOGY

The largest geographical community of living organisms and their nonliving environment is known as the *biome*. It is regional in extent and characterized by a life form (grass, deciduous trees, etc.) that dominates. Of course, there are smaller units (ecosystems) within the region, some of whose communities are in stages of succession. However, in most cases a biome will be characterized with reference to the life form of the climax community for that particular climate. There are only a few major land biomes of the world, as explained below. An outline of these biomes may helpful before we discuss them:

I. Forest biomes
 A. Tropical (low latitude) forests
  1. Rain forest (including jungle, galeria, and mangrove)
  2. Tropical deciduous forest
  3. Scrub forest
 B. Nontropical (middle and high latitude) forests
  1. Chaparral (Mediterranean scrub forest)
  2. Temperate broadleaf (and broadleaf-coniferous) forest
  3. Coniferous forest (taiga)
II. Grassland biomes
 A. Tropical grassland (savanna)
 B. Temperate grassland (steppe and prairie)
III. Desert biomes
 A. Desert (dry desert)
 B. Tundra (frozen desert)

Tropical Rain Forest (Selva) is the Richest in Diversity

Trees are close together and tall in the *rain forest.* Their density results in a close canopy that allows very little sunlight to reach the forest floor. Weak penetration of light, in turn, makes the trees have straight, branchless trunks and creates an unfavorable situation for the growth of underbrush. Tree trunks are commonly strengthened with buttresses; this is advantageous in providing additional support, since the roots are too shallow to give the best kind of anchorage. The trees have broad, leathery, evergreen leaves. There are many species of trees, none of which occur in any sizable stands. The absence of pure stands makes lumbering more difficult and expensive, since it has to be done selectively. Ebony, mahogany, and teak are three of the valuable lumber trees. Some other trees are sources of products like chocolate, rubber, camphor, gums, resins, and rattan. Associated with the trees is a profusion of vines (lianas), epiphytes, and parasites. Decay organisms favored by high temperatures and abundant moisture permit little or no humus to accumulate.

Animals are predominantly arboreal (tree-dwelling). Insects, birds, and reptiles are especially abundant. Some of the mammals are the bat, tapir, opossum, peccary, jaguar, armadillo, anteater, and monkey. Many of the animals are brilliantly colored.

Rain forests are located along the Amazon River, equatorial west Africa, on the islands between Asia and Australia, on the east coast of Madagascar, and in smaller places in the equatorial belt.

A variation of the rain forest is the *jungle,* which differs primarily in having a dense tangle of undergrowth. Undergrowth cannot develop in the shade of the rain forest but does develop whenever sunlight can penetrate to the forest floor. Penetration of sunlight is possible under three conditions—first, in areas ravaged by humans or fire; second, on steep slopes; third, along the banks of wide streams, lakes, or coasts where the overhead canopy is broken.

Jungle along broad river banks seldom extends more than a few hundred meters from the water.

Another variation of the rain forest is the *galeria* (gallery) *forest*. It is really a narrow extension of the rain forest along streams into areas where the climate is much drier and where vegetation is lighter. Streams are narrow enough for the tree crowns to meet overhead. In effect, the stream flows through a tunnel of green.

A third variation of the rain forest is the *mangrove forest*. It grows in muddy tidal zones where there are no strong tides or currents. In some places it may extend a considerable distance from the shore at high tide. This type of environment supports a forest of trees that stand on stilt-like, interlocking prop roots.

The rain forests of the world are the sites of earth's greatest diversity of organisms. It is estimated that 10 percent or fewer of rain forest insects have been scientifically identified. Efforts are underway to conserve the remaining rain forests, which are mere remnants of those existing a few hundred years ago. Human agricultural, lumbering, and mining activities are the main causes of the drastic shrinking of this biome in recent times.

Tropical Deciduous Forest Grows Where Wet and Dry Seasons Alternate

The *deciduous forest* grows in tropical regions having a definite wet and dry season. Many of the trees are the same species as the evergreen trees of the rain forest, but most of them shed their leaves during the dry periods. Commonly growing in pure stands, the trees are smaller and more widely spaced than in the rain forest. Their smaller size and occurrence in pure stands make the forest easier to lumber. The result is that much of the forest has been disturbed by human activity. The wider spacing of trees breaks the density of the canopy and allows enough light penetration to cause undergrowth.

Ground animals are numerous. Insects, birds, and reptiles are more abundant here than in the rain forest. In addition, large herbivores, like the elephant and giraffe, find shelter in the margins where the forest adjoins grassland. It is also the home of lions and tigers.

The best development of this forest is in southern Asia, where the monsoon winds blow inland from across the equator. Jungle, galeria, and mangrove forests are associated with the tropical deciduous forest just as they are with the rain forest.

The Tropical Scrub Forest Is a Transition Biome

The *scrub forest* is a zone of transition between the rain forest or tropical deciduous forest, where there is more rainfall, and the tropical grassland, where rain is scarce. Rainfall is low and the dry season longer. Vegetation supported by such conditions is deciduous and shrubby. In wetter places the vegetation looks like dense, shrubby second growth; in dryer places the shrubby growth is scattered in a semi-grassland. The shrubs have such xerophytic characteristics as small leaves, thick bark, and thorns.

Animals are abundant, especially insects, birds, and reptiles. Certain of them are even more abundant than in the rain forest. The scrub cover adjacent to grassland is the ideal place for ground animals like the carnivorous tigers and lions and the herbivorous elephants and giraffes.

Large areas of the tropical scrub forest are in South America, southern Africa, southern Asia, and northern Australia. Smaller areas are in Mexico and Central America. Jungle, galeria, and mangrove forests may also be associated with the scrub forest.

Chaparral Is Semitropical but Outside of Equatorial Latitudes

Unlike the preceding scrub forest, the *chaparral* (or *Mediterranean scrub forest*) is in the middle latitudes. It is, however, semitropical. The dominant plants of the forest are xerophytic shrubs and trees. They are predominantly broadleaved evergreens with small, thick, waxy leaves; but some conifers may be scattered among them. The widely spaced evergreens are usually twisted and low. A shrubby ground cover blankets much of the soil beneath them. The shrubby growth, either mixed with trees or in a pure cover, is called chaparral in California and Arizona and maquis in southern Europe. In drier or disturbed areas the chaparral is entirely treeless. During the summer, which is the dry season, the plants become drab and semidormant; during the winter, which is the rainy season, they are green and growing. This forest has been much disturbed by humans. It is the region where cork oaks, wine grapes, and olives are grown commercially.

In terms of worldwide distribution, the chaparral is limited in extent. It is confined primarily to a fringe around the Mediterranean Sea, two areas in southern Australia, parts of the southwestern United States, the southern tip of Africa, and central Chile. Rainfall varies from 15 to 40 centimeters per year.

Temperate Forests Are the Broad-leaved Forests of Hot Summer and Cold Winter

Dominant plants in the *temperate forests* nearest the tropics are evergreen. Here climate is warmer and winter and summer seasons are less distinct. The forest somewhat resembles the tropical rain forest, but the species are different. The trees are not as close, and undergrowth is thicker. In poorer, dryer places within the general area conifers replace the broad-leaved species. In swampy places, cypress dominates.

Farther poleward, winter and summer seasons are distinct. Winter temperatures are low enough to force plant dormancy during that season. Here the evergreen broad-leaved trees give way to the deciduous broad-leaved trees. The number of species is smaller, and they often grow in almost pure stands. Examples of some of the trees are oaks, hickories, beeches, and maples.

Broad-leaved forests have been so extensively exploited by humans over the centuries that virgin (never lumbered) areas are seldom seen. Despite the frequent almost pure stands, there is a tremendous diversity in vegetation and environmental conditions. Conifers are scattered among the hardwoods (deciduous trees) and sometimes grow in pure stands on poorer soils. Other local conditions support meadows, moors, and grass or heath balds.

Animal life is abundant and varied. Fish, amphibians, reptiles, birds, and mammals are well represented.

The temperate forest is confined to the northern hemisphere, except for a part of the southern tip of South America, the east coast of Australia, and islands adjacent to the latter. It is found in eastern North America below the taiga, most of Europe, and eastern Asia.

Coniferous Forest (Taiga) Occurs in Harsher Climates

Coniferous forests are able to survive under conditions that are too adverse for broad-leaved trees. Wherever they grow in association with the broad-leaved trees, they are on drier, rockier sites. At the risk of overgeneralizing, it may be predicted that coniferous forests will be found at higher altitudes and latitudes than deciduous broad-leaved forests.

High latitude forests, generally called *taiga* or *boreal forest*, are dominated by cone-bearing softwoods such as pines, firs, spruces, hemlocks, and cedars. Some deciduous broad-leaved trees may be present, but the weather is too cold for them to do well. Near the poleward limits of the forest, trees become fewer as well as smaller. The region occupied by the taiga was once glaciated and is dotted with swamps, lakes, and bare rocks. The swampy places that interrupt the continuity of the forest contain stunted trees, grasses, sedges, and mosses. The assemblage of swamp plants is called *muskeg*. Human development of the coniferous forest area is seriously hindered by the shallow rocky soil and exposed bedrock which interfere with agriculture, and the swamps and lakes, which make transportation difficult. The trees have been exploited for pulpwood. New growth is slow to replace what is removed because the climate is unfavorable.

Large-animal life is abundant in the taiga, making the northern coniferous forest a sort of sportsman's paradise. In America bear, deer, fox, mink, moose, rabbit, and weasel are some of the hunted mammals. Freshwater fish abound in streams and lakes. Birds are also plentiful. Insects are a real nuisance to humans and other mammals; in some seasons mosquitoes make life almost intolerable for them.

The coniferous forest just described occurs primarily in the northern hemisphere: in Canada, northern Europe, and Asiatic Siberia. A tiny coniferous forest area is in South America.

Another kind of coniferous forest occurs in the coastal plain of the southern United States. Tall, widely spaced pines dominate. The forest floor is covered with a moderate to scarce shrubby undergrowth or grass; in some places the forest looks almost park-like. The climate is, of course, quite different from that of the northern regions. This coniferous forest can be accounted for on the basis of soil conditions and/or fire.

Tropical Grassland (Savanna) Occurs in Drier Tropics

Tropical grasslands (also called *savannas*) do not get enough rainfall to support many trees. Dry tropical areas are located near the poleward boundary of the tropics or inland from the east coasts from which the rain-bringing winds blow. The dry season is so long and the rainfall so scant that the maximum vegetation they can support is a luxuriant, coarse growth of grass sometimes up to four meters tall. The grass is scattered so that bare ground is exposed among the plants. In some places isolated or small clumps of low trees may be scattered here and there, but the aspect is grassland.

Tropical grasslands are rather inhospitable to human habitation. They are, however, ideal for elephants, giraffes, zebras, lions, tigers, reptiles, termites, tsetse flies, and mosquitoes.

About half of Africa and extensive areas of South America are grass-covered. Other areas are in northern Australia and the West Indies. In the United States the nearest similarity to the tropical grassland is the everglades in southern Florida.

Temperate Grassland (Steppe and Prairie) Has Rich Soil

Both *steppe* and *prairie grasslands* occur in areas where rainfall is not sufficient to support trees. They frequently occur side by side, with the steppe adjacent to deserts and the prairie adjacent to forests. Thus, the difference is determined by available water. A fair generalization to make is that steppes are supported by 25 to 50 centimeters of rain annually and prairies by 50 to 75 centimeters, plus or minus. Of course, a sharp line between the two types is usually nonexistent.

As a natural habitat, a grassland can best support herbivorous animals. In early America it was the home of the buffalo. Small ground animals are the most abundant types. Some common animals are burrowing owls, hawks and other soaring birds, prairie chickens, badgers, prairie dogs, coyotes, ground squirrels, rabbits, skunks, and wolves.

Steppe grasses form an almost continuous mat from 10 to 20 centimeters high. Now and then stunted trees and xerophytic shrubs break the monotonous continuity of grasses. By now, much of the world's steppes have been altered by humans, either by cultivation or by the grazing of cattle.

Steppe areas are in Asia, Australia, North America (mostly the United States), and northern Africa just above the Sahara. In North America it is called the short grass plains or the Great Plains.

Prairie grasses form an almost continuous mat averaging about 60 centimeters tall, but sometimes growing to nearly 2 meters. Occasionally trees or shrubs grow where they can get water, along streams for instance. Because rainfall is enough to support grain and some other crops, most of the prairie of North America is in cultivation. It is the grain belt or breadbasket of the

United States and Canada; the same is true in Argentina, Australia, and Ukraine.

The largest prairies of the world are in North America and in South America (including the Argentine pampas). Smaller prairies are in Europe (Russia), Manchuria, and southern Africa.

Desert (Dry Desert) Is an Area of Few Plants and Little Water

Dry deserts are areas of very low rainfall (under 25 centimeters annually) located in the low and middle latitudes, usually in coastal areas away from the direction of the prevailing winds or in inland areas of continental masses. The daily range of temperatures may be extremely great; summer temperatures are hot. The plants are relatively sparse, and some areas may be totally barren. When plants are present, which is usually the case, they are commonly widely spaced xerophytic shrubs, bunches of short grass, or the succulent type of plants.

Animals are common but not abundant. Reptiles, birds, rodents, and small carnivores are typical. Fish are scarce, confined to irrigation canals, water holes, and a few streams. There are also a few amphibians where local conditions permit. Whatever animals are present usually burrow into the ground or hide under rocks and come out to forage only at night or in the cooler parts of the day. Their water requirements are low, or they have adaptations for conserving water.

The largest dry desert areas are in north Africa (Sahara), inland Asia, and central Australia. Smaller deserts are in western North America, southwestern Africa, and southern South America (primarily Argentina).

Tundra Is a Frozen Desert

For most of the year the *tundra* is a frozen desert. In fact, the subsoil is permanently frozen almost everywhere. Under the severe weather conditions where tundra exists, organic materials decompose slowly. Swamps and ponds abound because drainage is poor. Plant life consists of lichens, peat mosses, grasses, sedges, and small herbs. There are a few dwarfed trees with trunks up to one centimeter in diameter.

Typical animals are birds, caribou, musk ox, barren-ground grizzly bears, lemming, arctic fox, arctic hare, and arctic wolf.

Tundra is almost altogether in the northern hemisphere, just north of the taiga and south of the polar ice. It also occurs at high elevations, where it is called *alpine tundra.*

Mountain Regions Have Biomes of Several Types

Mountain vegetation is not a different kind of community but is composed of communities already mentioned. High mountains have a wide range of climatic conditions, especially temperature, because of altitudinal differences.

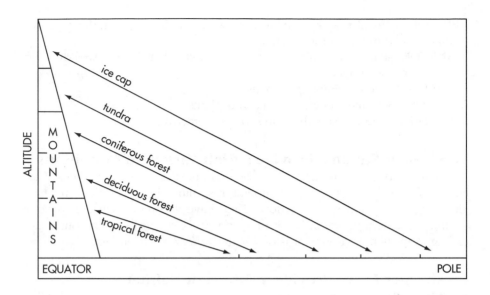

FIGURE 17.2. *Corresponding altitudinal and latitudinal zonation.*

The community at the base of the mountain is the same kind of community as the climax community typical of the general area surrounding its base. Upon ascending the mountain, one passes through a sequence of altitudinal zones corresponding to the biomes one would find in going toward the poles (Fig. 17.2). This is a generalized observation, subject to modifications caused by local conditions.

Communities Are Usually in Transition

Communities are seldom absolutely uniform throughout their extent or sharply delimited from adjacent communities. Climate, soil conditions, and other factors that support them intergrade with other conditions in adjacent areas. The absence of distinct environmental boundaries results in transition zones or merging zones between clearly definable communities.

AQUATIC LIFE OCCURS IN A VARIETY OF HABITATS

Organisms live in a multitude of aquatic habitats of the *hydrosphere*, which comprises about 75 percent of the earth's surface. Their adaptations are legion, but from one ecological point of view they may be reduced to five categories that are based on their mode of life:

NEUSTON—Organisms like duckweeds, water striders, or whirligig beetles that float or move on the surface.

PLANKTON—Plants and animals, larvae or adults, that float or weakly swim at or near the surface.

NEKTON—Actively swimming animals.

BENTHOS—Bottom-dwelling animals and plants.

PERIPHYTON—Organisms that cling to submerged vegetation.

Freshwater Streams Provide a Rich Variety of Conditions

Streams vary so much that they provide an unusual variety of habitats. They have different depths, turbidity, temperatures, mineral content, bottoms, marginal vegetation, and speed of flow. They have a common characteristic in having movement of water, but it may vary from sluggish to vigorously cascading. Any such environment is designated a *lotic environment.*

Freshwater Lakes Provide a Different Habitat by Their Slower Currents

Lakes, which by contrast to streams have still water, are known as *lentic environments.* The kind of zonation that develops is closely related to the size of the lake. In any case, the shallow zone around the margin where rooted plants commonly grow is the *littoral zone.* Lakes that are deep enough also have a horizontal stratification based on degree of light penetration. The upper stratum, the *limnetic zone,* is the area where enough light can penetrate to cause photosynthesis to proceed at a rate that is at least equal to the rate of respiration. The deep water zone (including bottom) below the limnetic zone is the *profundal zone.*

Another kind of stratification based on temperature differences is possible in deep lakes at certain times of the year. During the summer the upper layer, the *epilimnion,* is circulated as wind blows along its surface. It is distinctly warmer than the bottom. Beneath it is a layer known as the *thermocline,* which is a zone where temperatures decrease about one degree centigrade for every meter increase in depth. Beneath the thermocline is the *hypolimnion,* which is cold and relatively undisturbed.

In temperate regions seasonal upheavals may occur during which time the stratification is destroyed and all the water circulates. Some lakes have two turnovers a year, one in the fall and one in the spring. With the approach of cold weather the surface water cools until it is approximately the same temperature as the deeper water. Wind blowing against the surface can then circulate the entire lake in the same way that it circulated the epilimnion during the summer. During the winter the surface temperature drops below 4°C at which point water is at its greatest density. A further drop causes the water to expand. This produces a lighter surface stratum above the denser water below. Ordinarily the surface also freezes. The spring turnover occurs when the surface ice melts and the temperature rises to about 4°C. Again the wind can circulate the entire lake because the surface water is approximately the same density as the water below.

The Oceans Support a Tremendous Volume of Organisms

Marine habitats (zones) may be classified in several ways, depending on what one has in mind. A composite classification outline, taking into account several factors, follows this paragraph. Most of the categories are adjectives that may be used preceding the word zone, habitat, or type of life. For example, one can say littoral zone, littoral habitat, littoral animal, or littoral plant. Some of the categories overlap. For instance, littoral zones are also photic zones (penetrated by light).

I. Littoral (also photic*)—Between high and low tide lines (alternately exposed and flooded).
II. Nonlittoral—Beyond low tide line.
  A. Pelagic—Of the open sea (from low tide line on out).
    1. Neritic (also photic*)—Over the continental shelves (up to about 200 meters deep), subject to wave action.
    2. Oceanic—Over deep part of ocean.
      (a) Photic*—Where photosynthesis can occur (down to about 200 meters), subject to wave action.
      (b) Aphotic*—Little or no light penetration.
        Bathyal—Deep (from about 200 to 4,000 meters), currents but no wave action.
        Abyssal—Very deep, water permanently cold.
        Hadal—In deepest trenches.
  B. Benthic—Of the bottom.
    1. Sublittoral (also photic*)—Under neritic zone of continental shelf.
    2. Bathyal—From margin of continental shelf to 4,000 meters deep.
    3. Abyssal—Under deep water.
    4. Hadal—Under water of trenches.

(*Not descriptive of type of life)

COOPERATIVE LIVING IS NOT UNCOMMON AMONG ORGANISMS

Despite the tremendous struggles for survival among living organisms, there are innumerable adaptations of a cooperative nature between two different kinds of organisms. For example, insects pollinate plants and get food in return. Also nitrogen-fixing bacteria in the roots of leguminous plants provide nitrogen and obtain protection and food.

A number of animals cooperate with other members of their own species. Some commonly live as mates and share responsibilities. Some live in colonies and perform highly specialized obligations. The largest are among the insects, where a colony of bees may contain 75,000 members or a colony of tropical termites may contain three million members.

HUMAN ACTIVITIES HAVE PROFOUND INFLUENCE UPON THE ENVIRONMENT

In the approximately 200,000 years of the existence of *Homo sapiens*, there has been an almost unparalleled success of the species. We now number approximately 6 billion individuals and inhabit or have impact upon every biome. We have a strong propensity to change any environment that we enter and can do so with great effect because of the technology that we have developed. This has led to serious trouble.

Biodiversity Loss Is Mostly Caused by Humans

The fossil record indicates that the average extinction rate over many millions of years has been about one species per year (assuming the present degree of diversity). It is estimated that the earth is currently experiencing an extinction rate of somewhere between 400 and 4,000 species per year. This staggeringly larger than average loss rate is likely attributable to human intervention.

Most species live in tropical areas where human societies have been rapidly expanding for the last several hundred years. This has accelerated in the past few decades, with tropical forests being logged at an alarming rate and agricultural fields replacing them.

Other human activities leading to extinctions include accidental or purposeful introduction of non-native species, which then compete too well with native species. Alteration of aquatic environments by filling in and by thermal and chemical pollution have had well-documented negative effects, especially in species-rich marsh lands.

Atmospheric Pollutants May Change Global Temperature

The so-called greenhouse gases (carbon dioxide, ozone, nitrous oxide, and others) that have been released by technological activity are accumulating at alarming rates in the atmosphere. They are so called because they retard the normal loss of heat from the atmosphere into space, just as glass does in a greenhouse. Although scientists disagree over the rate, there seems to be consensus that the earth's surface will be gradually warming as a result of these accumulations.

The effects upon the ecology of all biomes would be profound. Patterns of plant growth would shift, and animal populations would have to follow the plants or go extinct. The current best areas for agriculture would shift just like the natural regions, and areas currently supporting many people would become less productive. Sea level would rise as polar ice melted into the oceans, with disastrous results for human populations living and growing crops near the coasts. Patterns of rainfall would change as average temperature rose, with some currently moist regions becoming dry deserts and vice-versa.

Organisms would survive global warming; it has happened cyclically before without human cause. But the world would be profoundly different and living through the changes would not be enjoyable for large segments of the population.

The Protective Ozone Layer Is in Jeopardy

Recent trends in distribution of the gas ozone in the atmosphere have been disturbing. Ozone (O_3) is paradoxically a pollutant that is among the greenhouse gases and also a natural protectant against the effect of ultraviolet light. It has been known since the 1920s that certain wavelengths of ultraviolet radiation can damage DNA; i.e., cause mutations. Some ultraviolet light constantly hits earth and its inhabitants, but a significant proportion of that which streams from the sun is absorbed high in the atmosphere by a layer of ozone.

Since the mid-1980s, satellite reconnaissance has shown annual loss of ozone from the high-atmosphere layer. Although there is still controversy on this matter, it appears that the loss is attributable to human-released chemicals that react with ozone, particularly a category called chlorofluorocarbons (CFCs). These have been used in recent decades as propellants in aerosol cans, as coolants in air-conditioning units, and in other industrial applications. Legislation in several countries now makes release of CFCs into the atmosphere illegal. It is difficult to estimate whether this is the only reason for recent ozone layer depletion and whether we can reverse the trend.

Ozone released as a pollutant cannot be relied upon to replenish what is being lost. That which is released at the planet's surface apparently does not rise to the stratosphere where it would help block ultraviolet light.

People Must Be Active in Protecting Life on Earth

The examples cited above could be joined by many others. The lesson from them is simple: humans, by our technological capability and expanding population, threaten the welfare of other forms of life on earth. In so doing, we threaten ourselves, because the study of ecology teaches us that all organisms are intimately linked in many ways. We are largely ignorant of how the loss of any particular species has impact upon others in a community.

Extinctions and their consequences have been in the nature of things since life began. Humans are in the unique situation of being responsible for far more than the average number of extinctions and at the same time being the only organisms that can contemplate the consequences of those extinctions. It is fervently hoped that our intellects, our desire for self-preservation, and our compassion can reverse the trend for the betterment of all life on earth.

STUDY QUESTIONS

1. Discuss this statement, "Humans are both generalists and specialists."

2. Distinguish between habitat and niche.

3. Why is it said that parasites should not be too successful in reproduction and finding hosts?

4. In what major way do plants differ from animals in their mode of dispersal? What are some obstacles to the free movement of organisms?

5. What is ecological succession and what makes it possible?

6. What is the ecological climax? How can one tell if the life present in a particular place is the climax or a stage of succession?

7. Discuss succession in fresh water lakes as revealed in their zonation.

8. Explain how specialization and generalization are related to survival.

9. Describe some types of mimicry.

10. What structural modifications do hydrophytes have that adapt them to aquatic environments?

11. How are xerophytes adapted to conserve water? Name a xerophyte.

12. Define community as used in ecology. Define biome.

13. Briefly describe these biomes: tundra, taiga, temperate forest.

14. What environmental factor is necessary for the development of a jungle? What three conditions make possible the availability of that factor?

15. Where is the chaparral found in the United States?

16. Why is a temperate forest more economical to log than a rain forest?

17. Based on their mode of life, into what five categories may aquatic organisms be placed?

18. Contrast lotic and lentic environments.

19. Discuss vertical stratification in deep lakes. What is its cause? Under what conditions may the stratification be destroyed?

20. What change is it that may be causing global warming?

21. Since this book was written, has there been any change in the trend toward loss of biodiversity? Toward our understanding of global warming? Toward loss of the protective ozone layer?

GLOSSARY

A

abscisic acid. A plant growth regulator that inhibits growth of buds and prepares them for winter dormancy.

abscission layer. A layer along which a leaf or fruit naturally separates from the stem.

absorption. The process of bringing digested molecules into cells.

acetylcholine. A neurotransmitter released at neuromuscular junctions and at all synapses outside the brain.

acid. A molecule or ion capable of releasing ionic hydrogen.

acrosome. A membranous bag lying over the nucleus of an animal sperm cell, carrying enzymes needed for successful fertilization.

action potential. The temporary reversal in charge across a neuron's plasma membrane that is the message sent through a nervous system.

activation energy. The energy that must be provided before a chemical reaction can occur.

active transport. Process of molecules or ions moving across membrane against a concentration gradient, an energy-consuming process.

actin. The protein that forms microfilaments; together with myosin, a contractile portion of muscle.

ADP, adenosine diphosphate. Adenosine with two phosphate groups; converted to *ATP* by the addition of another phosphate group bound by a high-energy bond.

aerobic respiration. Respiration in which gaseous oxygen (O_2) is used as a hydrogen acceptor.

aggregate fruit. A fruit derived from several carpels of a single flower, e.g., strawberry.

alga. A member of Kingdom Protista that can carry on photosynthesis.

allele. One of two or more contrasting forms of a gene.

allergy. An inappropriate production of unusual antibodies that irritate one's body cells; a set of symptoms that occur because of this irritation, such as swelling, drippy nose, watery eyes.

allopatric. Not overlapping in geographic range.

alveolus. An air pocket of the vertebrate lung through whose wall gases are exchanged.

amnion. The sac immediately surrounding an embryo (or fetus).

amylase. A starch-digesting enzyme.

anaerobic respiration. The form of respiration during which gaseous oxygen is not used.

analogous. Superficially similar because of similar adaptation; not homologous.

anaphase. That phase of mitosis or meiosis when chromosomes are moving toward the poles of the spindle.

animal. A multicellular heterotrophic organism carrying on internal digestion of food; a member of Kingdom Animalia.

anisogamy. The situation in which the male gametes of a species are visually distinguishable from the female gametes.

anther. The distal part of a stamen that produces pollen.

antheridium. A male gamete-forming structure in many kinds of plants.

antibody. A protein whose shape enables it to attach to an antigen; part of the humoral immune system.

anticodon. A three-base sequence of tRNA that can base-pair with a codon of mRNA.

antigen. A molecule (usually a macromolecule such as a protein) that elicits an immune response because it is recognizable as foreign to the body.

anus. The exit of a digestive system.

aorta. The large artery leaving the heart and supplying blood to all parts of the body.

apical meristem. Actively growing tissue at tips of roots and stems.

Archaea. The kingdom of bacteria-like organisms called archaebacteria, when a six-kingdom classification system is used.

archegonium. An egg-producing organ in many plants; consisting of a base, venter, and neck.

archenteron. The cavity of the gastrula stage of an embryo; becomes the gut.

Archezoic. The first era of geologic time in which there was evidence of life.

arteriole. A small artery.

ascus. A sac-like container of spores in fungi of division Ascomycota.

asexual. Sexless; able to reproduce without joining genetic material from two individuals.

association neuron. A nerve cell between two other nerve cells. Also called *interneuron*.

assortative mating. The phenomenon of sexual mates choosing each other on the basis of phenotypic similarity (or dissimilarity) to each other; a type of nonrandom mating.

aster. A centriole with its radiating microtubules (including spindle fibers) in a dividing animal cell.

ATP, adenosine triphosphate. Adenosine with three phosphate groups; converted to *ADP* by the loss of one phosphate group; the most-used molecule for storing a small amount of energy for a short time.

atrium. A compartment of the heart receiving blood from the body.

autoimmunity. An inappropriate attack by the immune system upon one's own cells or molecules.

autonomic nervous system. That part of the nervous system regulating involuntary reactions of many organs of the body, especially the digestive tract and associated structures.

autosome. A chromosome that is not a sex chromosome.

autotroph. An organism that produces its own food.

auxins. Plant growth regulators produced by actively growing tissues; primary effect is cell elongation.

axon. An elongated portion of a neuron conducting impulses away from the body of the neuron.

B

bacillus. A rod-shaped bacterium.

backcross. Crossing a dominant phenotype with a homozygous recessive to determine if the phenotype is homozygous or heterozygous.

bacteria. Prokaryotic organisms of Kingdom Bacteria (six-kingdom system) or Kingdom Monera (five-kingdom system).

Bacteria. The kingdom containing true bacteria, when a six-kingdom classification system is used.

bacteriophage. A virus that parasitizes bacteria.

bark. All tissues outside of the vascular cambium of roots and stems.

basal body. A centriole-like object at the base of a cilium or flagellum.

base. A molecule or ion capable of accepting ionic hydrogen.

basidium. A club-shaped spore-producing structure in club fungi.

Batesian mimicry. The phenomenon of a defenseless organism taking the shape and/or color of an obnoxious one, thus gaining protection.

benthos. Bottom-dwelling organisms of the sea.

binomial. A two-part name given to each organism, consisting of its genus and species names.

biome. A large biotic community of wide geographical extent characterized by a dominating life form; e.g., desert, tundra, rain forest.

blastula. An early stage of an animal embryo consisting of a hollow ball of cells.

bond. The linkage between any two atoms in a molecule.

boreal forest. A high latitude coniferous forest; taiga.

Bowman's capsule. The bulbous part of a kidney tubule surrounding the glomerulus.

bronchiole. A small air passage in the lung.

bronchus. An air passageway leading from the trachea into a lung.

budding. An asexual reproductive process in which a fragment is separated from the body and grows into a new organism; a form of grafting in which a bud is used as the scion.

C

calorie. A unit of heat; the amount of heat required to raise the temperature of one gram of water one degree Celsius.

Calvin cycle. A series of reactions taking place during photosynthesis, during which carbon, hydrogen, oxygen, and energy are combined to make a stable organic molecule that can be used as food; also called carbon fixation or the dark reactions.

cambium. A meristematic tissue occurring in layers in roots and stems and producing secondary tissues.

cancer. A complex of diseases characterized by uncontrolled cell reproduction and disruptive invasion of cells into normal tissues.

capillary. A small blood vessel consisting of only one layer of endothelium and usually connecting arteries to veins.

carbohydrate. An organic molecule composed of carbon, hydrogen, and oxygen in which hydrogen and oxygen occur in the same proportion as in water; sugars and their polymers.

carbon fixation. See *Calvin cycle*.

carcinogen. A cancer-causing agent.

carnivore. An organism that ingests heterotrophs (usually animals) as food.

carpel. The female part of a flower, composed of one or more megasporophylls.

Casparian strip. A thin stripe of wax deposited on walls of endodermis in roots.

cell center. A structure near the nucleus of a nondividing cell, including a pair of centrioles.

cell membrane. The outer boundary of the cell's cytoplasm, often called the plasma membrane.

cellular immune system. The portion of a vertebrate animal's immune system that depends upon action of T cells.

cellulose. A polysaccharide making up the framework of the walls of plant cells.

Cenozoic. The most recent geologic era, known as the age of mammals, birds, and angiosperms.

centriole. An arrangement of microtubules associated with organizing a spindle during cell reproduction.

centromere. The point on a chromosome where a spindle fiber is attached during cell reproduction.

cephalothorax. The fused head and thorax of animals like crustaceans.

chaparral. A shrubby type of vegetation in semitropical areas where the rainy season is in the winter; Mediterranean scrub forest.

chemiosmosis. A complex process that transfers the energy of electron transport for the building of ATP; occurs in mitochondria and chloroplasts.

chitin. The horny skeletal material of Arthropods, a polysaccharide.

chlorenchyma. A plant tissue composed of thin-walled cells that contain chlorophyll.

chlorophyll a. The most common type of green photosynthetic pigment in most plants and many algae.

chloroplast. A plastid containing chlorophyll and other pigments.

chorion. The outer membrane surrounding a mammalian embryo, providing direct connection with the uterine wall before the placenta is fully developed.

chromatid. One of two identical portions of a chromosome during the early portions of cell reproduction.

chromosome. A physical structure holding genetic information in a cell; composed of DNA and (sometimes) associated proteins.

cilia. Small projections from cells, containing microtubules and used to circulate materials over the surface of stationary cells or to propel some single-celled organisms.

citric acid cycle. See *Krebs cycle.*

cleavage. One of the early cell divisions of an embryo.

climax community. A group of species living in harmonious balance so that succession is unlikely to occur.

clitoris. The female structure homologous to the tip of the penis; a center of sexual sensations during intercourse.

cnidocyte. A stinging cell characteristic of animals in phylum Cnidaria.

coccus. A spherical bacterium.

codon. A sequence of three bases (of either DNA or RNA) that signifies a single event in the translation process; most signify the placement of specific amino acids.

coelom (true coelom). A body cavity lined with cells of mesodermal origin.

coenzyme. An organic substance, often a vitamin or vitamin derivative, necessary for an enzyme to function.

collenchyma. A simple plant tissue composed of cells with some thickening of the walls, especially in the corners.

colon. The large intestine.

commensalism. A close relationship (symbiotic) between two organisms where one member benefits and the other is neither harmed nor helped.

community. A group of species living together.

companion cells. Small phloem cells adjacent to sieve tubes.

complete flower. A flower with all kinds of floral parts of both sexes.

compound. A molecule composed of two or more different elements.

contractile vacuole. A vacuole in freshwater protozoa that periodically expels excess water to the outside.

convergence. The evolving of organisms from dissimilar groups so that they become similarly adapted to similar environments.

consumer. A heterotrophic organism.

cork. Tissue replacing epidermis in older stems and roots; dead when mature.

corm. A modified stem similar in general appearance to a bulb but solid like a potato.

corona radiata. A layer of follicle cells adhering to an ovulated mammalian egg.

corpus luteum. In the ovary, yellow cells filling the follicular space after ovulation and producing progesterone and estradiol.

cortex. A region in the bark of roots and stems internal to the epidermis or cork and primarily composed of parenchymatous cells; also the outer part of organs such as the kidney or adrenal gland.

cotyledon. The leaf on an embryonic seed plant; the seed leaf.

covalent bond. A bond between atoms in which one or more electrons are shared.

crossing-over. An exchange of genes on homologous chromosomes resulting from their becoming entangled during synapsis of meiosis; results in recombination.

cutin. A water-proofing material associated with the epidermis of plants.

cyclic AMP (cAMP). A modified nucleotide, cyclic adenosine monophosphate; acting as a "second messenger" for hormonal systems within target cells.

cytokinesis. Cytoplasmic division after chromosomal movements of mitosis.

cytokinins. Plant growth regulators that induce cell division.

cytoplasm. The materials of a cell outside of the nucleus; non-nuclear organelles and the fluid in which they are suspended.

cytosol. The semiliquid portion of cytoplasm.

D

decomposer. An organism that extracts its food from materials of dead bodies; a saprobe.

dehydration. The chemical separation of water from a molecule, an energy-releasing step of respiration; also the loss of water from cells, tissues, or organs.

dendrite. A portion of a neuron conducting impulses toward the cell body.

deoxyribose. A five-carbon sugar in *DNA*.

differentiation. Specialization of cells, usually occurring in embryonic stages.

diffusion. The net movement of molecules or ions from their higher to their lower concentration area.

digestion. The breaking of large food molecules into smaller molecules.

dihybrid cross. A genetic cross in which two characteristics are considered simultaneously.

diploblastic. Having two complete layers of embryonic cells: ectoderm and endoderm.

diploid. Containing two sets of chromosomes as in zygotes or most somatic cells.

disaccharide. A compound sugar composed of two joined monosaccharides.

divergence. The evolving of closely related species so that they become dissimilar when they adapt to different environments.

division. The plant kingdom taxonomic group equivalent to a phylum.

dizygotic twins. Siblings born nearly simultaneously and derived from different fertilized eggs.

DNA, deoxyribonucleic acid. A nucleic acid that comprises the genes of nearly all organisms.

dominant. Expressed; used to describe a gene whose characteristic hides the characteristic of a recessive allele when both alleles are present.

dorsal root. The dorsal branch of a spinal nerve connecting with the spinal cord; the pathway of sensory neurons entering the spinal cord.

drift. See *genetic drift*.

duodenum. The first part of the small intestine, the part into which the liver and pancreas discharge their secretions.

E

ectoderm. The outermost layer of an early stage of the embryo, after gastrulation.

ectotherm. An organism that passively takes on the temperature of its environment.

effector. The part of a body responding to a stimulus; a muscle or gland.

electron transport chains. Sets of membrane-bound molecules that can pass electrons; used in photosynthesis and aerobic respiration.

embryo. A multicellular organism between the time of fertilization and the organism's hatching, birth, or emergence from a seed.

embryo sac. The megagametophyte, or egg-producing phase, of flowering plants.

emulsification. The action of bile in breaking fats into small droplets.

endocarp. The inner layer of a fruit wall (pericarp).

endocrine glands. Glands that secretes hormones, generally into blood rather than into specific delivery tubes.

endocytosis. The process of bringing bulky objects (larger than individual molecules) into a cell.

endoderm. The innermost layer of an early stage of the embryo, after gastrulation.

endoplasmic reticulum. A system of cytoplasmic membranes closely associated with the nuclear membrane, plasma membrane, ribosomes, and Golgi apparatus.

endosperm. A food-storage tissue in seeds, located adjacent to the embryo.

endosperm nuclei. Two nuclei in the center of the plant embryo sac that fuse with a sperm to develop into a food-storage tissue called the endosperm.

endotherm. An organism capable of thermoregulating and therefore becoming somewhat independent of the environment's temperature.

enzyme. An organic molecule (almost always a protein) that increases the rate of a cellular reaction without itself being used up in the process.

epicotyl. The bud of an embryonic seed plant, located above the cotyledon(s).

epididymis. A tube carrying sperm cells from the seminiferous tubules to the vas deferens.

epinephrine. A hormone secreted by the medulla of the adrenal gland (also called adrenaline).

equational division. Meiosis II; the portion of meiosis in which cells remain haploid throughout.

erythrocyte. A red blood cell.

estradiol. The most common form of estrogen.

estrogen. An ovarian hormone that establishes and maintains female characteristics.

ethylene. A plant growth regulator whose functions include abscission of leaves and ripening of fruit.

eukaryote. An organism or cell that has a nucleus and membranous organelles.

evaporative cooling. A temperature regulating device in which water's evaporation from a body surface draws heat from an organism.

excretion. The discharge of metabolic waste.

exocarp. The outside layer of a fruit wall (pericarp).

exocrine glands. Tissues or organs that release products into tubes for delivery to specific places.

exocytosis. The process of expelling objects from a cell by the use of vesicles.

extracellular digestion. Digestion of foods outside of cells.

F

facilitated diffusion. Diffusion through a membrane that is enhanced by the presence of specific proteins in the membrane.

facultative anaerobes. Organisms that can exist either in the presence or absence of gaseous oxygen; contrast with obligate anaerobes.

Fallopian tube. An oviduct in humans.

fat. A food substance composed of carbon, hydrogen, and oxygen with oxygen being in a much smaller proportion than in carbohydrates; may be broken into one glycerol and three fatty acid molecules.

fermentation. The process of converting the end products of glycolysis to other molecules in the absence of gaseous oxygen.

fertilization. The fusion of gametes to produce a zygote.

fetus. As applied to humans, the developing organism from about eight weeks until the time of birth, having a definite human form.

filter feeder. An organism that captures food by the use of a net-like filtering anatomical device; e.g., a sponge.

fission. Asexual reproduction of unicellular organisms whereby the cell is divided into two cells of approximately equal size.

flagella. Whip-like projections used for locomotion by some motile cells; longer than cilia, but with similar internal microtubular arrangements.

fluid feeder. An organism that obtains its food by sucking liquid from another organism; e.g., a mosquito or an aphid.

follicle. An egg and the supporting cells around it, in an ovary.

food chain. A linear sequence of food relationships—from plant, to herbivore, to carnivore.

food web. Similar to a food chain, but a more complex interrelationship of producers, consumers, and decomposers.

frond. A pinnately compound leaf as in ferns.

fructose. One of the common monosaccharides, found in honey and many fruits.

fungus. A multicellular heterotrophic organism with cell walls and digesting food outside its body; a member of Kingdom Fungi.

G

gametangium. A plant organ producing gametes.

gamete. A sex cell, sperm or egg.

gametophyte. A haploid gamete-producing plant.

ganglion. A clump of neuron bodies.

gastrin. A stomach hormone that activates cells to produce hydrochloric acid.

gastrula. An early stage of the animal embryo consisting of three layers: ectoderm, mesoderm, and endoderm.

gene. Sufficient genetic material of a chromosome to hold the information for building a polypeptide.

gene pool. All of the alleles of all of the genes in a breeding population.

gene therapy. A set of techniques for inserting normal genes into cells to correct genetic defects.

generative cell. A cell in a pollen grain that divides into two male gametes.

genetic drift. Random evolutionary change that can be attributed to a population being small in size.

genotype. The genetic composition of an individual.

genotypic ratio. The proportion of genotypes resulting from a specific cross.

germination. The sprouting of a seed or spore.

germ layer. An early layer of the embryo from which certain tissues arise; the layers are ectoderm, mesoderm, and endoderm.

gestation. The prebirth period of an organism.

gibberellins. A group of plant growth regulators that stimulate stem elongation and other actions.

gill. The gas exchange device used most by aquatic animals.

glomerulus. A bundle of capillaries from which many substances leave the circulatory system and enter the kidney tubule.

glottis. The opening between the pharynx and larynx.

glucagon. A hormone produced by the pancreas and acting to increase the supply of glucose in blood.

glucose. A monosaccharide sugar; the most-used sugar for gaining energy by being broken down.

glycogen. A polysaccharide particularly characteristic of animals, consisting of many glucose molecules strung together.

glycolysis. The first phase of cellular respiration during which glucose is converted to pyruvic acid.

glycoprotein. A protein with an attached short chain of carbohydrates.

Golgi complex. A loose stack of platelike membranous structures in a cell; a center for producing vesicles.

gonad. A gamete-producing organ, testis or ovary.

gradualism. A proposed mechanism for speciation in which small changes accumulate over time until the population is different enough to be considered a new species.

grana. Concentrations of "stacked" thylakoids in chloroplasts.

guard cells. Pairs of cells forming the boundaries of stomata on leaves.

H

habitat. All of the surroundings with which an organism or population is in contact and in which it lives.

haploid. Containing one set of chromosomes, as in gametes.

Haversian canal. A small canal of bone tissue, containing a small artery and vein as well as a nerve supply and surrounded by several zones of bone cells.

herbivore. An organism that ingests plant materials as food.

hermaphroditic. Having both sexes in the same body (when referring to animals).

heterosporous. Producing two kinds of spores, microspores and megaspores.

heterotroph. An organism that must obtain its food from the environment, not capable of manufacturing food.

heterozygous. Adjective referring to an organism that has two different alleles of a gene.

homeobox. A segment of DNA within a homeotic gene; codes for part of a protein active in controlling gene transcription.

homeostasis. The condition of being able to maintain optimal conditions within the body.

homeotic genes. A class of genes that control major events in embryonic development.

homologous. Similar because of common derivation from a common ancestor.

homozygous. Adjective referring to an organism that has two identical alleles of a gene.

hormone. A potent secretion of some cells that produces profound effects, usually elsewhere in the organism.

humoral immune system. The portion of a vertebrate animal's immune system that involves the action of plasma cells in releasing antibody.

hyaline layer. Extracellular material binding adjacent cells in a tissue or organ.

hydrogen acceptor. Chemical that combines with hydrogen during respiration or photosynthesis; e.g., NAD or NADP.

hydrogen bond. A weak interaction between two atoms that carry slight (and opposite) charges.

hydrolysis. A chemical reaction during which a molecule of water is broken into its ionic components, H^+ and OH^-; occurs when complex food molecules are broken into smaller units.

hydrophyte. A plant adapted to live in very wet habitats.

hypha. The filament of a fungal organism.

hypocotyl. The stem of an embryonic plant, located below the cotyledon(s).

hypothalamus. A region of the brain, active in controlling several homeostatic activities such as temperature and osmotic regulation.

I

ileocolic valve. A valve between the small and large intestines.

ileum. The last and longest part of the small intestine, where absorption of digested food occurs.

imperfect flower. A flower possessing structures of one sex only.

inbreeding. The practice of mating between closely related organisms.

incomplete flower. A flower with one or more kinds of floral parts missing.

independent assortment. The phenomenon of genes being on different chromosomes and therefore moving independently of each other during meiosis.

induction. Directing a gene or set of genes to undergo transcription (genetics); directing a cell or group of cells to begin differentiating in specific ways (embryology).

inferior ovary. In plants, an ovary partially or completely embedded in the receptacle.

inner cell mass. The portion of an early mammalian embryo that becomes the body proper.

instinct. Animal behavior that is genetically determined rather than learned.

insulin. A hormone produced by the pancreas, acting to facilitate uptake of glucose into cells.

intermediate filament. A thin solid rod composed of proteins; supports animal cells internally.

interphase. The state of a cell when not dividing.

interstitial cells. Cells located among seminiferous tubules of the testis and secreting testosterone.

intracellular digestion. Digestion within a cell.

ion. An atom or group of atoms having lost or gained one or more electrons, thus electrically charged.

ionic bond. A bond between atoms in which one or more electrons are transferred.

islets of Langerhans. Patches of pancreatic cells that manufacture and release glucagon and insulin; the endocrine portions of the pancreas.

isogamy. The situation in which the gametes of a species are visibly alike but sexually different.

isotope. A variant of an element differing from other forms in having a different number of neutrons in the nuclei of its atoms.

J

jejunum. The second part of the small intestine, where most intestinal secretions are produced.

K

karyotype. An analysis or display of the number and kinds of chromosomes in a cell.

Krebs cycle. A cyclic portion of aerobic respiration during which acetate is broken down and its hydrogens are released to yield their energy; the citric acid cycle.

L

lactic acid. A product of fermentation in most organisms.

lactose. Milk sugar, a disaccharide.

lacuna. A cavity in which a bone or cartilage cell is located.

larva. An immature but free-living form of an animal, not possessing the shape of the adult.

larynx. The organ of voice in mammals; the modified upper part of the trachea.

leaf gap. A location devoid of conducting tissue, immediately above the position where phloem and xylem leave the vascular bundle of the stem to enter the leaf.

learning. A relatively long lasting change in an animal's behavior resulting from previous experiences.

lentic. Of or pertaining to lakes.

lenticel. A patch of loosely arranged cells through which gas exchange can occur, in the corky layer of plant stems.

leukocyte. A white blood cell.

lignin. A complex organic constituent of the walls of sclerenchyma.

linkage. The association of genes on the same chromosome, so that they do not independently assort during meiosis.

lipase. A fat-digesting enzyme.

lipid. A molecule composed mostly of carbon, hydrogen, and oxygen and soluble in ether but insoluble in water.

littoral zone. A shallow zone around the margin of lakes where rooted plants grow; in oceans, the zone between high and low tides.

long-day plant. A plant that flowers when exposed to short periods of darkness.

lotic. Of or pertaining to streams.

lung. An internal pouched structure used for gas exchange by many terrestrial animals.

lymph. Constituent of blood that seeps through capillary walls into tissue spaces, eventually returned to veins by way of the lymphatic system.

lymph nodes. Enlarged areas along lymphatic veins where filtration of bacteria and other foreign objects can occur.

lymphatic system. A system of tubes that collects lymph and returns it to veins.

lymphocyte. A type of white blood cell that is very active in producing and regulating the immune response.

lysosome. A vesicle containing digestive enzymes.

M

macroevolution. Evolution on a scale to produce new species or higher taxa.

macronucleus. A large nucleus in the Protistan organism *Paramecium* and other ciliates, concerned with regulating day-to-day activities by producing RNA.

Malpighian tubules. Excretory tubules in insects, emptying into the gut.

mammary gland. A milk-emitting gland of mammals.

mantle. A fold of skin that (usually) secretes the shell of a Mollusc.

medusa. A free-swimming form of a member of phylum Cnidaria; the jellyfish stage.

megagametophyte. The egg-producing phase of a plant life cycle.

megaspore. A large spore that grows into a female gametophytic plant.

meiosis. The steps that allow production of haploid reproductive cells from diploid cells.

memory cells. B cells and T cells that remain in the body after a primary immune response; responsible for the secondary immune response.

meristem. A plant tissue consisting of unspecialized cells that are capable of active cell division.

mesocarp. The middle layer of a fruit wall (pericarp).

mesoderm. The middle layer of an early embryo, after gastrulation.

mesoglea. A jelly-like layer between the two cellular layers of Cnidaria; the jelly of a jellyfish.

mesophyll. A tissue in leaves located between the two epidermal layers.

mesophyte. A plant living in intermediate situations with respect to available water supply.

Mesozoic. An era of geologic time known as the age of reptiles.

messenger RNA. RNA that carries the encoded messages from DNA to the ribosomes.

metabolism. The set of chemical reactions that cause energy to be made available to the cell.

metamorphosis. The phenomenon of passing through several distinctly different body stages between embryo and adult forms.

metaphase. That stage of mitosis or meiosis when chromosomes are aligned at the equator of the spindle.

metastasis. The ability of cancerous cells to break away from a tumor mass and settle into other regions of the body.

microevolution. Evolution within a species.

microfilament. A thin solid rod composed mostly of actin; involved in changing cell shape.

microgametophyte. The sperm-producing phase of a plant life cycle.

micronucleus. A small nucleus in the Protistan organism *Paramecium* and other ciliates, containing genetic material that can be exchanged with that of another individual during conjugation.

micropyle. A pore in a plant's ovule through which the pollen tube grows; an opening in a fish or insect egg's shell through which sperm can reach the egg plasma membrane.

microspore. A small spore that grows into a male gametophytic plant.

microtubule. A hollow rod composed of the protein tubulin; composes the spindle and other cellular structures.

middle lamella. The first layer of a cell wall deposited by the two daughter cells during cell division.

mitochondrion. A cellular organelle, the center of respiratory activity.

mitosis. The portion of asexual cell reproduction that involves movement of chromosomes.

molecule. A group of two or more atoms bonded together and stable.

Monera. A kingdom of single-celled organisms without nuclear membranes, including bacteria and archaebacteria; divided into Kingdoms Bacteria and Archaea in a six-kingdom scheme.

monocyte. A large white blood cell that matures into a nonspecific scavenger, the macrophage.

monohybrid. A genetic cross involving only one characteristic being examined.

monosaccharide. The simplest kind of sugar; e.g., glucose or fructose.

monozygotic twins. Siblings born nearly simultaneously and derived by the splitting of a single zygote or a single early embryo.

morphogenesis. The shaping of the body during development.

motor neuron. A neuron sending messages to a responsive organ (effector).

Mullerian mimicry. The phenomenon of an obnoxious organism taking the shape and color of another species that is also obnoxious, thus benefiting both species.

multiple fruit. A fruit derived from carpels of several to many flowers located close together, e.g., pineapple or fig.

muscle fiber. The functional unit of muscle; composed of many myofibrils bundled together by a common sarcolemma.

mutagen. A mutation-causing agent.

mutation. A change in the message carried by a gene.

mutualism. A symbiotic relationship between two organisms whereby both members are benefited.

mycelium. The vegetative hyphae of a fungus.

myelin sheath. A covering over some neuron processes, formed from Schwann cells.

myofibril. The contractile material within a muscle fiber; composed of repeating units called sarcomeres.

myofilaments. Proteinaceous structures within sarcomeres of muscle fiber; actin and myosin filaments.

N

NAD. Nicotinamide adenine dinucleotide. A carrier of hydrogen and electrons during respiration.

NADP. Nicotinamide adenine dinucleotide phosphate. A carrier of hydrogen and electrons during photosynthesis.

nekton. Swimming animals.

nematocyst. The harpoon-like stinging device in each cnidocyte of animals in phylum Cnidaria.

nephron. An excreting unit of the kidney.

nerve. A cord containing extensions of many neurons bound together with connective tissue.

neural tube. A hollow tube formed after vertebrate gastrulation; becomes the spinal cord.

neuron. A nerve cell.

neurotransmitter. Any of a number of molecules released into synapses and capable of either exciting or inhibiting neurons or effectors at the synapse.

neuston. Organisms that live on the surface of water.

niche. The position or status of an organism, within which it can maintain existence; composed of geographical and biological surroundings of the organism.

nitrogen fixation. The process of converting gaseous nitrogen into a form that can be used by green plants.

notochord. The dorsally located, gristle-like supporting rod of Chordates; usually lost after the embryonic stage.

nuclear envelope. The double membrane forming the limiting boundary of a nucleus.

nucleic acids. DNA and RNA, polymers composed of units called nucleotides.

nucleolus. A spherical body seen in the nucleus during interphase, composed of RNA.

nucleoplasm. The semiliquid material in a nucleus, in which chromosomes are suspended.

nucleotide. The monomeric unit of which polymeric nucleic acids are made; composed of a phosphate group, a five-carbon sugar, and a nitrogenous base.

nucleus. An organelle of the eukaryotic cell containing the chromosomes.

nymph. An intermediate developmental stage of some insects, resembling the adult but having disproportionate body parts.

O

obligate anaerobes. Anaerobic organisms that are killed by gaseous oxygen.

omnivore. An organism that is equipped to ingest both autotrophs and heterotrophs as its food.

oncogenes. A class of genes that control normal cell reproduction in embryos and can cause cancer when inappropriately operating after the embryonic period.

oogonium. An ovarian cell that is the precursor of an egg cell.

organelle. A distinguishable portion of a cell having a specific function.

osmoconformer. An organism whose salt and water concentrations passively conform to those of its environment; opposite of an osmoregulator.

osmoregulator. An organism capable of performing appropriate homeostatic activities enabling it to be at least partially independent of its environment in terms of salt and water concentrations.

osmosis. The diffusion of water molecules through a membrane.

ovary. In animals, the female gonad; in plants, the basal part of the carpel.

oviduct. A tube for conducting eggs.

ovulation. The discharge of eggs from the follicles of the ovary.

ovule. An immature seed containing embryo sac with egg.

oxidative phosphorylation. The process of converting ADP to ATP under aerobic conditions.

oxytocin. A hormone produced by the posterior lobe of the pituitary gland, causing contraction of smooth muscles such as those of the uterus during childbirth.

P

Paleozoic. An era of geologic time rich with marine life and during which arose the first vertebrates, land plants, amphibians, insects, reptiles, and conifers.

palisade mesophyll. Column-shaped cells beneath the upper epidermis of leaves.

palmate. Arranged around one point like leaflets of Virginia creeper.

parasitism. A symbiotic relationship in which an organism lives in or on another species and imparts harm to the host.

parasympathetic system. One of the two portions (with the sympathetic system) of the autonomic nervous system.

parathyroid hormone. A hormone produced by the parathyroid glands; regulates metabolism of calcium and phosphorus.

parenchyma. A simple plant tissue composed of thin-walled cells.

parthenogenesis. Development of an egg without fertilization.

pathogen. A disease-causing organism.

pelagic. Of or pertaining to the open sea.

penis. An organ used in depositing semen during internal fertilization by animals; may also carry the urethra to the outside.

pepsin. A stomach enzyme that acts upon proteins.

perfect flower. A flower having structures of both sexes.

pericarp. The wall of a fruit.

pericycle. The outer layer of a vascular cylinder, especially in roots.

peristalsis. Wave-like muscular contractions of a tubular organ such as an intestine.

peroxisome. A vesicle containing the enzyme peroxidase.

petiole. The stalk of a leaf.

PGAL, glyceraldehyde-3-phosphate. A three-carbon compound regarded as the direct organic compound synthesized in the Calvin cycle of photosynthesis.

pH. The negative logarithm of the concentration of hydrogen ions in a solution; a measure of the acidity or basicity of the solution.

phage. See *bacteriophage.*

phagocytosis. The process of cellular ingestion of materials by engulfment.

pharynx. A part of the digestive tract of many animals; in humans, the passage between the mouth and the stomach.

phenotype. The expressed hereditary characteristic of an organism.

phenotype ratio. The proportion of phenotypes resulting from a specific genetic cross.

pheromone. An animal secretion that influences the behavior of other members of the same species.

phloem. Plant tissue forming sieve tubes that carry food from place to place.

phospholipid. A lipid resembling a fat except that a phosphate group (and additional atoms) has been substituted for one fatty acid; a major component of membranes.

photolysis. The splitting of water into hydrogen and oxygen during the light-dependent phase of photosynthesis.

photoperiodism. The physiological response of an organism to the duration of light and/or dark that it perceives.

photophosphorylation. A process that converts ADP to ATP, occurring in chloroplasts and dependent upon photosynthesis.

photosynthesis. The synthesis of simple foods from carbon dioxide and water in which light is the source of energy.

photosystems. Complex sets of molecules embedded in thylakoids, including chlorophylls that capture the energy of light and direct it toward photosynthesis reactions.

phylum. A major category of classification in non-plant kingdoms, hierarchically located between the kingdom and the class.

phytochrome. A plant protein linked to regulating the internal clock.

pinnate. Arranged along both sides of a longitudinal axis.

pituitary. A composite organ of both neural and hormonal tissues at the base of the brain; releases hormones that control a number of endocrine glands and other organs in the body.

placenta. An organ of exchange between embryo and mother; in plants, the place where an ovule is attached to the ovary.

plankton. Organisms, larvae or adults, that float or weakly swim at or near the surface of water.

plant. A multicellular autotrophic organism; a member of Kingdom Plantae.

plasma membrane. The outer limiting membrane of a cell; the cell membrane.

plasmid. A circular piece of DNA sometimes inhabiting a bacterium; reproduces whenever the bacterium does.

plastid. A plant organelle usually containing pigments or stored food.

platelet. Cell fragment concerned with the clotting of blood.

polar bodies. Nonfunctional cells (containing little or no cytoplasm) produced along with functional eggs during meiosis in female animals.

pollen. Male gametophytes of seed plants.

pollination. The transfer of pollen to the carpel of a plant.

polymer. A large molecule composed of many identical (or nearly identical) subunits in a chain.

polymerase chain reaction (PCR). A technique for making many copies of DNA from one or a few initially available.

polynucleotide. A polymer of nucleotides; e.g., DNA, RNA.

polyp. A form of Cnidaria that is usually attached, such as the sea anemone.

polypeptide. A polymer of amino acids, each linked to the next by a peptide bond.

polyploidy. The condition of having more than two complete sets of chromosomes in a cell.

polysaccharide. Any compound carbohydrate that may be broken into six or more molecules of monosaccharides.

polyspermy. The condition of more than one sperm cell successfully fertilizing an egg; causes polyploidy of the resulting zygote.

prairie. A middle-latitude grassland, with tall grasses.

predation. The characteristic of devouring or destroying animals.

primary structure. The level of a protein's structure involving the sequence of its amino acids.

primary tissues. Tissues produced by apical meristems of plants.

primordial germ cells. Cells from which gametes are derived.

producer. An autotrophic organism.

progesterone. A hormone secreted by the corpus luteum; the pregnancy-maintaining hormone.

prokaryote. An organism or cell lacking a nucleus and other membranous organelles.

promoter. A region of DNA that can be recognized by RNA polymerase in initial steps of transcription.

pronucleus. A haploid nucleus of a gamete (egg or sperm), after fertilization but before fusion to make a single diploid nucleus in the zygote.

prophase. The earliest phase of mitosis or meiosis during which chromosomes develop to their shortest and thickest form.

prostate. An exocrine gland that produces seminal fluids.

protein. A polymer of many (usually considered 50 or more) amino acids linked by peptide bonds; a large polypeptide.

Proterozoic. An early era of geologic time represented in the fossil record by marine algae, sponges, and worms.

prothallus. The gametophyte of ferns and related plants.

Protista. A kingdom of unicellular eukaryotic organisms.

protonema. In mosses, a filamentous gametophytic structure arising from a spore and producing buds that grow into gametophytes.

pseudocoelom. A body cavity (other than the gut) not entirely lined by cells of mesodermal origin.

pseudopodium. A variable and temporary appendage extended from certain cells and used for locomotion or feeding.

punctuated equilibrium. A proposed mechanism for speciation in which some organisms of a species suddenly change enough to become a new species.

purine. A nitrogenous base of nucleic acids, either adenine or guanine.

putrefaction. The decay of proteins accompanied by the production of foul-smelling compounds.

pyrimidine. A nitrogenous base of nucleic acids, either cytosine, thymine, or uracil.

Q

quantitative inheritance. The form of inheritance in which a single phenotype is determined by the additive effects of two or more genes.

quaternary structure. The level of a protein's organization involving combination of two or more polypeptides.

R

radicle. The lower tip of an embryonic plant that grows into a primary root.

radiocarbon dating. A method of

estimating the age of an object such as a fossil, by measuring the amount of certain radioactive elements in the sample.

receptacle. The crown of a peduncle or pedicel where floral parts are attached.

recessive. Unexpressed; used to describe an allele whose phenotypic characteristic is hidden by the characteristic of a dominant allele.

recombinant DNA technology. Methods used to place genes into organisms or cells by human intervention.

recombination. Any action that causes new combinations of alleles to be placed together in a gamete (and therefore in organisms of the next generation).

reductional division. Meiosis I; the portion of meiosis in which a cell goes from diploid to haploid.

reflex. An unconscious response to a stimulus.

reflex arc. The pathway of impulse transmission in a reflex.

regeneration. The phenomenon of regaining lost body parts by growth and differentiation of cells near the damaged area.

rennin. A secretion of the stomach that curdles milk.

replication. The reproduction of a DNA molecule to produce two exact copies.

respiration. A metabolic process within cells that provides energy for cell activities.

restriction enzymes. Enzymes of bacteria that cut foreign DNA (thus inactivating it) at specific base combinations; useful tools in recombinant DNA work and in characterizing DNA.

rhizoid. A root-like structure of gametophytes, lacking conducting tissues.

rhizome. A horizontal underground stem.

ribose. A five-carbon sugar in RNA.

ribosome. A tiny cytoplasmic granule often attached to the endoplasmic reticulum and acting as a center of protein synthesis.

ribozyme. RNA that acts as a catalyst.

root hair. An epidermal outgrowth from the absorption zone of roots, used for absorbing water and other substances.

root pressure. Movement of water from roots to upper plant portions, powered by osmotic pressure at roots.

S

saprobe. An organism that extracts its food from dead bodies; a decomposer.

sarcolemma. The plasma membrane surrounding a muscle fiber.

sarcomere. A repeating unit with in muscle fiber's myofibrils; composed largely of actin and myosin proteins.

sarcoplasmic reticulum. An intricate network of membranous bags overlying myofibrils.

savanna. A tropical grassland.

sclerenchyma. A simple plant tissue composed of cells with thickened walls like fibers or stone cells.

secondary structure. The level of a protein's structure involving helical or sheet-like formations.

secondary tissues. Tissues like cork, phloem, and xylem that are produced by cambium.

selection. The mechanism of evolutionary change suggested by Darwin; the fittest variants survive in competition for limited necessities in the environment.

seminiferous tubules. Tubules of the testis whose cells undergo meiosis to become sperm.

sensory neuron. A neuron transmitting a stimulus from a receptor toward the central nervous system.

sepal. A part of the lowermost portion of most flowers; leaflike in appearance.

seta. A hair-like bristle along the body of an Annelid worm.

sex chromosome. A chromosome whose genetic information is involved in sex determination; in humans, the X and Y chromosomes.

sexual. Pertaining to reproduction that involves the uniting of cells from two organisms.

short-day plant. A plant that flowers when exposed to long periods of darkness.

sieve tube. A conducting tube in phloem composed of a row of cells whose end walls are perforated by pores.

simple fruit. A fruit derived from a single carpel of a single flower, e.g., tomato.

siphon. A jet-like device for moving water into or out of an animal's body, as in Molluscs.

speciation. The process of forming a new species.

species. A particular kind of organism, written as a binomial consisting of a generic and specific name; a group of natural populations of organisms that can successfully reproduce by sexual means.

spermatogonium. A testis cell that is the precursor of sperm cells.

sphincter. A band of muscles encircling an opening or passage and controlling the flow of materials through it.

spicules. Simple or branching needles of calcareous or silicious materials giving support in sponges.

spinal ganglion. A concentration of sensory cell bodies in the dorsal root of the spinal nerve.

spindle. A biconical fibrous structure seen in a dividing cell and functioning in movement of chromosomes.

spindle fibers. A set of microtubules forming the spindle of a reproducing cell.

spiracle. A pore opening to a trachea (breathing tube) in insects and some other Arthropods.

spirillum. A spiral-shaped bacterium.

sporangium. A spore container in plants.

spore. A reproductive plant cell not requiring fertilization; in life cycles, a haploid cell produced by the sporophyte.

spore mother cell. A diploid cell from which haploid spores are derived.

sporophyll. A spore-bearing leaf, sometimes considerably modified as in the case of a stamen.

sporophyte. A spore-producing phase of a plant life cycle.

sporulation. Production of spores.

stamen. The male part of the flower.

steppe. A middle latitude grassland, with short grasses.

steroid. A lipid with characteristic interlocking ring formations; e.g., cholesterol, estrogen.

stigma. The part of a carpel receptive to pollen.

stipules. Paired appendages sometimes occurring at the base of a leaf.

stoma. An epidermal pore in plants through which gases are exchanged with the environment.

substrate feeder. An organism that obtains its food by eating the materials upon (or in) which it lives; e.g., an earthworm.

succession. Predictable change of community in a region, leading to a climax community.

sucrose. Table sugar, a disaccharide.

symbiosis. Dissimilar organisms living in a close relationship.

sympathetic system. One of the two portions (with the parasympathetic system) of the autonomic nervous system.

synapse. The junction between two neurons or between a neuron and an effector such as a muscle.

synapsis. Pairing of homologous chromosomes during prophase I of meiosis.

syncytium. A multinucleate tissue with cells confluent, not separated by cell membranes.

sympatric. Overlapping in geographic range.

syrinx. The voice box of birds, located at the base of the trachea.

T

T tubules. System of membranous tubules that are extensions of a muscle fiber's sarcolemma, forming a conduit for action potentials to move to the fiber's interior.

tadpole. The aquatic larval stage of frogs and toads.

taiga. The boreal forest, a high latitude coniferous forest.

taxis. Directional orientation of unattached animals or motile reproductive cells.

taxon. A group of organisms in a classification category, such as bluebirds, ants, or legumes.

telophase. The last phase of mitosis or meiosis, when chromosomes reach the poles of the spindle and a new nucleus begins to form.

tertiary structure. The level of a protein's structure involving the specific three-dimensional twisting of the polypeptide(s).

testosterone. A steroid hormone responsible for male characteristics.

tetrad. Four spores sticking together, derived from one spore mother cell; also a group of four chromatids (of two homologous chromosomes) during meiotic synapsis.

tetrapod. An animal using four appendages for locomotion.

thallus. A simple plant body without conducting tissues.

thermocline. A layer in deep lakes between the epilimnion and hypolimnion where the temperature decreases about one degree Celsius for every increase of one meter in depth.

thylakoids. A system of stacked membranes in chloroplasts; the sites of chlorophyll and other photosynthetic molecules.

thyroxin. The hormone secreted by the thyroid gland, influencing the rate of metabolism.

totipotency. The ability of a single cell to have and use all of the necessary genetic information to make a complete and normal organism.

tracheae. Conducting tubes such as

air passages to lungs, air tubes in insects, or vessels in wood.

tracheophytes. Plants with conducting tissues, including ferns and seed plants.

transcription. The process of building RNA under the direction of DNA.

transfer RNA. Cytoplasmic RNA that holds specific amino acids, carrying them to ribosomes for construction of polypeptides.

transgenic organisms. Organisms carrying and expressing genes from other species.

translation. The process of building specific polypeptides in a cell.

translocation. The movement of foods from one place to another in plants.

transpiration. The evaporation of water from the leaves of plants.

transpirational pull. The movement of water through a plant against gravity, powered by transpiration at leaf surfaces.

transverse tubules. See *T tubules.*

triploblastic. Having three complete layers of embryonic cells: ectoderm, mesoderm, and endoderm.

trophoblast. The region of an early mammalian embryo that contributes to the making of the placenta and the amnion.

tropism. A directional growth response in plants.

tube cell. A part of a pollen grain that grows to the vicinity of the egg.

tuber. The enlarged tip of a rhizome, e.g., Irish potato.

tundra. A frozen desert (where subsoil is permanently frozen).

typhlosole. The dorsal fold of the intestine of an earthworm, providing increased surface area for absorption of digested food.

U

ureter. The tube conducting urine from the kidney to the urinary bladder.

urethra. The tube conducting urine from the urinary bladder to the outside.

urine. An aqueous fluid containing nitrogenous waste, made in the kidneys and excreted periodically.

uterus. The organ where the embryo or fetus develops; womb.

V

vacuole. A large cytoplasmic cavity filled with water and other materials.

valve. The half-shell of a Mollusc such as a clam or oyster, or the one-part shell of a snail.

vas deferens. The tube carrying sperm from each testis to the urethra.

vascular. Containing or using a system of tubes; e.g., vascular plants.

vascular bundle. In plants, a strand of conducting and supporting tissues.

ventral root. The ventral branch of a spinal nerve that connects to the spinal cord; the pathway of motor neurons.

ventricle. A compartment of the heart that pumps blood to the tissues.

venule. A small vein.

villus. A finger-like projection from the inner wall of the small intestine that increases the surface area for enhanced absorption of digested foods.

virus. An infective organism (or, arguably, particle) composed only of nucleic acid surrounded by a protein capsule, able to reproduce only within a host's cell.

visceral reflex. An unconscious response in the eye, internal organs, or blood vessels; its pathway is in the autonomic nervous system.

vitamins. Organic molecules needed in small amounts by an organism, but not synthesized by that organism; usually acting as coenzymes.

W

water vascular system. A unique system in Echinoderms used especially for locomotion, circulation, and food-getting.

X

xerophyte. A plant adapted for survival where water is scarce; a water conserving plant.

xylem. A complex plant tissue that conducts water and dissolved inorganic substances and provides support; the wood of plants.

Y

yolk. A complex of food materials (largely protein and lipid) stored in an egg for use by the embryo during development.

yolk sac. A sac extending from the ventral surface of an embryonic bird, reptile, or mammal embryo; holding stored food in the first two groups.

Z

zona pellucida. A jelly-like layer adhering to the plasma membrane of a mammalian egg.

zygote. The cell resulting from uniting of two gametes in fertilization; the first cell of an embryo.

INDEX